CIRP Design 2012

Amaresh Chakrabarti
Editor

CIRP Design 2012

Sustainable Product Development

Springer

Editor
Amaresh Chakrabarti
Centre for Product Design
and Manufacturing (CPDM)
Indian Institute of Science
Bangalore
India

ISBN 978-1-4471-4506-6 ISBN 978-1-4471-4507-3 (eBook)
DOI 10.1007/978-1-4471-4507-3
Springer London Heidelberg New York Dordrecht

Library of Congress Control Number: 2012949375

Printed on acid-free paper

Springer is part of Springer Science+Business Media (www.springer.com)

Contents

Part II Creative and Inventive Design (TRIZ)

Part III Enabling Technologies and Tools

Chapter 1
Introduction

Amaresh Chakrabarti

Abstract The collection of chapters in this book constitutes the proceedings of the 22nd CIRP Design Conference (CIRP Design 2012) held at the Indian Institute of Science in the city of Bangalore, India during 28–31 March 2012. This is the first time that this flagship design conference of the International Academy of Production Engineering (CIRP) has been held in India; the intent has been to bring together the international community from diverse areas of design practice, teaching and research, to share cutting edge research about design among the stakeholders, and provide a platform for its interaction and collaboration with the design research community in India. The conference is intended for all stakeholders of design, and in particular for its practitioners, researchers, teachers and students.

The collection of chapters in this book constitutes the proceedings of the 22nd CIRP Design Conference (CIRP Design 2012) held at the Indian Institute of Science in the city of Bangalore, India during 28–31 March 2012. This is the first time that this flagship design conference of the International Academy of Production Engineering (CIRP) has been held in India; the intent has been to bring together the international community from diverse areas of design practice, teaching and research, to share cutting edge research about design among the stakeholders, and provide a platform for its interaction and collaboration with the design research community in India. The conference is intended for all stakeholders of design, and in particular for its practitioners, researchers, teachers and students.

The theme of the 22nd CIRP Design is "Sustainable Product Development" —development that focuses on creating products that would sustain economy while ensuring sustenance of both society and ecology. This is a key challenge of

A. Chakrabarti (✉)
Centre for Product Design & Manufacturing, Indian Institute of Science,
Bangalore, India
e-mail: ac123@cpdm.iisc.ernet.in

A. Chakrabarti (ed.), *CIRP Design 2012*, DOI: 10.1007/978-1-4471-4507-3_1,

growing global importance that affects all stakeholders of a product's life cycle, both current and future. The conference provides a platform for exchange of ideas, discussion, and debates about the issues faced and resolution possible. The conference included original work on all aspects of design research, as reflected in the list of topics.

Fifty two full papers were submitted to the conference; these were reviewed by at least three reviewers from among 93 reviewers drawn from the CIRP Design 2012 International Committee comprising members from over 80 institutions or organisations from 23 countries spanning 3 continents. Finally, 40 full papers, authored by researchers from 32 institutions and organisations from 14 countries spanning 3 continents, have been selected for presentation at the conference and publication in this book. The topics span from those focusing on early stages such as creativity and synthesis, through those that are primarily considered in specific stages of the product life cycle, such as safety, reliability or manufacturability, to those that are relevant across the whole product life cycle, such as collaboration, communication, design management, knowledge management, cost, environment, and product life cycle management. Foundational issues such as the nature of design theory and research methodology is also a major area of focus. The papers span a variety of areas of application; of aerospace, healthcare, and automotive are but a few of these. A large number of papers embrace the theme of this year's conference—Sustainable Product Development; this reflects the growing importance of this theme within design research. The conference had 6 keynotes by eminent speakers from 6 countries, 2 invited papers and 6 podium sessions. Nine papers received certificate of merit.

This book has 40 chapters that are divided in six themes: Design Theory, Methodology and Research Methodology; Creative and Inventive Design (TRIZ); Enabling Technologies and Tools; Global Product Development and PLM; Design For X (Safety, Manufacture, Assembly, Cost, Risk, Reliability, Modularity, etc.); and, Sustainable Design and Manufacturing.

On behalf of the conference Organizing Committee, I would like to thank all the authors, reviewers, session chairs, keynote and invited speakers, organizations and delegates that participated in the conference, and the members of the conference programme committee for their help and support in organising CIRP Design 2012, reviewing the papers, and putting this book together. We are thankful to CIRP for its kind endorsement and support of the conference. We also thank Indian Institute of Science, Bangalore, India, and Mori Seiki, Japan—our two major sponsors—for their generous support of this event, and the Solid State and Structural Chemistry Unit of Indian Institute of Science, Bangalore, for providing the venue for the conference. I am thankful to all members of the Organizing Committee and the volunteers for their invaluable support in organising the conference, and making this publication possible. I also wish to place on record and acknowledge the effort and assistance of Mr. Ranjan B.S.C., Ms. S. Harivardhini, and Ms. Shakuntala Acharya Nair in the preparation of this book and the CD-ROM proceedings. Finally, I thank Mr. Anthony Doyle and Ms. Grace Quinn of Springer-Verlag, London for their sincere support in publication of this book.

Part I
Design Theory, Methodology and Research Methodology

Chapter 2
Design Framework for Micro and Nano-Scale Products

Sang-Gook Kim, Stephen Bathurst and Firas Sammoura

Abstract The design and manufacturing of high quality micro-electromechanical systems (MEMS) is becoming increasingly complex as the manufacturing tools available diversify. Design of nano-scale products is even more complex. Small-scale products tend to have highly coupled designs, decoupled at best, because of the serial nature of fabrication, material and process constraints and lack of adequate fabricating processes. This paper proposes a structured design framework, which enables designers to achieve the correct functionality by either reducing the complexity in multi-scale manufacturing or by developing a new manufacturing process to circumvent the existing constraints.

2.1 Introduction

Products have grown into ever-larger multi-scale systems while their components have shrunk to a smaller-scale (i.e., micro and nanostructures). The design and manufacture of multi-scale products with newly developed materials (such as carbon nanotubes and graphenes) has become increasingly complex. Small-scale products are not realizable when engineers make bad designs. This paper aims to establish a design framework for micro- and nano-scale products where no structured framework for design has been readily available.

S.-G. Kim (✉) · S. Bathurst · F. Sammoura
Massachusetts Institute of Technology, Cambridge, MA, USA
e-mail: sangkim@mit.edu

F. Sammoura
Masdar Institute of Technology, Abu Dhabi, UAE

A. Chakrabarti (ed.), *CIRP Design 2012*, DOI: 10.1007/978-1-4471-4507-3_2,

A micro system can be characterized by small dimensions, either of the system/component itself (one or more critical dimensions) or of the functional features of the system/component [1]. Generally, two categories of micro products are identified: (1) components with at least two critical dimensions in the sub-mm range, (2) relatively large components with functional features in the μm range. If the characteristic length in the above is in the order of 100 nm or smaller, they become nano systems.

Microelectromechanical Systems (MEMS) technology has shown the potential to revolutionize the production of sensors and actuators. However, many industrial efforts have experienced significant delay in delivering products in time for market success. This is somewhat due to poor design of the devices and production methods, but is mainly due to the lack of delivering the functions customers want. A structured design framework is necessary to effectively guide designers for small scale products.

We propose that small scale products can be designed with the following three steps: (1) *Functional Design with Axiomatic Design Framework*, (2) *Reduction of Complexity*, (3) *Multi-domain Design Mapping*. Regardless of the scale of products, a good design should clearly define the functionalities of the product ("what we want") at the top level of the design process and ensure that they can be attained during and after the product development. Without a structured design framework, functional requirements may be misinterpreted or lost. Therefore the first and the most important step to make good small-scale product design is to search and generate the functions of the product and maintain them throughout the design process. Axiomatic Design has been a useful design framework for generating functional requirements and mapping them across the design domains.

Design of small-scale products tends to be complex because micromachining and nano processes involve specialized top-down or bottom-up processes and the cost of prototyping (or make and see iterations) is enormous. In order to predict the performances of these devices before prototyping, domain-specific computer aided engineering (CAE) tools and solvers are necessary [2]. Unlike the narrow and deeply studied integrated circuit (IC) products, however, MEMS product design must account for the interaction between electrical, mechanical, fluidic, chemical and optical forces among others, which makes it practically difficult to develop general CAE tools for MEMS. Furthermore, CAE tools alone cannot help to reduce the complexity in coupled designs. Therefore, the second step of the small scale product design is to understand the nature of complexity in small-scale product design and to find ways to reduce it.

In evaluating the complexity of a MEMS device the designer is presented with the challenge of determining which manufacturing processes are best suited to meet the system's functional requirements and the sequence of those interdependent processes. Compounding the difficulty, MEMS design requires integrating an understanding of the coupled nature of the manufacturing tools into the design process. Design for manufacturing rules of thumb are commonly used to ensure manufacturability in macro scale products. These rules are often process or design specific such as radiusing internal corners on milled parts or depositing materials with higher processing temperature first in thin film deposition. It is unclear then, how a designer who is attempting to discover the best process or sequence would be able to make use of such rules. What is needed is a way for MEMS engineers to understand the impact of process changes on device functionality during the design phase. Axiomatic Design is a useful tool for this because it provides a structured basis for mapping design parameters to process variables as well as to understand the impact of the process variables on the functional requirements. Sometimes, this effort can lead to a novel manufacturing process, which uncouples the design couplings with the existing processes.

Three short case studies are presented to demonstrate how each step could be accomplished.

2.2 Functional Design is the First Step

A successful MEMS product, like macro-scale products, must align a functionally uncoupled or decoupled design with processes that ensure reliable manufacturing. Axiomatic Design is a useful tool for this because it provides a structured basis for generating functional requirements and mapping them across the design domains.

Axiomatic design was developed in the late 1970s by Prof. Nam Suh at MIT offering two axioms that provide a framework for the decision-making and the mapping between "what we want (function requirements)", and "how we can achieve them (design parameters)" [3, 4]. Originally the method was developed for managing the design process for complex engineering products and systems but is extended nowadays to a broad range of systems (for example in healthcare or lean systems applications [5]).

Axiomatic Design (AD) approach is based on the distinction of four different design domains: Consumer domain, Functional Domain, Physical Domain and Process Domain. The Consumer domain is where customer's needs reside. These customer's needs must be mapped into the functional domain and translated into a set of functional requirements. The functional requirements (FRs) are defined as the minimum set of independent requirements that characterize the design goals. These FRs are then mapped into the physical domain, where the design parameters (DPs) are chosen to satisfy those FRs. The AD process can be summarized as shown below:

1. *Conceive the top level FRs.*
2. *Map FRs to Design Parameters DPs at the same level.*
3. *Mapping process can be analysed/evaluated with the two design axioms to ensure a good design.*
4. *Above steps* (1–3) *are repeated top to down in a zig-zag manner until the physical solution can be conceived from the mapped DPs.*
5. *If all the FRs reach leaf nodes (where the conceived FR-DP is clear and no further decomposition is necessary), physical integration of them to a feasible solution will lead to the final design solution.*

The first axiom—the so-called Independence Axiom—requires an uncoupled or at least decoupled design, which guarantees independent control of the functionality of the product. The FRs must be translated into DPs without affecting other FRs. That means the set of DPs has to be chosen so that they satisfy the FRs as well as maintain their independence [4].

The information axiom (the second axiom) makes it possible to benchmark different design alternatives by comparing the overlap of the design range which is required to fulfill the product's functionality and the system range that the different design options offer. The design, which offers the lowest information content, that is the design with the highest probability of success, will be selected.

Here a case that demonstrates how a new MEMS product is designed by defining the top-level functional requirements at the early stage of design.

Case 1: Design of Piezoelectric Ultrasonic Micro Transducer

Advanced medical ultrasonic imaging and fast 3-D scanning systems, operating in both transmit and receive modes, can be achieved with tiny 2-D arrays of transducers with piezoelectric ceramics. However, the labor-intensive manufacturing processes, such as dicing, bonding, and delicate assembly of crystallized piezoelectric ceramic limit the production yield, rate, and quality. Piezoelectric Micromachined Ultrasound Transducers (PMUTs) would offer several advantages to the conventional bulk machined ultrasonic sensors, including batch fabrication, electronics integration, and better acoustic matching to the surrounding medium [6]. Most of all, it can be formed into a 2 dimensional (2D) array for 3 dimensional (3D) imaging which will enable advanced medical imaging capability such as intracardiac echocardiography. The goal of this design is a tiny 2-D array of piezoelectric ultrasound transducers that enable ultraportable and high power ultrasound imaging systems. Top-level functional requirements were searched to achieve this goal and the matching design parameters were explored to make the design uncoupled or at least decoupled.

A typical PMUT is composed of a suspended membrane, made up of a structural silicon layer and a piezoelectric layer sandwiched between two metal electrodes. The membrane is clamped at its edges, resulting in reduced device acoustic impedance. With the application of an AC voltage across the piezoelectric layer, the membrane vibrates as a result of the strain mismatch between the elastic layer and the piezoelectric layer. Most PMUTs operate with a pulse-echo in the range of 1–16 MHz, where the optimum operating frequency is chosen based on the required depth of penetration, resolution, and tissue composition. A typical ultrasonic device for medical applications operates at a resonant frequency of 3 MHz (FR1), which can be controlled by varying the membrane radius as well as the membrane thickness (DP1). It is not easy, however, to achieve the precise resonant frequency of the membrane due to manufacturing tolerances. Intentional residual stress on the membrane can be used tune the resonant frequency as well to correct for any manufacturing tolerance. Precise resonant frequency (FR2) is thus provided by the tuning ring on the membrane (DP2). The output acoustic pressure is directly proportional to the membrane displacement and deflection shape, where

Table 2.1 Top-level functional requirements (FR) and the design parameters (DP) of PMUT for medical imagining application

	Functional Requirements (FR)	Design Parameters (DP)
1	Resonant frequency of 3 MHz	Membrane thickness
2	Tunable center resonant frequency of 2 MHz	Voltage applied to an external tuning ring
3	Membrane Deflection: 0.1 μm	Membrane radius
4	Deflection Shape: piston-like	Corrugation edge design
5	Impedance matching to external environment	Membrane stiffness
6	Cross-talk between elements	Elastomeric material separating elements

a 10 nm average membrane displacement generates roughly 1 MPa. For high acoustic power applications such as brain imaging, about 100 nm membrane deflection is needed (FR3). The applied voltage level, the membrane radius, as well as the membrane thickness would control the membrane displacement. We chose membrane diameter as the DP3. The AC voltage that drives the membrane can be used to control the membrane displacement, whereas optimized values of the membrane radius and thickness can be used to control the resonant frequency in order to achieve a decoupled system. A piston-like membrane motion (FR4) is desired to increase the displaced volume, and is realized by attaching the membrane via a corrugation to the fixed support (DP4). In order to maximize the energy transfer to the surrounding, the membrane impedance (FR5) should match the acoustic impedance of the surrounding medium. Changing the membrane stiffness (DP5) or depositing a layer on top of the membrane can adjust the membrane impedance. Separating the elements (DP6) with a polymeric foundation will minimize the cross talk between the elements (FR6). Table 2.1 summarizes the functional requirements and the design parameters of the PMUT generated. The proposed top level FRs and DPs shown in Table 2.1 make an uncoupled system which resulted in a US patent pending design.

The process of codifying functional requirements and design parameters not only identifies possible couplings, but also ensures clear focus on successfully satisfying critical requirements. This way of expressing a design also serves as a basis for complexity analysis and multi-domain design. Defining a comprehensive set of independent functional requirements is therefore a key part of the proposed design framework.

2.3 Step Two: Reducing the Complexity

Nano science has found many interesting nano structures with novel potential functions in real world applications. However, nano products require an integration of nano-components to macro systems and the degree of complexity is very high when nano structures are integrated (or assembled) to macro products. Success of small-scale system design depends on better understanding complexity and subsequent reduction of complexity in assembling multi-scale components into a functioning system. There have been many different views and approaches to understanding complexity in the fields of information technology, system biology, mathematics, meteorology and economics among many fields of science. Algorithms to find out absolute measures of complexity, however, have not been well established. In this paper, complexity of small-scale system design is understood though the axiomatic design framework.

A relative measure of complexity has been introduced by Suh [7]. In his complexity theory, complexity is defined as a measure of uncertainty in satisfying the functional requirements (FRs) within the specified design range. There are subtle differences among the uncertainty, complexity and difficulty which all provide hardship in design. Since it is believed that design axioms have been always true and no counter examples have been observed yet, we may define complexity as a result of a design's failure to conform the design axioms. Complexity can be defined as a collective outcome when a design doesn't satisfy the design axioms. The four kinds of complexity can be explained by their causal nature with respect to the design axioms.

- *Time-independent real complexity*: when a design is coupled. (Independence axiom violation)
- *Time-dependent periodic complexity*: when the coupled nature of design is capsulated to prevent the propagation across the system.
- *Time-independent imaginary complexity*: when a design is decoupled and not solved in the particular sequence (lack of knowledge).
- *Time-dependent combinatory complexity*: when a design has many states (FRs, DPs), which are not at equilibrium and change as a function of time (non-equilibrium).

The above speculation about complexity can be applied to small-scale systems design in that a coupled design with the high scale order will become extremely complex. When the scale order is high, all of the four types of complexity become larger. Suh suggested functionally periodic systems have a smaller scale complexity when the complexity is divided and confined in functionally uncoupled spatial/temporal sub-domains. Functional periodicity by dividing the system into sub-domains can be applied to micro and nano product designs in order to reduce and confine complexity. A new assembly process for carbon nanotubes has been developed based on the above framework to reduce the system complexity.

Case 2: Reduction of Complexity by Reducing the Scale Order

Assembly of individual nanostructures in a deterministic manner has been a big challenge due to its complexity. This case demonstrates a new method of handling and locating individual carbon nanotubes (CNTs) and integrating them into microelectromechanical systems (MEMS). An effective method of CNT assembly was demonstrated by reducing the complexity of assembly. It provides MEMS designers with the ability to locate an individual CNT nearly arbitrarily for the first time [8].

The key idea of transplanting assembly is to grow the vertically aligned individual CNTs on a substrate at ideal growth conditions and to transfer well-grown CNT strands to the target locations via micro-scale CNT carriers. The major technical issues solved are: (1) how to grow vertically aligned single strand CNTs at predefined locations, (2) how to preserve/control the orientation/length of an individual CNT during transplanting processes, and (3) how to locate/release an individual CNT at the target location. This assembly concept transforms the scale of tools necessary for CNT assembly from nano-scale to micro-scale, which enables manual, automated or parallel assembly of individual CNTs to MEMS in a deterministic and reproducible way.

The first step in transplanting assembly is the vertical growth of CNTs, which requires seeding catalytic nickel nano-dots at predefined locations on the Si substrate. An array of Ni catalytic dots (100–200 nm in diameter and 30 nm in thickness) was defined using electron beam lithography followed by metal deposition and liftoff process. An array of vertically aligned CNTs was grown using a home-built plasma enhanced chemical vapor deposition machine. Each CNT strand was then embedded into a micro-polymer block, which serves as a CNT carrier. A double polymeric layer encapsulation process was used with SU-8 as the top layer and polymethylglutarimide (PMGI) underneath. The SU-8 block (15 μm in thickness and 20 μm in diameter) forms the body of the carrier, while the PMGI layer (1.5 μm in thickness) holds the body until the release of the carrier from the substrate and then is removed to expose the CNT tip. The orientation of the embedded CNT is near parallel to the axis of the SU-8 block. The diameter of CNTs matches the size of Ni dots, and the length is 5–10 μm with a uniform cylindrical shape. The thickness of the bottom PMGI layer was chosen to be 1.5 μm so that a target aspect ratio of the exposed CNT tip is about 10–1. The 20 μm SU-8 blocks can now be used to manipulate and move the CNT to the desired location.

This technique was used to manufacture a CNT tipped atomic force microscope (AFM) cantilever with a unique ability to resolve high aspect ratio features as shown in Fig. 2.1. It is shown in this case decomposing the system into smaller uncoupled micro-domains can reduce the complexity. The scale order reduction of a multi-scale system into smaller-scale periodic systems can reduce the complexity of the whole system and subsequently ensure a good design and manufacturing by generating functional periodicity as was proposed in the complexity theory.

2.4 Step Three: Multi-domain Mapping

MEMS designers face an increasingly complex set of choices when selecting manufacturing processes. Furthermore, because the manufacturing processes often drive the designs of micro-systems used, the impact of process selection must be understood early in the design process.

Mapping design parameters to process variables early on in the design phase can provide information about the processing tools likely to be used and their limitations. Often with MEMS design, the impact of manufacturing process selection is apparent after only a few levels of decomposition. With this information the designer can evaluate the impact of process decisions on FR-DP

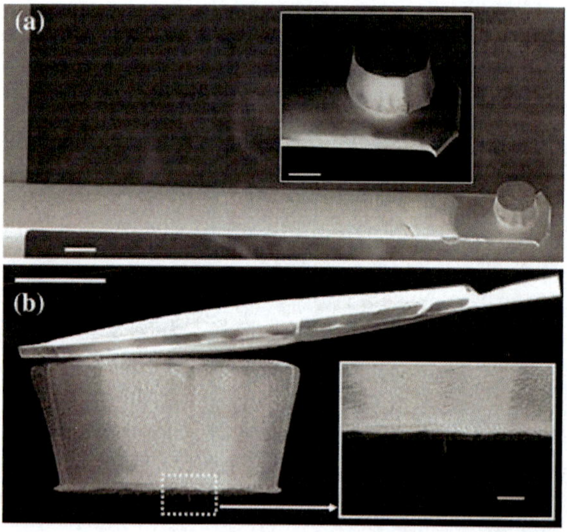

Fig. 2.1 CNT tipped AFM probes, scale bars 20 μm (**a**), 10 μm (**b**), 2 μm (inset) [10]

coupling. If processing constraints are not considered carefully, and instead DPs are selected based on intuition or previous experience, it is possible for functionality of the design to be sacrificed due to perceived limits on manufacturing processes. Using multi-domain design, it becomes apparent when a process improvement may enable a new DP and remove functional coupling. With this analysis method, the designer is able to evaluate whether functional coupling or process coupling is limiting the functionally of the product and to focus development effort appropriately.

Case 3: **Improving Process Flexibility by Ink Jet Printing**

Thin film lead-zirconate-titanate (PZT) is an attractive material for piezoelectric MEMS due to its high coupling coefficients. Manufacturing limitations have thus far limited wide commercial application of this material. For this case, we studied the couplings inherent to PZT deposition and tried to eliminate them.

Many methods of depositing thin film PZT have been demonstrated including sputtering and laser ablation. However, the most common PZT deposition method is sol–gel spin coating which achieves high quality films at low cost. Analysis of the forward and backward couplings of process variables helps identify what process changes might most significantly affect device performance [9]. Through our work spin coating PZT, we became aware of a significant, irresolvable coupling between the spin coating process and device structural layer.

The thermal processing of solution deposited PZT causes large tensile stresses in the final film. It is not possible to spin coat PZT over even small out of plane features without significant cracking. This causes current leakage or shortage and deteriorates performance. This forward coupling from device geometry limits device designs to planar geometries. One way to eliminate this coupling is to sequence the process such that the PZT is deposited before the formation of the structure. This has been demonstrated but requires either carrying the fragile PZT through a micromachining process, or using an additive process which requires compromising on material properties [10, 11]. A better, fully uncoupled, approach is to create a more flexible way of depositing the PZT precursor.

Recently, direct write methods have been shown to be a viable alternative to standard lithographic processing of MEMS [12]. Digital fabrication of solution-based PZT via drop-on-demand printing has the potential to remove many of the process constraints of spin coating, while maintaining all the advantages of a chemical solution deposition method. It provides for as deposited patterning, coating on or around arbitrarily out of plane features, and deterministic thickness control. A new process of depositing PZT, based on printing, was developed that may enable new process sequencing and therefore improved device performance.

Several experiments were conducted in order to realize this drop-on-demand based deposition method. Solutions were formed from a Mitsubshi 50/49 PZT sol–gel (E1) based on butyl alcohol and propylene glycol. As purchased, this sol is 85 %(wt) solvent and 15 %(wt) a mixture of lead, zirconium, and titanium metal oxides. Combinations of 2-methoxyethanol, isopropanol, and propylene glycol were added to the sol to create the diluted PZT inks. To prevent clogging of the printer nozzle and defects in the film, sources of particle contamination in the solution and in the environment were addressed and controlled. Throughout this work, over thirty ink chemistries, with dilution levels ranging from the as-purchased 15 %wt of metal oxides down to 2 %wt, were tried and the levels of dilution require for stable printing were observed. Preventing clogging also requires controlling the evaporation rate of the ink. If the solvent evaporates too quickly, metal organic molecules are built up inside the nozzle and firing chamber and concentrations that exceed stable printing requirements result. Maintaining a flow through the nozzle at all times prevents this type of clogging.

Controlling the distribution of solute material on the substrate after drying is also critical for uniform deposition of high quality films. The diffusion of solutes towards the film edges during solvent evaporation known as the coffee stain effect can lead to significant non-uniformity. In order to overcome this non-uniformity a study was conducted to determine the ink volatility and substrate temperature required to achieve the optimum level of spreading and diffusion.

After optimization of solution chemistry and substrate temperature, PZT films between 100 and 500 nm with less than 40 nm variation could be printed with a droplet size of 80 pl. Spot sizes as small as 43 μm were achieved with a 10 pl droplet of PZT ink deposited on a platinum substrate. The edge variation of printed lines was controlled within ± 10 μm. The ability of this new process to impact process sequencing was demonstrated through the deposition of PZT on pre released test cantilevers (Fig. 2.2). Analysis of forward and backwards couplings that exist due to process limitations highlighted the improved functionality that could be achieved by eliminating those couplings. When a MEMS designer properly understands the level and direction of process couplings, effort may be focused on process improvements that relive these couplings. As a result new geometry and material DPs become available and improved device performance may result.

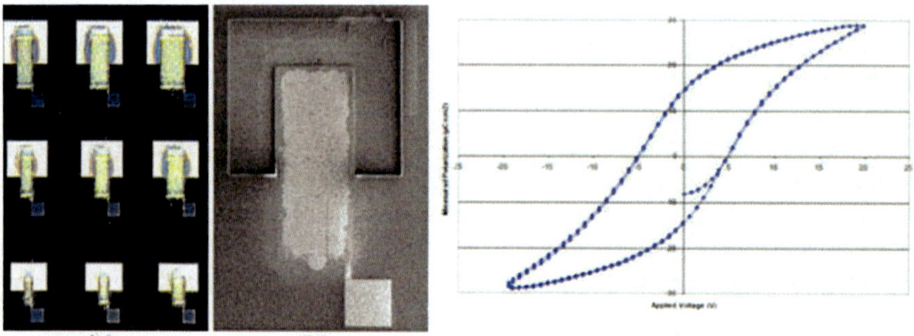
(a) (b) (c)

Fig. 2.2 **a** An array of PZT cantilever resonators (PZT printed after geometry formation). **b** Detail of a single PZT cantilever. **c** Polarization vs. voltage data from a printed PZT film

2.5 Conclusion

The complexity of small-scale products can originate from the functionally coupled designs and/or the high scale order of their manufacturing processes. A three-step design framework is proposed to cope with the complexity and to minimize the costly iterations in micro/nano product development. Three case studies demonstrate how this framework can be applied. Top-level functional requirements in developing a tiny 2-D array of piezoelectric ultrasound transducers were explored and investigated to make the design uncoupled or at least decoupled. A new nano-assembly process for carbon nanotubes has been developed by periodically dividing the complexity associated with the scale order. Finally a multi-domain design analysis approach was used to show how process decisions might impact device functional requirements.

Authors are grateful to DARPA (Grant HR0011-06-1-0045), Hewlett Packard and the MIT-MASDAR Program for the support of this study.

References

1. Hansen HN (2001) Approaches to design of micro mechanical systems. CIRP design conference Daejon, Korea
2. Mukherjee T and Fedder GK (1997) Structured design of microelectromechanical systems. Proceedings of 34th design automation conference pp. 680–685
3. Suh NP, Bell AC, Gossard DC (1977) On an axiomatic approach to manufacturing and manufacturing systems. Transactions of the ASME 77-WA/PROD-14
4. Suh NP (1990) Principles of design. Oxford University Press, New York
5. Peck J, Kim SG (2008) Improving emergency department patient flow through optima fast track usage. Ann emergency med 52(4) Supplement 1: S88
6. Kino GS (1987) Acoustic waves: devices, imaging, and analog signal processing. Prentice Hall, New Jersey
7. Suh NP (2005) Complexity: theory and applications. Oxford University Press, New York

8. Kim SH, Lee HW, Kim SG (2009) Transplanting assembly of CNT-tipped AFM probes. Appl Phys Lett 94(19):193102
9. Kim SG, Koo MK (2000) Design of a micro-actuator array against the coupled nature of microelectromechanical systems (MEMS) processes. Annals of the CIRP (Int'l Academy for Production Engineering) 49(1):101–104
10. Conway NJ, Traina ZT, Kim SG (2007) A strain amplifying piezoelectric MEMS actuator. J Micromech Microeng 17:781–787
11. Hajati A, Bathurst SP, Lee H, Kim SG (2011) Design and fabrication of a nonlinear resonator for ultra wide-bandwidth energy harvesting applications. IEEE international conference on micro electro mechanical systems Cancun, Mexico
12. Fuller SB, Wilhelm EJ and Jacobson JM (2002) Ink-jet printed nano particle microelectromechanical systems. J Microelectromech Sys 11(1):54–60

Chapter 3
Towards an Ontology of Engineering Design Using SAPPhIRE Model

V. Srinivasan, Amaresh Chakrabarti and Udo Lindemann

Abstract In this paper, a first version of an ontology for early phases of engineering design is developed using the SAPPhIRE model. A representation is empirically developed for the constructs of the SAPPhIRE model. Based on this representation, an ontology is developed by building clusters of nouns, verbs, adjectives, adverbs and mathematical equations from earlier work based on the SAPPhIRE model. Relationships are identified between the clusters and the constructs of the model. The developed ontology is validated by comparing it against existing ontologies to assess similarities and differences. Potential applications and means for expanding the ontology are also outlined.

3.1 Introduction

An ontology is defined as an explicit specification of a conceptualization; an ontology specifies entities and their relationships that are required for describing a domain [1]. Ontologies are important in engineering design for various reasons—indexing information in unstructured information repositories for

V. Srinivasan (✉) · U. Lindemann
Technical University of Munich, Boltzmannstr. 15,
85748 Garching, Germany
e-mail: srinivasan.venkataraman@pe.mw.tum.de

U. Lindemann
e-mail: udo.lindemann@pe.mw.tum.de

A. Chakrabarti
Indian Institute of Science, Bangalore 560 012, India
e-mail: ac123@cpdm.iisc.ernet.in

A. Chakrabarti (ed.), *CIRP Design 2012*, DOI: 10.1007/978-1-4471-4507-3_3,

better information retrieval [2], reducing ambiguity in engineering design [3], increasing uniformity within different functional design models [4], etc. Ontologies in engineering design are developed for better retrieval and reuse of information from unstructured engineering documents [2], modelling functions in engineering design [4], supporting bio-mimetic design [5], etc.

The SAPPhIRE (*S*tate Change, *A*ction, *P*art, *Ph*enomenon, *I*nput, o*R*gan, *E*ffect) model of causality is developed to explain the working of biological and engineered systems [6]. An integrated model of designing, GEMS of SAPPhIRE as req-sol, is developed and validated in terms of activities, outcomes, requirements and solutions, to describe the early phases of engineering design [7]. The SAPPhIRE model can describe outcomes at various levels of abstraction at the early phases of engineering design for both, biological and engineered systems [6, 7]. Idea-Inspire and a catalogue of physical laws and effects are both developed using the model [6, 8]. However, no ontology has been developed using this model. Due to the potential generality and empirical underpinning of the SAPPhIRE model, this research is undertaken to develop an ontology using the model. The development of the first version of the ontology is reported in this paper.

3.2 Literature Review and Objective

In this section, the SAPPhIRE model and, the existing taxonomies and ontologies are reviewed. The SAPPhIRE model provides a richer description of function (via action, state change and input), behavior (via phenomenon and effect), and structure (via organ and part) [6]. Phenomenon is an interaction between a system and its environment; effect is a principle underlying an interaction; state change is a change in a property of a system (and its environment) due to an interaction; action is an abstract interpretation of an interaction; input is a physical quantity that comes from outside the system boundary that is required for an interaction; organ is a set of properties and conditions of a system and its environment that are also required for an interaction; part is a set of components and interfaces that make a system and its environment [7]. Examples for these constructs are shown in Sect. 3.4. The SAPPhIRE model of causality is explained as follows. Parts create organs which along with inputs activate effects; effects create phenomena which in turn create state changes; state changes are interpreted at a higher abstraction level as actions [6]. The model is empirically found to be able to describe outcomes at different levels of abstraction in engineering design [7]. Idea-Inspire, a computer-based tool to support ideation by providing analogous stimuli from a database of biological and engineered systems, uses the SAPPhIRE model to describe entries in the database and assists search for stimuli [6]. In Idea-Inspire, actions, state changes, phenomena, inputs and parts; effects; and organs are represented in machine-understandable form using verbs, nouns and adjectives; name of effect; and phrases, respectively. Phrases are also used to represent all the constructs in the human-understandable form. In the catalogue of physical laws and effects, developed as a support for variety and novelty, the constructs of the model are represented using phrases and mathematical equations [8].

In the following paragraph, the existing ontologies are reviewed. Overall-function and sub-functions of engineering systems are represented using a combination of generally valid functions (verbs) and flows (nouns) at higher abstraction levels [9]. Five types of generic functions (change, vary, connect, channel, store) and three kinds of flows (material, energy, signal) are listed.

An ontology of functional concepts of artefacts is proposed in [10]. This ontology consists of four spaces—meta-function, function, behaviour, structure—organised using *is-a* and *part-of* hierarchies. Each of these spaces is further detailed by the granularity size and relationships between granularities of the same size. The entities in meta-function and function spaces are represented using verbs and nouns; entities in behaviour and structure layer are represented using nouns. This ontology is intended to help organise domain knowledge, provide vocabulary for explanation, explain rationales, enable redesigning, and help automatic identification of function structure.

A function-based taxonomy for mechanical design consists of four basic types of functions related to motion, power/matter, control and enclosure [11]. Each type of function is represented using verbs and adjectives. This taxonomy is intended to provide a common function language for designers to avoid semantic difficulties, used as a pedagogical tool, and used as a basis for developing more complete ontologies.

An initial version of an ontology of functions and flows—Functional Basis—is developed for formal function representation, description and comparison of products [12]. The representation for functions and flows is based on the earlier work in [9]. In this ontology, flows are categorized into material, signal or energy, detailed further into three levels; functions are categorized into eight classes with two or more detailed levels.

Generic, domain independent taxonomies of engineering functions and associated flows are developed as part of research on generic product knowledge representation in [13]. The flow taxonomy is divided into material, energy and signal. The list of functions and flows are categorized into different levels based on hierarchy. The authors in [13] argue that various categorizations are possible and each represents a different view. This taxonomy contains about 130 functions and 100 flows.

The initial version of this taxonomy is applied and evolved as a part of a method to identify modular product architectures [14]. The existing three classes of flows and eight classes of functions are specified in greater detail by including more levels to decrease the degree of abstraction. Due to a high degree of research similarity between the efforts in [12–14], in spite of being carried out independently, a reconciled Functional Basis is developed to reconcile and integrate the differences between them, and to provide a more comprehensive ontology, as reported in [4]. The function and flow terms from the earlier work are reconciled and categorized into three levels (class, secondary and tertiary), and a fourth level (correspondents) is also introduced. The different levels here are linked hierarchically through lexical relationships. An engineering ontology is developed as a part of a computational framework for information retrieval and extraction from unstructured engineering documents for the domains of designing and manufacturing [2]. This ontology consists of ten taxonomies—environment, function, device, material, process, value type, measurement unit, property, shape feature, standard—and entities are identified under each taxonomy. Twelve relationships—is-a, has-part, has-function, interface-with/interact-with, has-material, has-process, use-material, has-property, has-measurement, has-value, has-feature, has-standard—are established between the taxonomies. This ontology is linked to an engineering lexicon which contains a list of lexical terms that are used to match with words in documents or queries. Each lexical term is a representation (morphology, abbreviation, acronym or synonym) of an entity in the ontology.

An ontology is developed for describing technical solutions in the domain of packaging industry to retrieve similar solutions [15]. The entities and their relations in this ontology are explained as follows. A *technical solution* realizes a *function* which is executed by the *function owner* and is used in an *industrial sector*. A *function* is represented using an *operation* which has an *operation goal*, and an *object* which owns a *property*. *Operation* and *object* are represented using verb and noun, respectively.

To bridge the gap between the biological and engineering domains, a thesaurus with biologically connotative words related to engineering flows and functions is developed in [5]. This thesaurus is merged with the reconciled Functional Basis. The thesaurus is intended to help engineering designers understand biological phenomena and organisms, search for biological inspiration, model biological systems and promote collaboration between engineering and biology researchers, with potential chances for enhancing creativity and discovery.

From the literature review, it is found that the SAPPhIRE model can describe outcomes from engineering and biology, at various levels of abstraction. However, no explicit efforts are taken to develop a representation for the constructs of the SAPPhIRE model and an ontology based on the model. Therefore, the objective of this research is to develop a representation for the constructs of the model and to use this representation to develop an ontology of engineering design. The first version of the ontology is reported in this paper.

3.3 Research Methodology

The following research methodology is used in this paper. A general representation is developed for the constructs of the SAPPhIRE model by using the model to describe the working of existing systems; four systems are shown in this paper. Based on the developed representation, an ontology consisting of clusters of verbs, nouns, adjectives, adverbs and mathematical equations, is developed. These clusters are identified from Idea-Inspire and the catalogue of physical laws and effects. The developed ontology is evaluated by comparing it against the existing ontologies, to assess its similarities and differences.

3.4 Representation of SAPPhIRE Constructs

In this section, working of four systems is described using the SAPPhIRE model. A horizontal transport of an object is explained in the first example. Consider an object lying on a surface (P). The object is initially at rest (at time, t = 0), has degrees of freedom to move in certain directions and obeys Newtonian conditions (R). If a horizontal acceleration, a (I) is applied for a time interval, t, to this object in a direction in which the object has a degree of freedom, then the acceleration and the organ activate the second equation of motion, $x = 0.5 \times a \times t^2$(E) ($x$: displacement). The effect creates a horizontal sliding (Ph) of the object and this creates a change in the object's horizontal position (S). This state change is interpreted as horizontal transport of object (A) based on a premise that the position changes with respect to a fixed reference frame.

A generation of small voltage is described as follows. Two different wires are joined at their ends (P). The wires have thermoelectric properties (constant Seebeck coefficient, S) and one of the ends is maintained at a constant temperature (R). If a different temperature (I) is applied at the other end of the wires, the input and the organ activate the Seebeck effect, $V = S \times (T_A - T_B)$ (E) (V: voltage induced, S: Seebeck coefficient for the two materials, T_A, T_B: temperature at the two ends of the joined wires). This effect induces a small voltage (Ph), creating a small change in voltage (S), and this can be interpreted as generation of small voltage (A) on a premise that there was no voltage earlier.

In the third example, cooling of a body is explained. Consider a body held in a fluid environment (P). The constant surface area of the body and the heat transfer coefficient between the body and the surrounding fluid medium are the properties (R). Let the body be at a higher temperature; the temperature difference between the body and the surroundings is the input. This input and the organs activate the convective heat transfer effect, $Q = h \times A \times \left(T_b - T_f\right)$ (E) (Q: rate of heat transfer, h: heat transfer coefficient, A: surface area of the body, T_b, T_f: temperatures of the body and the surrounding fluid medium). The effect creates a heat flow from the body to the surrounding fluid medium (Ph), causing a decrease in the heat energy in the body (S). This is interpreted as rapid cooling of the body (A) on an assumption that a decrease in heat energy decreases temperature, causing cooling.

In the fourth example, a quick flow of a liquid is explained. A liquid on a surface is considered (P). The constant dynamic viscosity of the liquid, constant temperature of the liquid and the stationary surface are the properties (R). When a high shear stress (I) is applied on the surface of the liquid, the input and the organs activate the Newton's law of viscosity, $du/dy = \tau/\mu$ (E) (du/dy: velocity profile in a direction normal to shear stress, τ: shear stress, μ: co-efficient of dynamic viscosity). The effect creates a rapid deformation of the liquid profile (Ph), and this creates a large change in the velocity profile (S). This state change is interpreted as a quick flow of the liquid (A) on a premise that a change in the velocity profile causes relative motion among the liquid layers, creating a flow.

Based on the above descriptions, concise descriptions and observations are made in Table 3.1. Actions can be represented using verb (v), noun (n) and adjective/adverb (adj/adv). E.g. "transport object horizontally" (v: transport; n: object; adv: horizontally). Actions can also be represented without adjectives or adverbs. E.g. "cool body". State changes can be represented using verb, noun and adjective/adverb. E.g. "change horizontal position" (v: change; adj: horizontal; n: position). Phenomena can be represented using verb, noun and adjective/adverb. E.g. deform liquid rapidly (v: deform, adv: rapidly, n: liquid). Effects are represented using mathematical equations. E.g. $Q = h \times A \times \left(T_b - T_f\right)$. The development of the catalogue of physical laws and effects using the SAPPhIRE model in [8] showed that effects can be represented using mathematical equations or phrases. Inputs can be represented using verb, noun and adjective/adverb. E.g. "apply high shear-stress on liquid-layer" (v: apply; adj: high; n: shear-stress, liquid-layer). Organs can also be represented using combinations of verb, noun and adjective/adverb. E.g. stationary object (n: object; adj: stationary). Parts are specified by the system, its environment and their relationship. E.g. object lying on ground (system: object; environment: ground; relationship: lying-on). The system and the environment are represented as nouns, and relationships between them using verbs. E.g., lie, join, held, etc. This section shows that the constructs of the SAPPhIRE model can be represented using verbs, nouns, adjectives/adverbs and mathematical equations. Adjectives and adverbs enable a higher degree of specification of the constructs.

3.5 Development of Ontology

To build an ontology using the SAPPhIRE model, verbs, nouns, adjectives, adverbs and mathematical equations are identified from the catalogue of physical laws and effects, and the general verb, noun and adjective clusters of Idea-Inspire.

A category is defined as verbs, nouns or adjectives/adverbs. A cluster of a category is defined as a group of words which have the same meaning. In this research, a cluster is represented by {word1, word2,…}. The identified verbs, nouns and adjectives are grouped into clusters. For example, {join, attach, connect, yoke, bond}, {material, object, matter, substance, body, matter, member, part} and {fast, quick, rapid, instant} are examples of clusters of verb, noun and adjective/adverb, respectively. A word can have many meanings but in this research only the meaning of the word used in the context of description is used to identify the word's cluster. In [16], a portion of the identified clusters of verbs, nouns, adjectives and adverbs is shown; adjectives and adverbs are grouped in the same category.

The following observations are made from the developed clusters. The identified nouns are of two types: objects and physical quantities. The object noun clusters include different kinds of systems and environments. E.g. {material,

Table 3.1 Description of engineering causality using the SAPPhIRE model

	Example 1	Example 2	Example 3	Example 4
A	Transport object horizontally	Generate small voltage	Cool body	Flow liquid quickly
S	Change horizontal position of object	Small increase in voltage	Decrease in heat-energy	Large change in velocity-profile
Ph	Slide object horizontally	Induce small voltage	Flow heat	Deform liquid rapidly
E	$x = u \times t + 0.5 \times a \times t^2$	$V = S \times (T_A - T_B)$	$Q = h \times A \times \left(T_b - T_f\right)$	$du/dy = \tau/\mu$
I	Apply horizontal acceleration on object	Input temperature at a junction	Input temperature-difference between body and fluid-medium	Apply high shear stress on liquid-layer
R	Stationary object; acceleration in direction of dof; Newtonian object	Constant temperature at another-junction; constant Seebeck coefficient; different materials	Constant surface- area; constant heat-transfer-coefficient	Constant dynamic-viscosity; constant temperature of liquid and surface; stationary surface
P	Object lying-on surface	Two wires joined-at their ends	Body held-in fluid environment	Liquid lying-on surface

object, matter, substance, body, matter, member, part, system, medium}, {electric-field}, {magnetic-field}, etc. The physical quantity noun cluster includes different forms of energy, physical variables and properties. E.g. {sound-energy, sonic-energy}, {force, thrust, load}, {modulus-of-elasticity, Young's-modulus}, etc. A "type of" relationship is observed among the noun clusters of objects and physical quantities. E.g. {solid}, {liquid, fluid}, {gas, fluid} and {plasma} are types of {material, object, substance, body, matter, member, part, medium, system}; {weight}, {impact-load}, {gradually-applied-load}, {tensile-load}, {compressive-load} and {shear-load} are types of {force, thrust, load}.

Multiple hierarchies of lexical "type-of" relationships are observed among verb clusters. These hierarchies are represented as different levels (1, 2, 3 and 4) in the cluster; degree of abstraction reduces as levels increase. E.g. {screw, rivet, bolt, latch, clamp, nail, chain, brooch, buckle} is a type of {hold, cling, grasp, grip, lock, fasten, mate, mesh} which is a type of {join, attach, connect, yoke, bond}; {brush}, {cut, divide, chop, slice}, {tear, shear}, {bomb, blast, burst, explode} and {unscrew, uncock, unfold} are types of {dismantle, disengage, detach, disconnect} which is a type of {separate, remove, release, free}. Another kind of "type of" relationship among verb clusters can be developed using assumptions. For example {increase, magnify, gain, amplify, maximize} and {decrease, minimize, reduce, lose, diminish, shorten} are types of {change, vary, modify, transit} on the assumption that change involves some form of increase or decrease. {flush, pump,

Table 3.2 Relationship between SAPPhIRE representation and clusters

	V	N	ADJ/ADV	ME
A	v(1)	n(obj/pq)	adj/adv	–
S	v(change)	n(pq)	adj/adv	–
Ph	v(2, 3, 4)	n(obj/pq)	adj/adv	–
E	–	–	–	me
I	v(apply)	n(pq)	adj/adv	–
R	v(2, 3, 4)	n(pq)	adj/adv	–
P	v(connect)	n(obj)	–	–

v(1): verb-clusters from level 1; v(change): verb-clusters related to {change}; v(2, 3, 4): verb-clusters from levels 2-4; v(apply): verb-clusters related to {apply}; v(connect): verb-clusters related to {connect}; n(obj): noun-clusters under objects; n(pq): noun-clusters under physical quantities; n(obj/pq): noun-clusters under objects or physical quantities; adj/adv: adjective and adverb-clusters; me: mathematical equation clusters

blow} and {launch, catapult} are types of {move, transport, transfer, shift, transmit, travel, go, advance, displace, navigate, pass, propagate} under the assumption that the first two clusters are a means for achieving the last cluster. Verbs from the biological domain are also included under verb clusters. E.g., swim, paddle, hop, leap, camouflage, see, touch, etc.

Verbs to denote actions and phenomena have lexical or assumption-based relationships between them. Verbs to represent state changes are associated with the clusters: {increase, magnify, gain, amplify, maximize}, {decrease, minimize, reduce, lose, diminish, shorten} and {change, vary, modify, transit}. Verbs to describe inputs are associated with the cluster: {give, supply, provide, input, exert, apply, impart, direct}. Verbs to specify parts are associated with the cluster: {join, attach, connect, yoke, bond}. Nouns to denote state changes, inputs and organs belong to the clusters under physical quantities. Nouns to describe organs pertain to the properties in physical quantities. Nouns to represent parts are types of object noun clusters. The relationships between the clusters and the constructs of the SAPPhIRE model are shown in Table 3.2. This table shows the type of cluster used in the representation of the SAPPhIRE constructs. For instance, v(1) in Table 3.2 shows that for representing verb in action, verb clusters from level 1 are used.

3.6 Comparison of Proposed Ontology With Existing Ontologies

In the existing function-based ontologies [4, 5, 9, 12–15], flows are categorized as material, energy or signal, and functions are represented using verbs; in [10] functions are represented using verbs and nouns. In the proposed ontology, physical quantities and object clusters together comprise the energy, signal and material categories of the earlier work. The verb clusters in the proposed ontology resemble the functions from the earlier ontologies. The physical quantities and objects in the noun clusters of the proposed ontology resemble the nouns in the behavior and structure layers in [10]. Among the earlier work, only in [11] and [17] adjectives and adverbs are used to represent functions and flows but the work in [17] does not extend the representation to ontology development. Among the earlier ontologies, only [2] addresses artifact-related properties like viscosity, resistivity, conductivity, etc. In the proposed ontology, properties are included in the noun clusters. The earlier ontologies do not include physical laws and effects as an entity, although catalogues and databases of laws and effects were developed earlier. In the proposed ontology, this gap is addressed using mathematical equations. The earlier ontologies and the one proposed in this paper map only the domain of product-knowledge but the ontology in [2] also maps the domain of process-knowledge in designing and manufacturing. The lexical relationships among verb clusters as observed in the proposed ontology have been identified earlier among flows in [4, 5, 10, 12–15]. However, assumption-based relationships among verb clusters have not been explored earlier. The proposed ontology also uses synonyms in the clusters. The function-based ontologies [4, 5, 10, 12–15] have deliberately left out synonyms but the lexicon in [2] uses synonyms as well as abbreviations and acronyms. The use of biological words in the verb clusters in the proposed ontology enables integration of, and analogies between, biological systems and engineering systems. This has been attempted earlier for functions and flows in [5]. Since the SAPPIRE model provides a richer description of function, behavior and structure by accommodating their different kinds in the form of the constructs of the model, an ontology based on this model can potentially provide richer descriptions. The proposed ontology is a recent work in progress while the ontologies reviewed have evolved over a number of years. So, no comparison is made between the sizes of these ontologies.

3.7 Applications, Future Work and Summary

The following are the potential applications of the proposed ontology. The ontology can interactively assist in creating SAPPhIRE model-descriptions of existing engineered and biological systems from natural language description. The ontology can also be a source of stimuli during ideation. The integration of biological and engineering terms under same clusters can assist engineering designers to search and use biological stimuli, at different levels of abstraction of the SAPPhIRE model. The ontology can help reduce ambiguity and enhance shared understanding in the usage of the terms across engineering design domains.

The current version of the ontology includes the entities from the catalogue of physical laws and effects, and the general clusters of verbs, nouns and adjectives of Idea-Inspire. The clusters in the ontology can be expanded by investigating each entry in the database of Idea-Inspire. The number of clusters and the size of each cluster can also be expanded by adding synonyms and antonyms of the words in the clusters. Fantoni et al. [18] demonstrate the benefits of using synonyms and antonyms as stimuli for generation of novel ideas.

In this paper, a representation for the constructs of the SAPPhIRE model and a first version of an ontology for early phases of engineering design based on the model, are proposed. A general representation for the constructs of the model is developed by describing existing systems using the model. This representation uses verb, noun, adjective, adverb and mathematical equations. Based on this representation, an ontology is developed by identifying the clusters of verbs, nouns and adjectives/adverbs from earlier work. Relationships among clusters and their relations to the constructs of the model are made explicit.

References

1. Gruber TR (1995) Toward principles for the design of ontologies used for knowledge sharing. Intl J Hum Comput Stud 43(5–6):907–928
2. Li Z, Raskin V, Ramani K (2008) Developing engineering ontology for information retrieval. JCISE 8(1):21–33
3. Chakrabarti A, Murdoch T, Wallace K (1995) Towards a glossary of engineering design terms. ICED, Czech Republic 1:185–186
4. Hirtz J, Stone R, Mcadams D, Szykman S, Wood K (2002) A functional basis for engineering design: reconciling and evolving previous efforts. RED 13(2):65–82
5. Nagel J, Stone R, McAdams D (2010) An engineering-to-biology thesaurus for engineering design. IDETC/CIE DTM, Canada
6. Chakrabarti A, Sarkar P, Leelavathamma B, Nataraju BS (2005) A functional representation for aiding biomimetic and artificial inspiration of new ideas. AI EDAM 19(2):113–132
7. Srinivasan V, Chakrabarti A (2010) An integrated model of designing, special issue on knowledge based design. JCISE. doi:10.1115/1.3467011
8. Srinivasan V, Chakrabarti A (2010) Development of a catalogue of physical laws and effects using SAPPhIRE model, ICDC2010, Japan
9. Pahl G, Beitz W (1988) Engineering design: a systematic approach. Springer, Berlin
10. Kitamura Y, Mizoguchi R (1999) An ontology of functional concepts of artifacts, AI Technical Report 99-01, I.S.I.R., Osaka University
11. Kirschman C, Fadel G (1998) Classifying functions for mechanical design. JMD 120(3):475–482
12. Little A, Wood K, McAdams D (1997) Functional analysis: a fundamental empirical study for reverse engineering, benchmarking and redesign. IDETC/CIE DTM, USA
13. Szykman S, Racz J, Sriram R (1999) The representation of function in computer-based design. IDETC/CIE DTM, USA
14. Stone R, Wood K (2000) Development of a functional basis for design. ASME JMD 122(4):359–370
15. Gaag A, Kohn A, Lindemann U (2009) Function-based solution retrieval and semantic search in mechanical engineering. ICED, USA
16. http://tinyurl.com/7h897dr
17. Umeda Y, Ishii M, Yoshioka M, Tomiyama T (1996) Supporting conceptual design based on the function-behavior-state modeller. AI EDAM 10(4):275–288
18. Fantoni G, Taviani C, Santoro R (2007) Design by functional synonyms and antonyms: a structured creative technique based on functional analysis. I Mech E Part B: J Eng Manuf 221(6):673–683

Chapter 4
A Behavioural Design Approach to Improving Engineering Design

Huichao Sun, Rémy Houssin, Mickael Gardoni and Jean Renaud

Abstract The complexity of products is increasing as more and more functions and structures are included in them, leading to higher requirements on better understand of users' behaviour in term of using the product during the early engineering design phase. Although, industry and academia agree that human aspects are important for the success of the product, there are few methods that support the designers concerning these factors in the synthesis part of the design works. This paper covers the multi-trade engineering design, and deals with the development of a behavioural design approach (BDA) to help designers to optimize the product performance in the early design phase through taking into account utilization conditions and requirements. Finally, a software application is in development to support and allow a systematic utilization of the "behavioural design approach" by integrating it into the daily work of the designer.

H. Sun (✉) · R. Houssin · M. Gardoni · J. Renaud
LGECO/INSA Strasbourg, 24 Boulevard de la Victoire, 67084 Strasbourg, France
e-mail: huichao.sun@insa-strasbourg.fr

R. Houssin
e-mail: remy.houssin@insa-strasbourg.fr

M. Gardoni
e-mail: mickael.gardoni@insa-strasbourg.fr

J. Renaud
e-mail: jean.renaud@insa-strasbourg.fr

H. Sun
College of Mechanical Science and Engineering, Jilin University, Jilin, China

R. Houssin
UFR Physique et Ingénierie, Université de Strasbourg, Strasbourg, France

M. Gardoni
ÉTS/LIPPS, Montréal, QC H3C 1K3, Canada

A. Chakrabarti (ed.), *CIRP Design 2012*, DOI: 10.1007/978-1-4471-4507-3_4,

4.1 Introduction

Engineering design processes are often technology-centered and fail to integrate user's behaviour in term of using the product adequately. This problem is encountered along the whole life cycle of a project, and is especially noticeable during the early design phase. These behaviours take place all over the product lifecycle. In order to improve product performance, our research carefully thinks out a piece of research linking user centered and functional engineering design approached into an integrated package, and aims to better integrate product and user behaviour during the early design phase. Designers have been obliged to set aside their dreams of a 100 % machine due to the vital requirement of the user to perform some definite tasks with machines. While machine productivity and utilization conditions are the main reasons for automating production systems, human intervention on such systems remains a critical need and the tasks performed by the user remain poorly defined at the early design stage.

In traditional engineering design, designers normally take into consideration product functions and structures, while users' behaviours in terms of using the system are generally not fully considered during the early design phase. A product's behaviour is studied only from a technical point of view in order to verify its reliability and potential problems in the detailed design phase. However, this behaviour is neither characterised nor studied from a utilization point of view. Nowadays, although designers do increasingly have some understanding of user behaviour, they rarely pay much attention to the behaviour which derives from the structure (how the structure will move to fulfil the function), and behaviour which is fulfilled by the user (how the user will react to the machine).

The objective of this paper is to introduce an approach to help the designer optimize product performance from the early design phase, taking into account utilization conditions and requirements. This approach is based on a Task Model and the fact that the behavioural system (system and end-user) must be studied and defined from the early design phase. We focus on a production system design, and so, to complete the mechanical system design method, we propose a global view of the behavioural design approach (BDA) in the early design stage.

4.2 Industry Observations

Here, two examples observed in real companies are quoted.

1. Example one: Designers want to fulfil the function of transferring a movement. The structure which fulfils this function could be two rollers (as in a gearbox). And the behaviour of this structure is characterised by the two rollers turning in opposite directions. It shows that the user's hand may be jammed between these two rollers when he operates the structure. The problem is as follows: when the user opens the door of this gearbox to be changed to intervene in the system, his

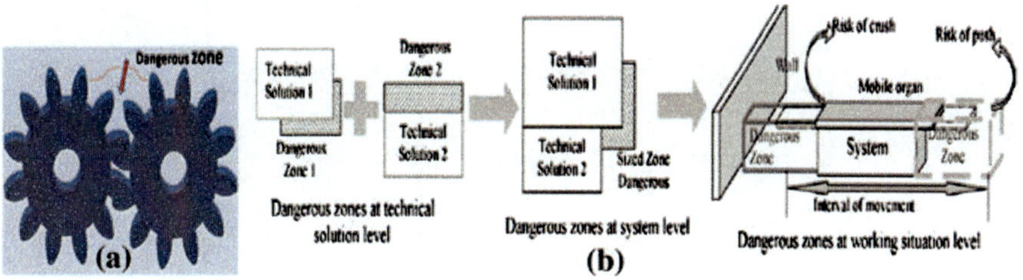

Fig. 4.1 **a** Dangerous zones between two rollers; **b** the dangerous zones sized at the various levels after [13]

hand might be close to the dangerous zone, as shown in Fig. 4.1a. Designers therefore build the cogs into the box. However, they do not research how the component (door) will guarantee the performance of the system (if the user opens the door, the system will be stopped, which decrease the system availability, and its performance). Also, designers do not analyze the tasks when the users intervene on the system.

2. Example two: At a Printer Manufacturer's, in order to respect legislation, designers try to reduce the dangerous phenomenon. Usually the potentially dangerous phenomenon is bounded according to its nature in a zone. The concept of the dangerous zone defines any zone inside and/or around a system in which a person is exposed to a risk of hurt or health damage. In our research, the dangerous zone can be situated at three levels, as shown in Fig. 4.1b.

   - Technical solution level, in which could be engendered a dangerous phenomenon.
   - System level, in which we represent the dimension 'assembly' of the technical solutions. In this level, the dangerous zones defined in the first level may be modified or even disappear.
   - Working situation level, in which a dangerous zone does not exist before system installation but results from its integration on using site.

4.3 Behavioural Design Approach

According to the two real industry examples discussed above, we conclude that there are two aspects to the concept of behaviour. The first involves behaviour which is carried out by the system according to the technical viewpoints. The second involves behaviour which is carried out by the users of the system or the correlative working team.

We herein propose a behavioural design approach to integrate user behaviours and structure behaviours from the early design phase. Behavioural design is a

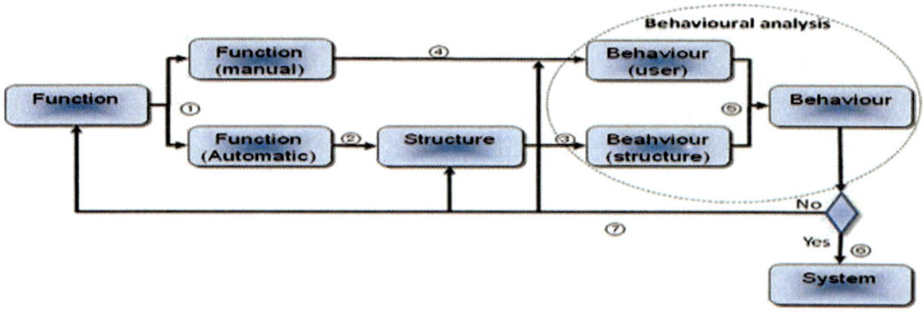

Fig. 4.2 Global view of the behavioural design approach [1]

mechanical system design method based on multidisciplinary knowledge that takes into account, from its preliminary phases, the analysis and the specification of utilization tasks necessary for accomplishing the functions [1]. We propose below the global view of the behavioural design approach. It represents a set of steps linking together the three concepts (function, structure, and behaviour), as shown in Fig. 4.2.

Seven steps are listed here to describe the modelling procedure:

Step 1: According to the functional analysis and requirements specification, we can divide the function into two parts. The first is the automatic function realized by technical solutions; the second is the manual function fulfilled by the user, because of the cost or the difficulties related to automation.
Step 2: According to some methods, such as FAST [2, 3], Axiomatic Design [4], we could find the necessary structure to carry out the function.
Step 3: According to structure decomposition, we can obtain the behaviour of structure tasks (operation, motion, etc.) that the structure has to perform to achieve the function.
Step 4: Manual functions will be carried out by the user. Thus, in this step we propose identifying and studying the tasks performed by the user to fulfil manual functions.
Step 5: To improve the performance of the system, we propose that the interaction between the structure's behaviour as well as the user's be analysed.
Step 6: If the structure's behaviour meets the performance criteria (functionality, productivity, safety, cost, quality, etc.), designers can continue to develop the system.
Step 7: Where the interaction between the user's behaviour and that of the structure does not ensure the needed performance, we have to change user's tasks, or go back to the structure level to modify the structure or go back to the function level to modify or change the function decomposition. We could also change the task performed by the user, which means changing the user's behaviour.

The BDA enters from these design steps to help designers to classify the manual function and automatic function, as shown in Fig. 4.3. And then our software help designer to obtain the structure behaviour which are derived from

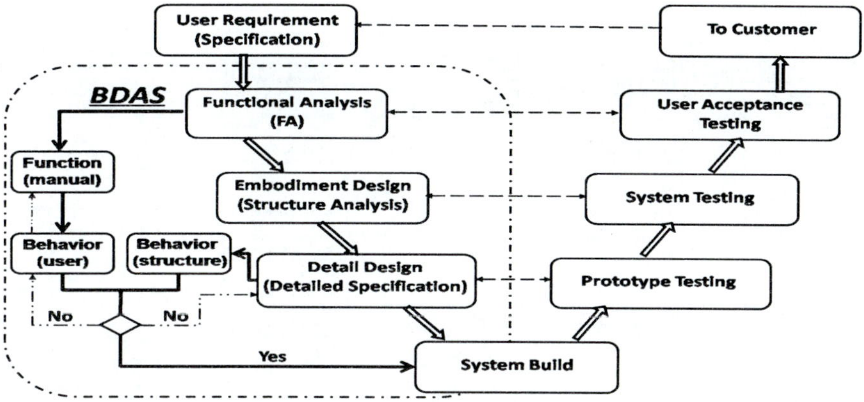

Fig. 4.3 BDA in the engineering design process

embodiment design (structure analysis). The BDA also can analyze the user behaviour which is derived from the manual function. And then in the detail design, it includes specifying the materials, the sizes and so on. All these factors are determined by both the technical solutions and the socio-technical solutions which are influenced by the integration of structure behaviour and user behaviour. The BDA will finally determine the integration of these two types of behaviour which can aid the designer to finish the system build. The BDA can help designers to find the potential dangerous factors before the manufacturing phase which approves the performance and reduce the cost of the redesign.

In order to implement our behavioural design approach at the early design stage, we introduce the task model (to be performed either by the product itself or the user). We adopt the definition proposed by Crandall [5]: the task is a goal to achieve, which involves a determined change of an object's state. In other words, the behaviour of the product presents all the tasks to be performed by these products. Moreover, we take into account those tasks to be performed by the end-user of the product. These tasks take into account the analysis and specification of the utilization conditions; that is to say, maintainability, user's safety [6], reliability and ways of system usage. It is based on a "Task model" integrated into the Function Analysis (FA). During the early phase of the design process, although system models are often primarily limited to geometrical aspects representing product-dimensioning and the associated functional surface qualities, they hardly or never take into account their behaviour and that of the future end-user and their interaction. In the following section, we present our task model which will be integrated in our behavioural design approach.

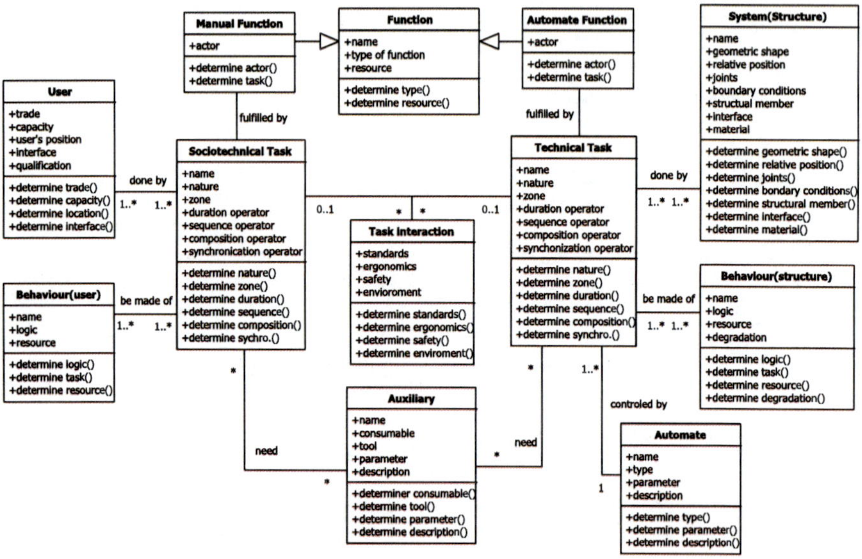

Fig. 4.4 Global view of task model

4.4 Task Model

In engineering design, the task is not very well defined and is used differently in different contexts. In our research, it encompasses two concepts. Firstly, it is used to refer to purposeful activities performed by users; such activities may involve a general class of activities, or a specific type of activity. Secondly, it refers to activities performed by one structure or a series of structures. Tasks arise from the relevance between behaviours delivered by a design system and a principle used in the system. We use a tier of levels to explain the relationship between behaviour and task. The highest level is behaviour, which is composed of a concrete task; and the concrete task is composed of an action; and the action is composed of the concrete operations shown by the image.

We present our contribution and the conceptual foundations and structure of the task model exemplified in Fig. 4.4 using the Unified Model Language (UML) method [7].

In this figure, we present details pertaining only to common concepts; other concepts had been detailed in [8]. This model supports most of the parameters linked to the environment and use parameters. Identified concepts presented in this model are the results of analysing real situations. In this model, we gather the parameters of use conditions from a socio-technical viewpoint (cognitive, social, etc.). To make the integration of our behavioural design approach easier, we used some concepts already used in FA [9].

4.5 Implementation of Behavioural Design Approach Software

Moreover, there is no computer-aided software that permits combining all these aspects which we discussed above into the design. We develop the software based on the behavioural design approach which integrates with task model and knowledge bases. The software aims to illustrate the practicality, applicability and validity of behavioural design approach. And it consists of four different knowledge bases, such as ontology base, structure base, task base and behaviour base. They store all the previously data capitalized by the designer and provide communication between designers and users in order to improve product performance.

Comparing with common engineering design system, our system first carefully thinks out piece of research linking user cantered and functional engineering design approaches into an integrated package. Designers can input the new user's task through fulfilling various manual tasks to improve the product performance. In order to appropriately the use knowledge for supporting BDA system, it is critical to identify the context of user's task and structure's task and to verify adaptive usage of certain knowledge [10]. After identify the context of the tasks, the designers are encouraged to evaluate and comment values of the task interaction according to the results of their interaction. Simultaneously, the task and knowledge usage are automatically recorded in the knowledge base which improves the traceability and trustworthiness of the knowledge elements of the BDA system. In addition to the above functional requirements of the BDA system, other aided systems are also considered, such as PLM system and CAD system, which are compatible with and collaborate with existing information infrastructure.

Based on the behavioural design model, the BDA software framework is designed by adopting the intelligent agent [11, 12] as shown in Fig. 4.5.

On the product design platform, the user requirements are analyzed into the design targets that are assigned to the collaborative working team. The designers collaborate with each other to fulfill the assigned design targets. Through a user machine interface with the support of CAD and BDA system, each designer uses knowledge and task information. The information agent provides BDA supporting programs such as accessing, searching, compiling, task comparisons and visualizing structures which derive from the CAD system. It follows the designer's operations and communications with other agents in the agent integration with respect to the agent requirements. The knowledge base accesses the agent information for storing and iterating relevant knowledge.

This software assists the designer to take into account systematic design standards, and to respect safety and ergonomics legislations. At the first step, designer calls the structure from CAD (for example: Solid works) system; the Information Agents receive the CAD files and then transfer the information into structure base; BDA system analyzes obtained structure and divides them into existed structures and new structures; Information Agents receive the feedback of structure's

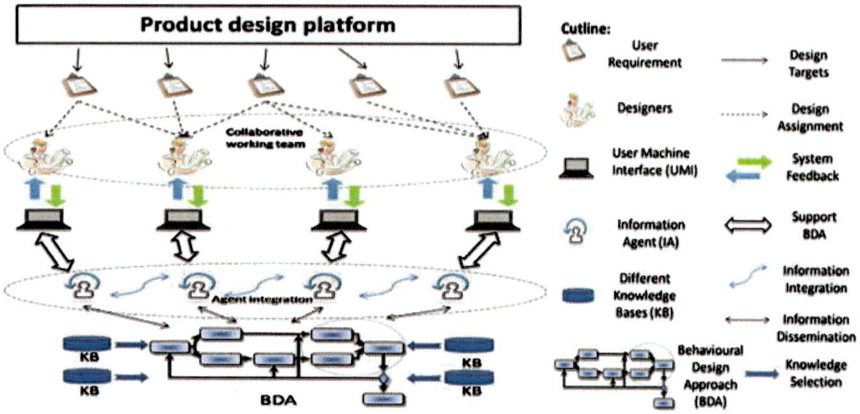

Fig. 4.5 Framework of distributed BDA software for engineering design

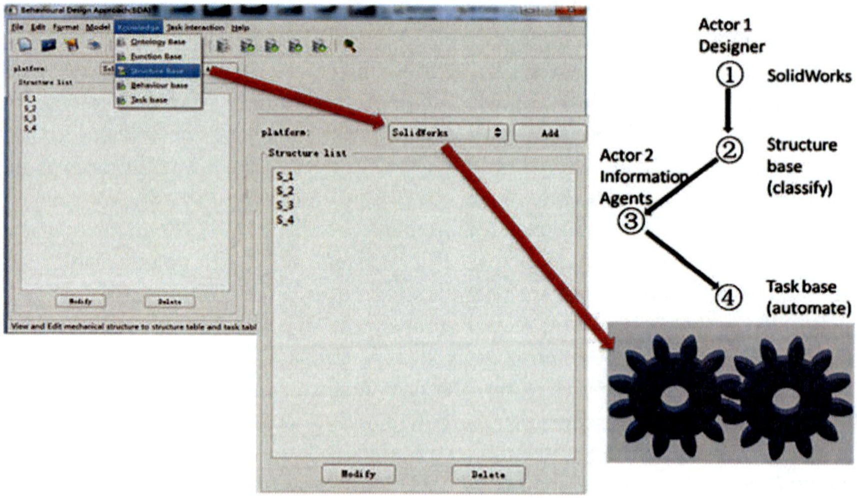

Fig. 4.6 Analysis of structure derived from the solid works (structure task)

analysis and realize an analogy between Structure Base and Technical Task Base (structure), as shown in Fig. 4.6.

In the next step, designer inputs manual functions which derive from the FA; Information Agents receive the information and then transfer them into Function Base; BDA system make an analogy between Function Base (manual) and Social Task Base (user); the task are divided into new and existed tasks; as shown in Fig. 4.7.

After that, Information Agents receive the two results: analysis of Structure Base (structure task) and analysis of Function Base (user task); and then the evaluation step occurs in the Task comparison; as shown in Fig. 4.8.

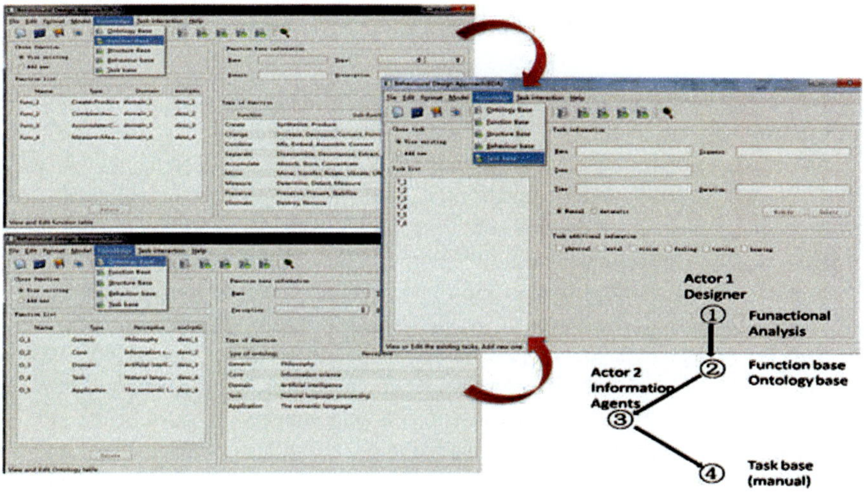

Fig. 4.7 Analysis of manual function (user task)

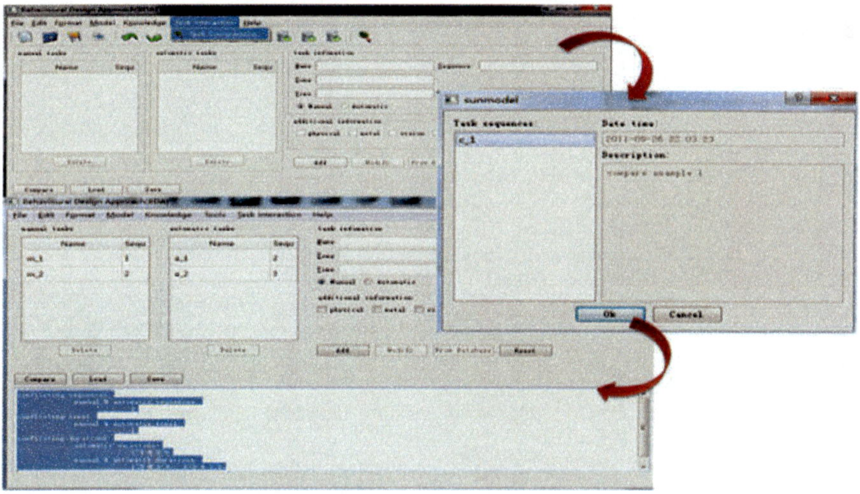

Fig. 4.8 The evaluation of task comparison

Finally, the Information Agents receive the feedback of result of Task Comparison; if the result is not acceptable, the software will propose designer to modify the solution, task, structure, etc. in order to eliminate or decrease the dangerous phenomenon and engendered hazard; if all these are not possible, the software proposes the new behavioural analysis; a step of new behavioural analysis to assure the modifications cannot lead any other accessibility or performance problem.

4.6 Conclusion and Perceptive

In this paper, we present the behavioural design approach which integrates the utilization tasks of products, and more particularly user behaviour from the early design phase. We propose a conceptual model to help the designer define each required task to fulfil system functions in order to improve its performance. We noticed that the requirements of human tasks are considered as the opportunities instead of the constraints. Thus, we propose that functional specifications are completed by behavioural ones. In the last section, an application is proposed to exemplify the practical aspects of the behavioural design approach.

The BDA system is developed based on the model to support and allow a systematic utilization of the behavioural design approach by integrating it into the daily work of the designer. Although considerable endeavours have been made in this study to improve product performance in the early design phase, there are still many problems in further research. Because of the time span of evaluation, only the simple evaluation is presented here. Further application and case studies are essential to totally establish the evaluation system and to quantify the more detailed evaluation of our approach.

References

1. Sun H, Houssin R et al. (2010) La conception comportementale : une nouvelle approche pour améliorer la performance de produit. 17ème Congrès de Maîtrise des Risques et de Sûreté de Fonctionnement Gentilly, France
2. Conway JB (1990) A course in functional analysis. Springer, New York
3. Wixson JR (1999) Function analysis and decomposition using function analysis systems technique, Citeseer
4. Suh NP (2001) Axiomatic design. Oxford university press, Oxford
5. Crandall B, Klein GA et al. (2006) Working minds: a practitioner's guide to cognitive task analysis. The MIT Press, Cambridge
6. Coulibaly A, Houssin R et al (2008) Maintainability and safety indicators at design stage for mechanical products. Comput Ind 59(5):438–449
7. Booch G, Rumbaugh J et al. (2005) Unified modeling language user guide, the (addison-wesley object technology series). Addison-Wesley Professional, Boston
8. Sun H, Houssin R et al. (2011) Improving product performance with integration of using tasks during the design phase: a behavioural design approach. In: International conference on industrial engineering and systems management, METZ FRANCE
9. Blanchard BS (2004) System engineering management. Wiley, New York
10. XU J, Houssin R, Bernard A, Caillaud E. Systemic modelling of knowledge for innovation in design. CIRP J Manuf Sci Technol. http://www.sciencedirect.com/science/article/pii/S1755581712000545
11. Franklin S, Graesser A (1997) Is it an agent, or just a program?: a taxonomy for autonomous agents. Intelligent agents III agent theories, architectures, and languages, pp 21–35
12. Russell SJ, Norvig P (2010). Artificial intelligence: a modern approach. Prentice hall, Englewood Cliffs
13. Houssin R, Bernard A et al (2006) Information system based on a working situation model for a new design approach in concurrent engineering. J Eng Des 17(1):35–54

Chapter 5
Systematic Sustainable Design in Architecture and the Need to Mimic Nature

Abraham George and Susan Abraham

Abstract Need for finding long-term sustainable design and development solutions; *which meets the needs of the present without compromising the ability of future generations to meet their own needs that* warrant continuing human existence and well-being is far more compelling in these days. Sustainable architecture reiterates the fact that humans receive what they need from the universe. Sustainable design is a positive response to awareness that nature provides, not a prescriptive formula for survival. A conceptual framework and strategies for sustainable coexistence, based on the three fundamental concepts; *objectives, means, and end*, is attempted. Limitations of the traditional design models are brought-out and the need to search for mimicking nature is established. Biomimicing thought process, is illustrated and application of Biomimicing in architectural design is illustrated along with a list of recommendations.

5.1 Concepts of Sustainability

Sustainable design and development is defined as; "design and development that meets the needs of the present without compromising the ability of future generations to meet their own needs" [1]. The need for finding long-term solutions that

A. George (✉)
Department of Architecture and Regional Planning, Indian Institute of Technology, Kharagpur, West Bengal 721302, India
e-mail: profageorge@gmail.com

S. Abraham
Research and Dam Safety Organization, Kerala State Electricity Board, Pallom, Kottayam, Kerala 686007, India
e-mail: sabraham007@gmail.com

A. Chakrabarti (ed.), *CIRP Design 2012*, DOI: 10.1007/978-1-4471-4507-3_5,

warrant continuing human existence and well-being is far more compelling in these days of depleting resources and catastrophic climate change, than in the former days [2]. Though, it is Green and sustainable that are the catch words of design, in the contemporary helpless climatic scenario which is worsening over the passage of time, these have also created ample ambivalence and confusion. Thanks to the urgency of the situation, debates on the terms green, sustainable or ecological architecture has become almost meaningless, but the kernel matter lies with addressing the needs of harmonic and ecologically sustaining design and development that would assure the future of the earth with all its myriad of living and nonliving systems. During a building's actualization process, its construction affects the local and global environments by way of interconnected human activities and natural processes. Needless to state that manufacture, transport and procurement of materials at various stages never fail to leave their impact on the global environment. Completed buildings require resources and energy in various forms for their useful performance. These, in turn, give rise to pollution and add to the mammoth problem of waste and environmental degradation which inflicts long-term impact on the environment. For instance, fuels and water used by its inhabitants produce toxic gases and sewage. Similarly, the process of extracting, refining, and transporting all the resources used in building operation and maintenance also has negative impacts on environment. Global ecosystem is essentially made up of three group namely inorganic substances, living organisms, and human beings [3]. Built forms contribute to the compounded impact of architecture on global ecosystems. It is therefore, important to study the impact of built forms on the totality of the environment, throughout various stages. Developments should be facilitated taking into consideration the entirety of the systems*; resource, energy and transport etc. and the myriad of population of the Nation,* along with the specific characteristics of the region being developed. Though, it is *green* and *sustainable* that are the catch words, an examination of the meaning of 'sustainable' is required to avoid the avoidable confusion these words tend to generate, knowingly or otherwise [4]. Sustainable architecture implies that we receive what we need, from the nature. Sustainable architecture, then, is *a farsighted positive response to awareness that everything we need is received from nature*, not a prescriptive formula just for our survival [5]. The goal of sustainable design is to find architectural solutions that warrant the well-being and coexistence of constituent groups [6]. Therefore, a conceptual approach to framework is to be developed in order to meet the goal of well-being and coexistence in an effort to attain sustainability. Three fundamental concepts of the framework proposed are; *Objectives, Strategies and Achievement*. These relate to the environmental responsibilities, creating environmental awareness, explaining the building ecosystem and designing sustainable built forms for future [3].

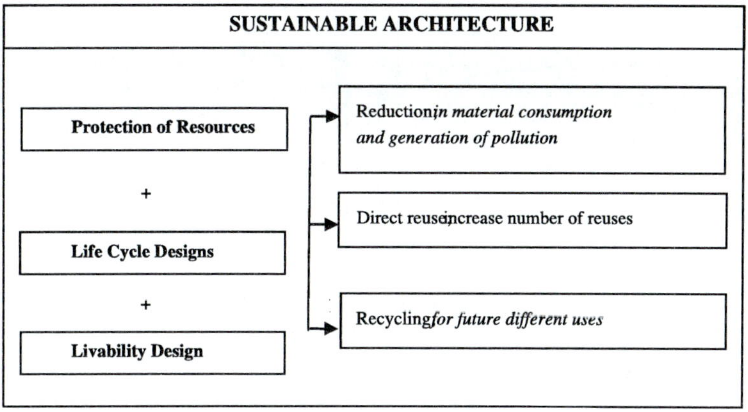

Fig. 5.1 Framework of concepts and strategies for sustainable architecture

5.2 Why Sustainable Architecture

Sustainable architecture aims at the *Protection of Resources—(PR)*, is the primary response to the awareness which influences all the following stages. *Life Cycle Design—(LCD)* and *Livability Design—(LD)* facilitate healthy habitation for humans. Protection of natural resources is proposed at the inception stage of building process, to be achieved through the reduction and reuse; *direct reuse or recycling,* of the physical resources involved [3]. While *Life Cycle Designs* provide a methodology for analyzing the building process and its impact on the environment in an effort to decide on the effectiveness of designer's choices, *Livability Design* focuses on the interactions between human beings and the natural environment [7] (Fig. 5.1).

A sympathetic attitude from architects and an understanding of the above objectives which embodies a unique set of intentions is important to develop a more thorough understanding of the designer's positive interaction with the environment. The genesis of a project leading to its geographic location is one of the extremely important phases in the effort to reduce the load on infrastructure and consumption of resources and generation of pollution. Locating a sustainable built form far away from the supporting or depending facilities would generate unnecessary traffic perils associated with commutation which is avoidable. Moreover, developing countries like India has the major part of 60 % of its population living in rural areas. Appropriately rated economic magnets like Special Economic Zone—(SEZ), IT parks, industries or the like may be effectively used in order to achieve balanced development in rural areas which would retard the unhealthy migration to already congested urban areas. Moreover, as a strategy, develop automobile free, *'walk to work'* rural or peri-urban communities, self-sufficient in water and energy requirements, equipped with appropriate waste management systems. Such sustainable communities may be the hubs that are effectively connected to others by means of high speed, ecologically friendly mass transit systems.

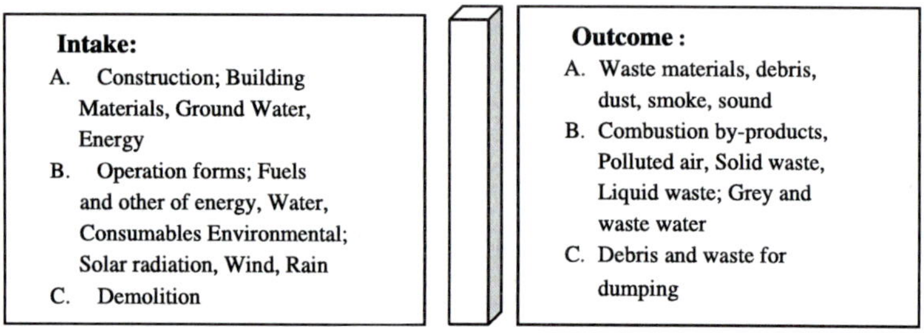

Fig. 5.2 The input and output streams of resource flow

5.3 Protection of Resources

It is the responsibility of an architect to reduce the use of nonrenewable resources in the construction and operation process of buildings in an effort to protect the resources and to preserve these for the future generations [8]. Natural and manufactured resources, as is seen, are in a continuous flow in and out of any building which begins with the production of building materials, continues throughout the building's effective life sustaining intended functions. There are two essential streams of resource flow as shown in Fig. 5.2.

Intake resources flow into buildings as input to the building ecosystem while *Outcome* is resources that flow out of the building to the ecosystem [6]. Strategies of protection of resources are multi thronged and must be attempted through ***Energy conservation, Water conservation and Material conservation.*** It is worthwhile considering appropriate legislation for levying *green* tax on buildings that exceeds the material and energy limits prescribed. The tax is to be stipulated for unit area of foot print on site. Further, those designs that comply with the norms shall be encouraged.

5.4 Life Cycle Design

Figure 5.3 illustrates the linear life cycle process of a building which consists of three major stages [9]. Each of these stages calls for sustainable approaches and strategies in an effort to achieve the goals envisioned.

Life cycle design (LCD) is not prescriptive, but suggestive in nature. However, LCD can contribute information and facilitate the effective decision making process. This approach accounts for the environmental consequences during the entire life cycle of construction materials; *from procurement to return to nature* [10]. Life cycle of a building can be brought into three phases namely *Pre-building, Building, and Post-building*. The phases can be developed into LCD means that focus on minimizing the environmental impact of a building. The procurement of building materials impacts the environment in various ways; *unscrupulous felling*

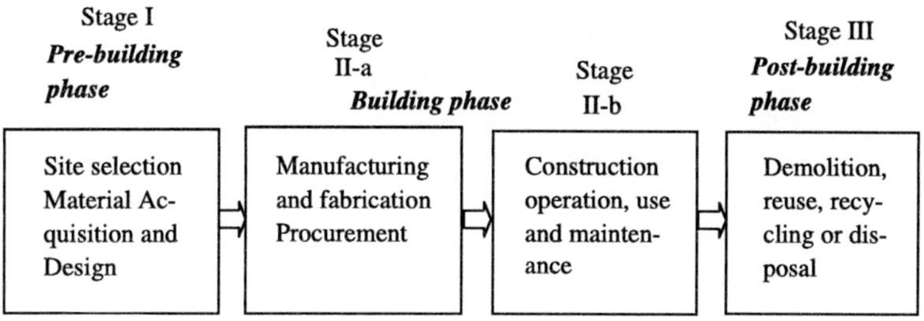

Fig. 5.3 The life cycle process of building

of trees leads to deforestation, mining mineral resources disturbs the nature and creates environmental pollution or the like. Building phase refers to the stage in the life cycle of a building when it is physically constructed and operated. In the sustainable design, the construction and operation processes shall embrace means to reduce environmental impact, resource consumption and sick-building syndrome. Post-building phase refers to the stage which begins when the useful life of a building has ended and its building materials are turned into resources for other buildings or waste to be recycled or returned to nature. The strategy is to reduce construction waste by recycling and reusing buildings and building materials.

Further,

Lowered consumption in building	Lowered production of waste, Need for production
Optimization of space requirement	Material Requirement, Lowered operational costs

5.5 Livable Design

Livable Design refers to the livability of all constituent spaces in built forms and spaces that form various groups in the global ecosystem [11]. It is concerned about the healthy coexistence among buildings, their environment and their respective occupants. Its broad objectives are intended to preserve the elements of the ecosystems in an effort to facilitate human survival.

Built forms are intended to provide safe and healthy environments that are comfortable for their occupants which in turn enhance their satisfaction and productivity. Hence, Livable Designs could be evaluated under *Generation and sustenance of natural conditions, Creation of satisfactory urban design and site, Generation of human comfort.* Needless to state that sustainable design shall offer human comfort; both internal and external, in the interest of individuals and the nation at large.

5.6 Need for Alternative Concepts of Sustainability

Traditional design models have limitations to generate unique built forms that meet the requirements of sustainable designs. Mostly, the inherent inability in the traditional models is manifest by way of stereotyped thinking which leads to no atypical designs. It is worthwhile to ponder the words of Albert Einstein "*We cannot solve the problems by the same thinking that created them*". Such designs, due to the lack of novelty in conceptualization, offer very little scope for optimization and lowering of energy and resource consumption. This problem of stereotype could be resolved creatively by a search for alternate models in designs. Nature, at this juncture, presents itself with harmonious designs that are sustainable, self-supporting and self organizing. Solutions that are found in the harmonious natural systems are always in evolution, perfecting and adapting to their contexts. Thus, what is seen today has been working over billions of years for evolving a reliable and sustainable model. Adoption of these evolved models in human designs would facilitate the making of future systems better sustainable; environmentally, ecologically and economically. Hence, Biomimicing reveals itself as a fine model to follow in the generation of alternative sustainable design solutions.

5.7 Biomimicry

Biomimicry is a new science that studies nature's best ideas and principles and imitates these designs and processes to solve human problems. In other words Biomimicry leads to *innovations inspired by nature* (Biomimicry Institute n.d.). Though some of nature's basic configurations and designs can be copied, most ideas from nature are best adapted when they serve as *inspiration for human-made designs and productions* [12]. Adaptation of natural systems and organisms has facilitated better understanding of related phenomena and principles in the design of novel designs, devices with better features and capability. For example, the cell-based structure that is the building block of biological systems has the ability to grow with fault-tolerance and self-repair. With the adaptation of Biomimic structures based on nano-technologies, such designs and devices are possible in human-made designs, but not with traditional materials and processes. On a different level, there exists the evident, inspirational link between the design of tongs and bird's beaks. The same inspiration is evident in the foldable hand-held fan design and the peacock feather display; *a magnificent attempt to impress the female.* One of the important features of nature is its evolution by responding to the system needs and generating solutions that work. Nature remains in an open, dynamic system establishing balance and continuous refinement in all its productions. Each of the successful natural creation that passes to the following generation has to withstand the test of survival, establishing the best fit for the following generation. Nature's laboratory through evolution generates information that is coded in genes and transferred to the following generation through the process of self replication. Nature thus, is perfecting models worth copying and inspiring novel engineering methods, processes, materials, algorithms, and designs. In a similar way production of designs and the elements and their organization in the design produced shall remain in a continuum of evolutionary changes, permitting adaptation and attainment of the best fit. Mimicking of nature may be done at various levels beginning with the full and complete appearance of the natural system to its every system detail in part or full. On the other extreme, natural models are interpreted and transformed in the making of human-made designs. Mimicking of life-systems demands the full capacity and intelligence of humans.

5.8 Biomimic Thought Process

- *Identify the real challenge*
  - What do you want to "do" (not make)? Be open, rational and creative. Learn inquisitiveness from kitten!
- *Interpret*
  - Identify the functions/purpose
  - How nature does perform function?
- *Discover Nature's Genius*
  - Go for a walk outside and observe and brainstorm. Look for the precious stones!
- *Abstract*
  - What patterns and principles work for your problem? Be creative and prudent.
- *Emulate or Imitate*
  - Play and design
  - Brainstorm and converse
- *Evaluate*
  - Revaluate and Re-Imagine the design deeper and rigorous each time with holistic thinking in order to solve the entire problem. It might be necessary to redefine to solve the problem as a whole, not in parts.

5.9 Formula for Sustainable Future

Intellectual Capital + Nature's Genius = Innovative, Sustainable solutions. If we are limited, it is by our own dreams! Therefore, dream great and be a B_2; *Beautiful Biomimic*.

5.10 Application

Biomimic thought process is adopted in the design studio of the final year Bachelor of Architecture, IIT Kharagpur (2011) under the guidance of the author, in creating sustainable design for a project on school of engineering at the Campus. The students explored various biological examples like honeybee, acorn, anthill, and palm tree. Students were asked to study the natural system along with the specific subsystem details, how these worked with an intention to *Identify, Interpret, and discover* in order to *Abstract and Emulate*. One such design had its inspiration from the palm leaf as a subsystem, how it's support and circulatory systems in the foliage facilitated by the form of the leaf, could lead to an interesting architectural product and palm leaf is used to articulate the design pattern, by its function and form. The design and modelling are done using *ArchiCAD* and *Google Sketchup*.

Aerodynamic behaviour of the flexible palm leaves is interesting. They get blown into a streamlined to any strong wind, which results in wind resistance and the needed strength to withstand the wind. This concept is used in the design of the overhead palm leaf structure of the proposed built form. The fronds; *divisions of the leaf*, of palms are connected to the exposed spine by means of fixed diagonal supports. Fronds of the palm leaf shaped structure have solar panels fixed on them. These panels can rotate along their axial supports, thereby, changing its direction with respect to the sun and wind. They are designed to control illumination, glare and ventilation in each enclosed space within. These panels, by rotating more than 90° from the normal, can also act as shading devices and the building is protected from direct solar radiation. Also they provide an ideal method for wind to permeate, reducing the wind forces excerted on the building. One of the two other palm leaf structures acts as a cover to the Open-air theatre, controlling the illumination and glare during performances. These fronds can also vary the aesthetic appeal of the entire setting, if required, for unique performances. Another one covers the informal meeting places provided in the site. Thus the palm leaf structures divide the site into functional zones. Adult palm trees have a system of concentric rings of fluid conduits that not only show an enormous resistance to flexion, but also behave like 'air mattresses', providing defense against fire; *preventing fire from destroying the entire tree*. This concept is appropriated by providing water channels, which flow through the building structure preventing the destruction of the entire building, in case of fire. The channels would be covered suitably by structural glass and are walkable. These water channels along with the

Fig. 5.4 Exterior views of the School of Engineering, Thulasi B. Arch 2011

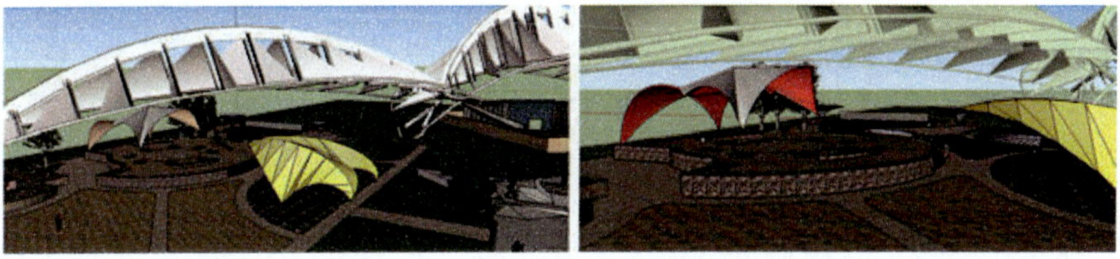

Fig. 5.5 Interior views, School of Engineering, Thulasi B. Arch 2011

landscaping inside the building provide an ideal setting and an inspirational atmosphere of a *'closer to nature'* academic environment. Further, the presence of these water channels regulates the circulation pattern and humidity levels inside the building (Figs. 5.4, 5.5).

5.11 Conclusion

Concepts and strategies aim to generate sustainable architecture which is the need of the hour and vital to the existence of life forms on earth. The scientific and technical complexity involved along with the commercial and political interests amalgamated to its objectives make 'defining sustainable architecture' a delicate issue. These approaches lack a real concern for the unique survival and special needs of the context to which such prescriptions are applied. Further, it is important to understand the limitation of traditional design models and look for better alternatives including Biomimicry. Space and area optimization warrant an effective strategy in the achievement of reduction of the need which lowers the generation of pollution. Architects have a greater role to play, therefore, at this important issue. Engineers have an equally important role in the design and selection of optimized systems; *mechanical, electrical, transport and disposal*, which are vital for the efficient performance of the built forms. It is important to consider necessary legislation to impose green tax on buildings that exceed the prescribed limits.

Education in architecture and allied fields shall impart rigor and soundness to design professionals engaged in the design, production, operation and reuse of built forms [11]. It is also important to accept the relationships and interconnectedness within the ecosystem and the built forms that designers develop. It is the vital responsibility of architects to find creative design solutions that facilitate well-being and harmonious coexistence of organic and inorganic groups.

There is inherent inability in the traditional models and stereotyped thinking leads to no atypical designs. The problem of stereotype could be resolved creatively by alternate models wherein mimicking natural systems holds great potential. Natural systems are always in evolution, perfecting and adapting to their contexts over billions of years. Adopting the natural models through Biomimicing facilitate the making of future systems better sustainable. Further, the Biomimic thought process is illustrated in evolving a formula for sustainable future. The Biomimic thought process is validated in a built form design thus illustrated its creative and environmental sustainable potentials.

References

1. Development WC (1987) Our common future. Oxford University Press, London
2. Papanek V (1995) The green imperative ecology and ethics in design and architecture. Thames and Hudson, Singapore
3. Celeb G (2003) Environmental discourse and conceptual framework for sustainable architecture. GÜ-J Sci 16(1):205–216
4. Andrew S (1992) Sourcebook of sustainable design. Wiley, New York
5. Bergen SB (1997) Ecological engineering: design based on ecological principles. ASAE Annual International Meeting, USA
6. Kim J-JR (2000) Sustainable architecture module: introduction to sustainable design
7. American Institute of Architects (1996) Environmental resource guide. Wiley, New York
8. Kua HL (2002) Demonstration intelligent building—a methodology for the promotion of total sustainability in the built environment. Build Environ 37(3):231–240
9. Curran MA (1996) Environmental life cycle assessment. McGraw-Hill, New York
10. Burall P (1996) Product development and environment. The Design Council, London
11. Celebi GA (2001) Architectural responsibilities within the context of sustainability. In: Proceedings of Trabzon: livable environments & architecture international congress. 4–7 July 2001
12. Bar-Cohen Y (2006) BIOMIMETICS-biologically inspired technologies. 6000 Broken Sound Parkway NW, Suite 300, Boca Raton, FL 33487-2742: CRC Press, Taylor & Francis Group

Chapter 6
Computational Models of Tacit Knowledge

Madan Dabbeeru and Amitabha Mukerjee

Abstract When an expert designer uses a term such as "interference fit" or "H7-r6", they effortlessly invoke a rich set of associations across a wide range of experience. While at one level, the meaning of a term such as H7 is formally specified, many of these associations are implicit and hard to characterize formally. The explicit concepts build on layers of implicit abstraction; e.g. the concept of fit would be difficult to achieve without the commonsense notion of "tight", discriminated by human infants from five month onwards. We propose that such ubiquitous expertise may be acquired as functionally relevant low-dimensional chunks in an experiential space, which are then stabilized through language. The technical terms of design build on these everyday concepts by mechanisms such as extension or narrowing of their semantics. We suggest a two-stage computational analog of this process: (a) the baby designer stage learns elementary concepts as tacit patterns on an input space; and (b) the novice designer stage relates these early concepts to explicitly defined design terms to arrive at a grounded semantics for the new symbols. We illustrate the process through the development of concepts such as interference fit.

> Now we see tacit knowledge opposed to explicit knowledge; but these two are not sharply divided. While tacit knowledge can be possessed by itself, explicit knowledge must rely on being tacitly understood and applied. Hence all knowledge is either tacit or rooted in tacit knowledge. A wholly explicit knowledge is unthinkable—Michael Polanyi [1].

M. Dabbeeru (✉) · A. Mukerjee
Indian Institute of Technology Kanpur, Kanpur, India
e-mail: madan.dabbeeru@gmail.com

A. Mukerjee
e-mail: amit@cse.iitk.ac.in

A. Chakrabarti (ed.), *CIRP Design 2012*, DOI: 10.1007/978-1-4471-4507-3_6,

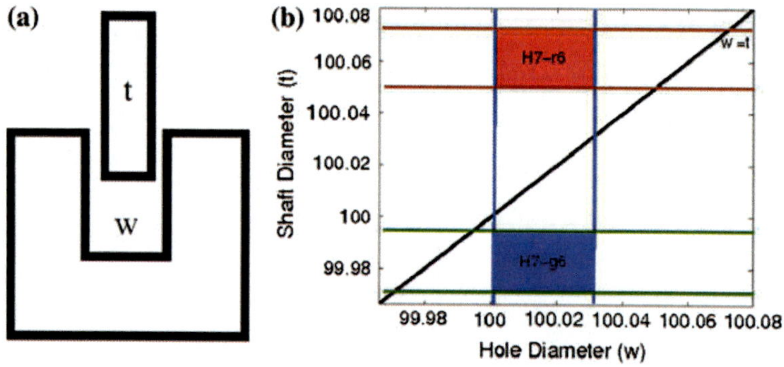

Fig. 6.1 H7 is a tolerance interval on the hole diameter w, and g6/r6 on the shaft diameter t. [H7-g6] represents a running fit (w–t > 0) and [H7-r6] is an interference fit (w–t < 0), **a** Peg-in-hole assembly **b** Tolerance intervals

6.1 Design Concepts Build on Everyday Experiences

Consider a designers' understanding of the terms "running fit" and "interference fit" in mechanical assemblies. A running fit, where an object slides smoothly with very little clearance, may have a tolerance of H7-g6 (e.g. in a bush bearing) [2]. On the other hand an interference fit that can transmit force (e.g. a gear on a shaft), may have a tolerance of H7-r6. Terms such as "H7" are formally defined as a band of tolerance for the hole, and "g6" or "r6", as a tolerance range on the shaft. A formal inference process can then combine these two definitions to determine that the combination "H7-g6" indicates an intersection of these two constraints, corresponding to a rectangular region in the space of hole and shaft diameters (Fig. 6.1).

However, for the experienced mechanical engineer, the meaning of these terms extends far beyond this formal definition—it includes a wide set of associated concepts and constraints—how they resist force or permit motions, the feel of trying to rotate a shaft in different fits and that the difficulty increases with tighter fits, the differing sounds a shaft makes when the fit changes due to component wear, and so on. Even if one could give a name for every sound and every sensory feel, it would not be possible to write down all rules related to all associations; thus much of this knowledge is tacit. Here, we use the term "tacit knowledge" knowledge of this kind, which we cannot explicate.

It has been suggested that tacit knowledge works together with the explicit, and may be difficult to separate out cleanly [1]. As one gains experience of a domain, one learns the stable patterns that lead to functional distinctions—these have been called chunks [3]. The structure of these chunks may be hidden even from one who knows it; it can be difficult to model. A novice relies on explicit knowledge-alone, and is unable to handle the high-dimensionality of the sensory data, and works falteringly. For an expert the scene resolves itself in a small number of chunks, each of which may encode patterns of considerable complexity. Thus, the dimensionality of the decision space is significantly reduced.

Despite the increasing awareness of the tacit component of design knowledge [4–6], most design theory formulations rely on quasi-logical systems that define one symbol only in terms of other symbols. But where does this process end, where can we relate the rules to an actual design? In practical implementations, this last link to real data is not defined formally, but is left to the implementation. This is why a common but frustrating experience in working with large symbolic design systems is that they seem to be working well for a suite of test problems, but suddenly fail on an unexpected situation. In the symbolic reasoning literature, the difficulty in specifying all possible contingencies is called the frame problem [7]. A more general view of this process is in terms of Kripke-Wittgenstein's "paradox of rules", which suggests that rules regress from one to the other, and in the end, fail to lead to action [8].

6.1.1 From Tacit to Explicit

We highlight the importance of tacit knowledge by considering an engineering student who somehow has not experienced assemblies, and is unable to distinguish tight from loose fits. Now, the professor, when encountering such a student, may feel that it would be very difficult for her to ever learn the concepts of "fit" competently. Many such prerequisites are assumed to be present even in the most novice apprentice; these are part of what has been called *ubiquitous expertise* [9]—a large set of concepts known to all members of a society.

Concepts relating to containment, such as the constraint that a large object cannot enter a small hole, are attested in human infants by the age of 3 months, and the tight-loose distinction by 5 months [10]. The tacit knowledge underlying such distinctions form the core of the eventual design concepts of fits and tolerances.

The importance of tacit knowledge in design has been widely acknowledged [4, 5, 11]. However, it has not been clear how such knowledge can be captured and related to the design process. Constructive approaches based on tacit knowledge have been criticized for being vague and not operationalizable [6]. It is our intention here to suggest a computational process for capturing tacit knowledge, which would then provide a mechanism for operationalizing this process. The approach extends our earlier notion of a *baby designer* to one that is able to deal with design symbols, whom we call a *novice designer*:

1. *baby designer:* discovers that certain patterns in its decision space result in consequences that have functional relevance (e.g. will an object enter the mouth or not?) This initial concept is coded as a low-dimensional chunk in this input space. Later it is associated with a linguistic label, and this helps align the

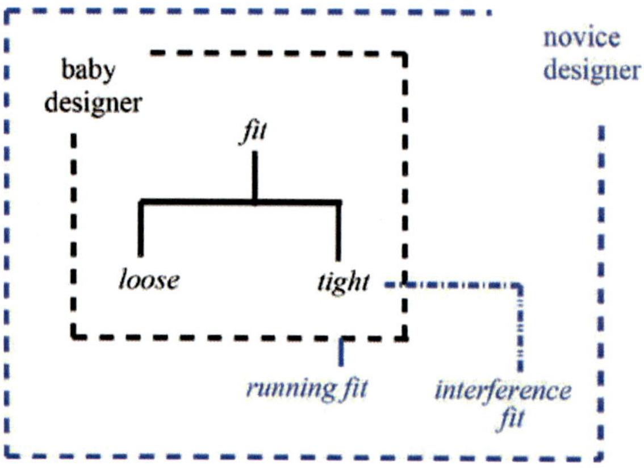

Fig. 6.2 Two-stage concept construction. First, the baby designer learns the initial concept-label pairs ([*TIGHT*]) and ([*LOOSE*]). The novice designer extends these to arrive at the semantics of design terms like [*RUNNING FIT*] or [*INTERFERENCE*]. For the expert designer, this semantics is further refined through a wide range of functional associations

concept to those prevailing in the society at large. This concept-label pair is the initial symbol.

2. *novice designer:* novice designer: is explicitly told the semantics of terms such as "interference fit". By associating these with existing symbols such as [TIGHT], forms an initial model of the concept.

Beyond the novice is the *expert designer*, who applies the concept in a wide range of applications, and learns various associations for it. This is beyond the scope of the present work. Here we limit ourselves to the mechanisms for learning the initial symbol, and how it is used to derive the meaning for additional design terminology. The process of defining a derivative symbol may involve two processes, both operating on its semantic space:

1. Narrowing or Broadening: Where the concept is either a specialization or a generalization of an earlier concept [12]. The space in which the concept is being defined remains the same.
2. Extension or Lateral transfer: Where the concept is an extension of a concept into a range of parameters or situations not available in the early models [13].

Here the space may have new linkages added to the original space. Thus, an initial concept such as tight may be narrowed (specialized), resulting in the more precise notion of running fit and even further refined into even more specific notions such as [H7-g6], etc. On the other hand, extending the initial concept into the previously forbidden zone where a peg is larger than its hole requires linkages with concepts such as expansion (either by thermal or mechanical stress), and enables the learning of concepts such as interference fit (Fig. 6.2). This is the process being elaborated in this paper, via the stage of symbol learning that we refer to as the *novice designer*.

We note that these derived symbols may be learned without direct experiential grounding but in terms of other symbols (e.g. in a classroom, or by reading). They inherit the semantics of the earlier symbols (including possible misconceptions)which is why it is crucial that they be directly experienced in terms of actual

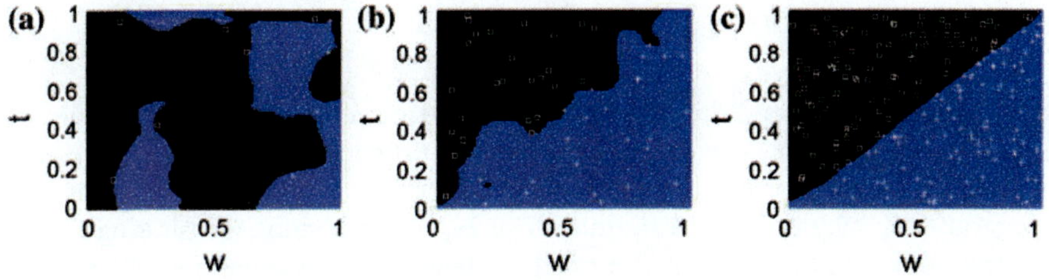

Fig. 6.3 Learning through experience that inserted-object-must-be-smaller-than-container (w > t). For the Fig. 6.1a, the object of thickness t goes into a hole of width w, the feasible solutions in the (w;t)-space are marked with "+" and failure regions as squares (□). **a** In very early stages of exploration (10 instances explored), the learned pattern is unsure of which t can go into which w. **b** After 50 instances, the pattern w > t is beginning to emerge and **c** is quite clear after 200 instances

designs. In the absence of such experience, the symbols are incomplete, and may contain many errors, which is why design education more than many others, insists on direct, hands-on exposure.

In the next section we review our model for the first stage, the *baby designer*, followed by the *novice designer* stage which is the main focus of this paper.

6.2 Stage 1: Baby Designer

Here we review earlier work on the baby designer [14, 15]. The baby designer is a system that has a built-in preference for compact descriptions of patterns in data. Thus, for the containment task (Fig. 6.3), a simple perceptron model is able to categorize the class of fits into feasible (below the w = t line) and infeasible (above). We note that as its experience increases (from 10 to 50 to 200 instances), the number of mis-categorized instances go down, resulting in an increased confidence in the discriminating function. The "compact description" that the baby designer seeks can be defined in terms of the number of parameters required to describe an input, or its dimensionality. We suggest that in many situations, functions defined on high-dimensional input spaces characterize a set of "good" solutions that lie on a lower-dimensional subspace or manifold. These lead to the chunks that are mapped to the initial (pre-linguistic) image schema. This is a tacit concept, learned through direct experience. However, if it is a concept that is referred to in language, it would be possible for the agent to learn a label for it. Subsequently, language can be used to align this concept to societal conventions.

6.2.1 Learning the SymbolTIGHT

During insertion tasks if the clearance is small, the inserted object may sometimes get stuck in the hole, when even very high insertion forces will not succeed in pushing it in (of course, a small wiggling motion or compliance may release the peg, but our Baby is yet to learn this). This type of situation is frustrating for the learner, and becomes a salient event that is attended to. The situations where this is more likely to occur may eventually get associated with the initial notion of tight. Similarly a loose clearance may involve some wiggle or play.

In the computational simulation, the learner explores many instances of tight fit and loose fit, and gradually comes to recognize the regions in the input space $< w, t >$ corresponding to these—we call these the functionally feasible regions or FFRs (Fig. 6.4a, b, FFRs in gray). Eventually, it is realized that for the "good" instances of tight, w and t appear to vary in a related manner; they lie on a 1-dimensional manifold in the 2-D input space (Fig. 6.4c, d).

One may abstract further patterns by considering lower-dimensional representations for the FFRs. In this situation, we may use linear dimensionality reduction (PCA) [16]. The dominant eigenvectors converge much faster than the decision boundary—thus, the first eigenvector after 20 samples is already roughly parallel to the $w = t$ boundary, and becomes more strongly so after a 100 samples (b, c above). This indicates that the concept of clearance lies along a 45° line in the w; t space (as shown in the bottom row). The invariant along either line is the quantity w–t, which becomes the learned chunk; its value eventually forms part of the semantics for the symbol [CLEARANCE].

The next step is to associate the frequently occurring chunks to linguistic labels. The availability of such labels will enable different instances of situations labeled as "tight" to be related to each other and inform the semantic model (Sect. 6.2.1).

The learning achieved up to this point is part of the ubiquitous expertise that all human adults would be expected to have. It is based on this knowledge that we expect the student to construct her learning of new symbols relating to tolerance fits (Sect. 6.3).

6.2.2 Language Label Learning

At this stage, our *Baby Designer* has an implicit notion of the categories TIGHT and LOOSE, in terms of low dimensional chunks. This is based on a measure of the probability of wedging, which may vary from person to person. Once the agent has a label with which this concept may be communicated across other agents, this concept will stabilize and acquire a richer meaning adapted to the social conventions.

To learn the label, we consider that the agent has the idea that sequences of sounds may refer to concepts. Then, when it is exposed to adult speaker narratives

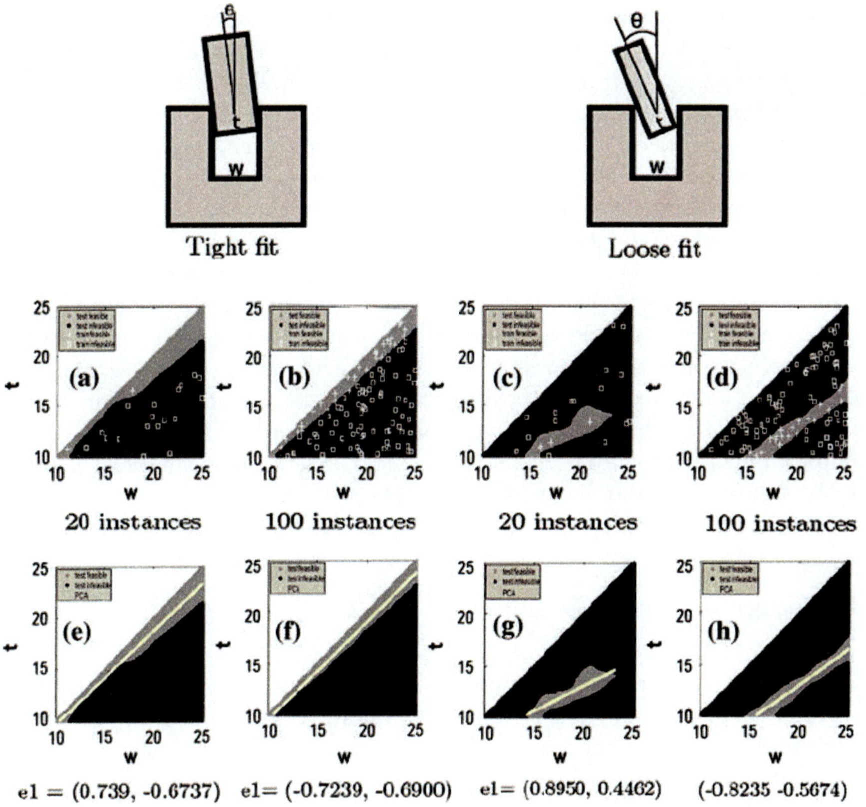

Fig. 6.4 Emergence *of chunks for fit: "tight" versus "loose":* Situations where t is close to w may cause wedging, which can occur when the insertion angle is tilted more than $= \frac{w-t}{w\mu}$. Tight situations are those where wedging is more likely (θ is low). Using this as a performance metric, we can learn the Functionally Feasible regions (FFRs) for both tight and loose. The FFRs (*gray*) learned after 20 instances (**a**, **c**), are poorer than after 100 instances (**b**, **d**). On reflective analysis, we find that these FFRs are well-approximated as lower-dimensional (1-D) linear manifolds (**e**, **f**, **g**, **h** *bottom row*). Using PCA, we discover the principal eigenvector to be dominant, and the number of parameters reduces from (w; t) in 2-D to an emergent 1-D chunk representing invariance in w–t. This process results in two initial chunks $TIGHT_i$and $LOOSE_i$—but we do not know that these are called "tight" or "loose" yet

that describe tight or loose situations, even without any knowledge of other words or grammar, we find that words like "tight" emerge as the most likely keyword for TIGHT and also that this can be done for any language [15].

Now, whenever tight and loose situations are observed, the learner may compute the probabilities of different words occurring in the two contexts. We find that even after a dozen or so exposures observes it is able to determine that the term "tight" occurs more frequently in the context of TIGHT and may form this label-meaning pair.

Once learned, the language label permits changes in the semantics. The learner observes other context where "tight" is being uttered, and modifies its own image schema to comply with the observation. If the observation matches the schema, then the learner gains confidence in the schema. If it does not match, but seems similar, the learner *extends* the semantics to include such cases. If its own usage fails to be understood, the interpretation may be *narrowed*. Other extensions, such as metaphorical extension to novel spaces etc. are not considered in this work.

Thus the language label serves as an index with which the meaning can be related to other concepts. The semantics of TIGHT, which in our simulation was initially correlated only with a range of insertion angles, can now be associated with other notions—e.g. fits have low clearances, the inserted object has less play, as w gets very near t, it may require some insertion force etc. As the learner finds that its image schema is able to cover most of the situations where the term is used, it transitions from the initial image schema $TIGHT_i$to a mature, linguistically-informed image schema TIGHT. It is now ready to learn more advanced concepts such as those related to fits and tolerances.

6.3 Stage 2: Novice Designer

At this point, the learner has a stabilized concept (the mature image schema tight), along with the linguistic label "tight", so it has a proper symbol [tight], with a semantics grounded in tacit knowledge. The learner can now extend this concept by being told about it, as opposed to having to encounter everything as a sensorimotor experience. Thus, the student in a design course may be told about fits and tolerances, and she can then construct these more complex concepts by building on the concept of tight. We consider the learner at this stage, when "running fit" is being introduced, where a shaft is located closely in a hole, but is free to move. The learner recognizes this as a low-clearance situation, a special case of tight. This enables many properties associated with tight—such as low degree of wiggle to be linked with this new concept. However, RUNNING FIT is a more specific concept than tight, since it applies to a very narrow band of dimensions of shaft and hole, and is also associated with relative motion. This results in a *narrowing* of the earlier relation.

Next, the novice designer is told about "interference fit" (in a course or from a book, say). It learns that fits requiring high torque transmission may be achieved by having shaft diameter slightly greater than the hole. Now, the concept of interference fit encroaches on a region ($w > t$) which was considered infeasible while learning the original $TIGHT_i$relation. However, since the interference is small, it may be considered an extension of the earlier concept.

The two new concepts are thus learned as a result of narrowing and extension of the concept of tight, operating on the same design space (w; t). We shall see that while running fit may be considered a sub-category; the concept of insertion fit is more of a lateral shift Fig. 6.2.

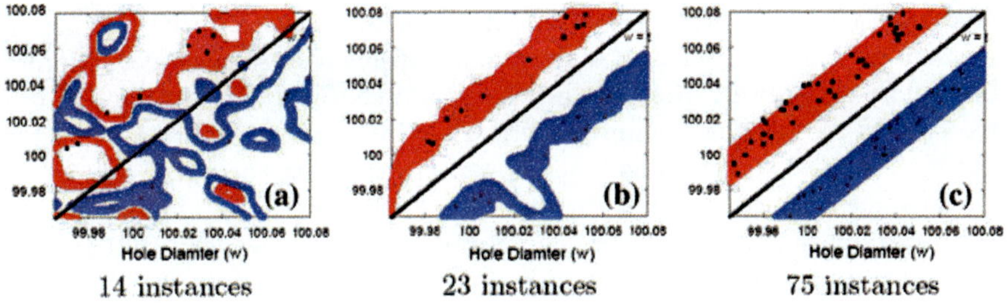

Fig. 6.5 **a** Learning the semantics for running fit H7-g6 (*blue*) and INTERFERENCE FIT H7-r6 (*red*). Though each specific instance is informed by explicit knowledge, the emergent patterns, especially in more complex situations, become part of the tacit knowledge gained by the designer through experience

In the problems the novice designer is asked to solve, it encounters some instances of interference fits and running fits. For example, in the interference fit situation, one may have a problem involving a gear-shaft assembly, where the functional criteria for interference fit may be explicitly defined in terms of the torque τ that the assembly must transfer. This torque is linked to the contact pressure pc through the explicit formula $\tau = 2\pi\mu r_i^2 L p_c$ where r_i is the nominal inside radius of the hole. The pressure $p_c = \frac{\delta_r E\left(r_o^2 - r_i^2\right)}{2rr_o^2}$, where $\delta_r = \delta_{rh} - \delta_{rs}^2$, where δ_r is the degree of inter-penetration, r_o the outer radius of the pinion, and E the modulus of elasticity. It may emerge that a certain range of δ_r would meet the desired torque requirement while remaining within the material strength constraints. Now, to manufacture parts that meet this interference criterion, each part would need to meet certain tolerances. All this is explicitly told to the novice designer, resulting in a shallow understanding of statements such as "H7-r6 is an interference fit".

Subsequently, as she begins practice, she encounters many actual design situations. This gives substance to this understanding, by encountering instances of the various tolerance classes and fits. Thus Fig. 6.5 shows the patterns that emerge for H7-g6 running fits and H7-r6 interference fits. The "good" designs within this range are sampled and the patterns that emerge after 14, 23 and 75 instances are shown in Fig. 6.5. Although each instance in this data is based on explicit knowledge, the finally emergent pattern may be implicit, especially in more complex situations. This gives the novice designer some tacit intuition for what it means to have a particular tolerance pattern. Also, once such a pattern emerges, one may become aware of it, a process known as reification, which would result in an explicit form of knowledge. Thus, explicit and implicit are in constant interplay in the designer's life. There has been much speculation on the nature of this interaction [16] but here we wish to provide an overview of the process.

Other parts already present in the mature tight concept are also enriched through this experience. For example, the fact that insertion forces may go upas w approaches or breaches the w = t boundary was already known, but is now reinforced along with explicit relations that relate this insertion force to the contact pressure pc. However, the tacit understanding of this process continues to inform the mature designer's expectation of insertion force. However, situations such as the process of assembly requiring thermal expansion/contraction are learned as specific only to insertion fit and are not part of the earlier concept.

Thus, interference fit is seen by the learner as being related to tight, but also extending it in several ways. The concept is constructed as a lateral shift on tight, rather than as a subcategory.

6.4 Conclusions

Here we have argued for a very ambitious approach to modeling design knowledge. While the process is demonstrated on a toy set of symbols from a single domain, there is reason to believe that the approach is scale-able. For one, beyond a knowledge of what constitutes an interesting functional distinction (e.g. inserting or not; wedging or non-wedging), we have used no domain knowledge at any step of the process. While the dimensionality reduction modality used here is linear, a number of non-linear manifold learning approaches are today feasible and may be applied to data which implicitly lies on a lower-dimensional space. Indeed, the power of such algorithms for discovering latent relationships in design remains to be explored.

The work has two main ramifications. One is in constructing a cognitive theory for the human design process. The concrete steps suggested in the theory overcome the obstacle of earlier suggestions of tacit knowledge that were considered vague and non-operationalizable. The key observation is based on the fact that chunks, incorporating often arise as a lower-dimensional pattern in high-dimensional design spaces is an important contribution of the present work to this literature.

6.4.1 Computational Tacit knowledge?

In addition to presenting an operationalizable approach to tacit knowledge learning, this work hints at a possibility of enormous potential that computational models may be tuned to gain tacit knowledge, that may lead to human-like flexible behaviours. This appears to be a contradiction in terms, since it is al-most axiomatic that computers work on representations that are precise and well-posed.

However, if we consider the patterns that the computers are learning—no the simple linear paradigms presented in this work, but also the general regions of a functional feasibility determination, such spaces can be quite complex. For example, most categorization algorithms in machine learning result in a capability to categorize (e.g. recognize objects) based on data without the programmer being able to articulate how exactly the algorithm is able to do the task.

This may open up a radically new approach to computational models for design knowledge, with far-reaching implications for maintaining design rationale, design repository management, and increasingly in design database exchange. However, at this point, we have a model that may be able to learn one symbol at a time. The composition of symbols often results in very different complexes, with large changes in semantics. Clearly, large scale simulations need to be runon different symbol classes to understand how their interactions may be acquired.

The methodology outlined here is a meager start in what may be a radically new direction in design cognition.

References

1. Polanyi M (1966) The logic of tacit inference. Philosophy 41:1–18
2. ISO-286-2 (1988) Bases of tolerances, deviations and fits, International Organization for Standardization
3. Chase W, Simon H (1973) The mind's eye in chess. In: visual information processing, vol 215. Academic Press 281
4. Schon DA (1988) Designing: Rules, types and words. Design studies 9:181–190
5. Visser W (2006) Designing as construction of representations: a dynamic viewpoint in cognitive design research. Human-Computer Interaction 21:103–152
6. Dennett DC (1984) Cognitive wheels: the frame problem of AI. In: Hookaway C (ed) Minds, Machines and Evolution. Cambridge University Press, Cambridge
7. Wittgenstei L (1953) Philosophical investigations. Blackwell, Oxford 1953/2001
8. Collins H, Evans R (2007) Rethinking expertise. University of Chicago Press, Chicago
9. Hespos S, Baillargeon R (2001) Reasoning about containment events in very young infants. Cognition 78:207–245
10. Bransford J (2000) How people learn: Brain, mind, experience, and school. National Academies Press, DC
11. Wilson D, Sperber D (2008) In: Relevance theory. Blackwell Publishing Ltd pp 606–632
12. Fauconnier G, Turner M (2003) The way we think: conceptual blending and the mind's hidden complexities. Basic books, NY
13. Dabbeeru MM, Mukerjee A (2011) Learning concepts and language for a baby designer. In: Fourth international conference on design computing and cognition, Springer, stuttgart pp 445–463
14. Mukerjee A, Dabbeeru MM (2010) Using emergent symbols to discover multi-lingual translations in design. In: Proceedings of DETC'10, 2010 ASME design engineering technical conferences
15. Bishop C (2006) Pattern recognition and machine learning. Springer, Berlin
16. Gourlay S (2006) Conceptualizing knowledge creation: a critique of nonaka's theory. J Manage Stud 43:1415–1436

Chapter 7
System-Environment View in Designing

B. S. C. Ranjan, V. Srinivasan and Amaresh Chakrabarti

Abstract A system interacts with its environment to satisfy requirements. Therefore, designing should involve developing the concept of both the system and its surrounding. A comprehensive review of literature on designing to analyse the use of system-environment view in designing revealed that while the concept of systems is used, implicitly or explicitly, by many design models, the concept of environment is rarely used as an evolvable construct in designing. Based on this, a system-environment view has been proposed in this paper that consists of: System, Subsystem, Elements, Environment and Relationships; each of these constructs is explicit and evolvable during design. The proposed system-environment view is empirically validated using protocol studies of design sessions, which were undertaken before this view was developed. The validation involved checking whether or not all the constructs in the system-environment view are naturally present, in these design sessions. An example of system-environment co-evolution during designing is also presented to show the importance of considering environment as an explicit, evolvable construct in models of designing.

7.1 Introduction

A system interacts with its environment in order to satisfy requirements of the system and its environment. Therefore, developing the concepts of both the system and its environment are important in designing. Various researchers, e.g. [1–4]

B. S. C. Ranjan · V. Srinivasan · A. Chakrabarti (✉)
Indian Institute of Science, Bangalore, Karnataka, India
e-mail: ac123@cpdm.iisc.ernet.in

A. Chakrabarti (ed.), *CIRP Design 2012*, DOI: 10.1007/978-1-4471-4507-3_7,

considered the interactions between system and environment as an important aspect of designing.

Asimow explained in detail the various interactions between engineering systems and their environment [1]. Deng et al. [2] showed that information of 'working environment' is useful for exploration of functional design solutions. According to Hall [3], physical systems not only exist in environment but they exist by means of an environment. Hubka and Eder [4], in 'Theory of Technical Systems', defined environment and discussed specifically about 'active environment', the portion of the environment that directly interacts with the system and plays an important role in the performance of the system.

While the primary focus of designing is to develop a system, environment must also be identified, specified, and variously modified, in order to ensure that they together are capable of fulfilling the requirements. Due to the primary focus of designing on the system, current literature on design theories, models and approaches either completely ignores this system-environment view—the perspective of taking both system and environment as explicit constructs—or focuses only on the system as the evolving construct during designing.

Henceforth in this paper, the system-environment view considers—system, subsystem, elements, relationships and environment, to help explain or design the interactions among them and the system-environment co-evolution.

The need for a system arises from its environment. As a system is developed, its environment may also get modified. Hence, any changes in a system may lead to change in its environment. There are various examples in literature on the evolution of products, where changes in a system have led to changes in its environment, and vice versa. An example is the historical evolution of writing devices from the likes of pen and inkpot, in which design efforts variously focused on: either the pen as the system to be designed with the inkpot being given and hence part of the environment, or the inkpot as the system to be designed, with the pen being part of the environment. Subsequently, the two were considered together, leading to design of integrated pen and inkpot systems, such as a fountain pen or a ballpoint pen [5]. Another example is design of material handling robots such as pick-and-place robots. For these robots, the environment should also be designed, such as the locations of pick-up and place-down, the path of the robot and so on, in addition to the design of the robot, in order to satisfy the task that the robot has to perform [6].

Therefore, we argue that a system-environment view must consist of explicit constructs to represent both the system and its environment, and must be incorporated into design theories, models and approaches as constructs that can evolve during design.

The specific objectives in this paper are:

1. To check if current literature on design theories, models and approaches uses the system-environment view in designing.
2. To propose a new, system-environment view.
3. To check if the constructs of the proposed system-environment view are used naturally in designing.

7.2 Literature Survey

Various design theories, models and approaches are reviewed to investigate whether and how these take into account a system-environment view.

In Cross [7], VDI 2221 [7], Visser [8], and Ulrich and Eppinger [9], problems are divided to sub-problems for which sub-solutions are found and are combined to produce solutions. This can be perceived as a system-environment view at the level of problems and solutions but without the environment and relationships.

The outcomes of designing in French [10] are selected schemes, general arrangement drawings, and part drawings. Selected schemes are at system level, general arrangement drawings are at subsystem level, and part drawings are at element level. In Cross [7] the outcomes are solution space, concept sketches, drawings, evaluated drawings, and final production documents. Solution space and concept sketches are at system level, drawings and evaluated drawings are at subsystem level, and final production documents are at element level. In Chakrabarti et al. [11], the outcomes are functions, solution-principles and embodiment. Functions are represented using input–output description, and solution-principles are represented by stringed laws and effects. The input–output descriptions are either at system or at element level. The individual laws and effects are at element level; these are strung together to form the system level description. Srinivasan and Chakrabarti [12] used the concept of system and environment to define the various constructs of the outcomes of the SAPPhIRE model. However, they did not explicitly use the concept of system-environment in their integrated model of designing.

Some researchers used system hierarchy structures to represent the system-environment view. Lossack [13] used task structure, physical principle structure and geometry structure. Pahl and Beitz [14] used function structure, working principle structure, assembly drawings and so on. VDI 2221 [7] used function structure, module structure and so on. Cross [7] suggested the use problem structure and decision trees. Similarly, Ulrich and Eppinger [9] suggested the development of a hierarchy of needs, in the concept development phase of design. In these researches, system-environment view has been used at several levels of abstraction, but without considering environment.

Hubka and Eder [4] developed a theory of technical systems. Their system-environment view consists of system, subsystem, elements, components, relationships, environment, and active environment. However, neither 'environment' nor 'active environment' is used as an evolvable construct in designing. They used system hierarchical structures at different levels of abstraction e.g. transformation process structure, function structure, organ structure, and component structure [15]. Hansen and Andreasen [16] developed the domain theory based on systems theory, and therefore implicitly considered system-environment view. In domain theory, there are three domains: transformation, organ and part. Domain theory and Hubka's function-means law were combined [16] and function-means law was modeled as a tree structure; it is set-up as a function/means tree which is a hierarchical arrangement of functions and means that are connected by causal relationships. The function/means tree represents the system view in this approach.

Hall [3] used a system-environment view that consists of system, subsystem, objects, relationships and environment. Hall considered environment as a major factor in the design process, and hence used 'initial environment' and 'final environment' in his model of the systems engineering process; however, he did not use 'environment' as an evolvable construct during the process.

Asimow [1] and INCOSE [17] used system-environment view without considering the environment. Asimow [1] in his morphology of design proposed using system, subsystem, components and parts. Although he considered the interactions between system and environment as important, environment was not used as an evolvable construct in his model of designing. INCOSE [17], for the systems engineering process, uses a view with these constructs: system, element or segment, subsystem, assembly, subassembly, components and parts. Ulrich and Eppinger [9] in the system level design explicitly used a system-environment view (without considering environment); in this view, a product is divided into functional and physical elements. Functional elements of a product consist of individual operations and transformations. Physical elements of a product are parts, components and subassemblies.

Howard et al. [18] explicitly used modular hierarchical structures to represent the system-environment view without considering the environment. Blessing [19] used product model (which consists of product, assemblies, components and standard components) to represent the system-environment view, without considering environment. Prabhakar and Goel [20] proposed Environmentally-driven Adaptive Modelling (EAM), which uses Environmentally-bound Structure-Behavior-Function (ESBF) model. ESBF model supports a hierarchy of interactions between a device and its environment. In Environment based Design (EBD) methodology, Zeng [21, 22] defines *product system* as "the structure of an object (Ω) including both a product (S) and its environment (E)". EBD uses a hierarchical representation in which system and environment are connected to each other via a set of objects. Zeng states that, "in the design process, any previously generated design concept can be treated as an environment component for the succeeding design, as a result, a new state of design can be defined as the structure of the old environment (Ei) and the newly generated design concept (Si), which is a partial design solution". This change in the state of the environment, where the new environment consists of the earlier environment plus the new design, is proposed as "evolution of environment". However, Zeng's work does not propose system-environment *co-evolution* in the same way as proposed in our paper. This is because Zeng considers system and environment to be mutually exclusive [22]; therefore, the new environment, created by adding a change in the system, amounts to changing either the system or its environment, but *not both and not together*.

Relationships—one of the constructs of the system-environment view—are considered explicitly by Hall [3], Hubka and Eder [4], Blessing [19], and Bhatta and Goel [23]. Hall [9] stressed that relationships is what "makes the notion of 'system' useful". Hubka and Eder [4], in their system-environment view, define relationships of various types e.g. analogy, homomorphy, isomorphy, equivalence, identity, causality, coupling, goal-means, spatial and logical. Blessing [19] uses a relationships model, which contains several relationships (e.g. spatial, functional and hierarchical). The structure of a device in the SBF model of Bhatta and Goel [23] is represented hierarchically in terms of its constituent structural elements and relations among them such as part-of, includes and parallely-connected. Lossack [13], and Pahl and Beitz [14] consider connections or interrelationships using system hierarchy structures. Chakrabarti et al. [24] developed the SAPPhIRE model of causality; according to this model: parts create organs. The relevant input and organs together activate a principle (effect), which in turn creates phenomenon. Phenomenon changes the state of the system and its environment (state change), which can be interpreted at a higher level of abstraction (action). Here creates, activates, changes, and interpreted, are the causal relationships among the outcomes.

The following points are concluded from the above review of literature:

1. Only some researchers consider the system-environment view. Very few consider the system-environment view explicitly (e.g. Hubka and Eder [4, 15], Hall [3]); others consider this view implicitly (e.g. Hansen and Andresen [16]). Most of these researchers do not consider environment.
2. Relationships are explicitly considered in very few theories, models and approaches (e.g. Hall [3], Hubka and Eder [4], Blessing [19], Bhatta and Goel [23]); the rest consider it implicitly (e.g. Lossack [13], Pahl and Beitz [14]).
3. Environment is explicitly considered in very few theories, models and approaches, e.g. Deng et al. [2] and Hubka and Eder [4].
4. None of the models consider 'environment' (that is, universe without the system) as an evolvable construct in designing. Consequently, none of the models note the occurrence of co-evolution of system and environment as a phenomenon during designing.

7.3 Proposed System-Environment View

We propose the following system-environment view, which consists of the constructs: system, sub-system, elements, relationships and environment. Here both system and environment are evolvable constructs in the process of designing. These constructs are defined as follows. A system is the overall product being designed, at any level of abstraction. A sub-system is a subset of a system that can be further divided. An element is a subset of a system or a sub-system, which cannot be further divided. An environment refers to all subsets of the universe apart from the system. The relationships are how system, environment, sub-systems, and elements are linked with one another. Elements (and sub-systems) combine together to comprise sub-systems. All sub-systems and elements combine together to comprise the system. System is characterized by a system-boundary

Table 7.1 Pattern of problem-solving

Problem brief	P1	P1	P2	P3
Designer	D1	D2	D3	D4

Table 7.2 Pattern of problem-solving

	Problem-briefs
P1	India has large number of people with transferable jobs. They need to shift frequently from one place to other (every 1–2 years). And often face problems transferring present types of furniture, which are bulky and heavy. It is not economical for them to buy furniture and sell it before shifting to other place. This furniture occupies lot of space and this is an additional problem since they live in small houses. It takes more time to pack the furniture and it damages during transport if it is not packed properly. Your task is to design portfolio of furniture which will help in sleeping and storing things while taking into account the above problems mentioned. Setup time and effort on the part of user should be minimal.
P2	India has large number of people with transferable jobs. They need to shift frequently from one place to other (every 1–2 years). And often face problems transferring present types of furniture, which are bulky and heavy. It is not economical for them to buy furniture and sell it before shifting to other place. This furniture occupies lot of space and this is an additional problem since they live in small houses. It takes more time to pack the furniture and it damages during transport if it is not packed properly. Your task is to design portfolio of furniture which will help in sit, write and eat while taking into account the above problems mentioned. Setup time and effort on the part of user should be minimal.
P3	There are many problems associated with the increase in temperature during summer. Huge numbers of people die because of heat waves. A number of products are available such as umbrella and hat to help alleviate part of these problems; however, there are useful only for blocking the direct sunlight. These are not able to protect a person from high temperature and heat waves. Air conditioners are available and solve this problem; however, they are expensive and work only in fixed setting. No mobile and portable equipment is available that can be carried around while in transit. There is a need for a product that will help in maintaining body temperature within a comfortable range. Your task is to design a product that will help in solving these problems. It should help the user in avoiding direct sunlight and maintaining body temperature. The user should be able to use it with-out any difficulty in setup and it should be portable.

that separates it from the environment. System needs an environment (which is outside the system-boundary) to satisfy its requirements.

These constructs are illustrated with a ballpoint pen, which is a system made up of a refill, a body, and a cap. The refill is a sub-system consisting of elements like nib, ink and ink reservoir. The body is another sub-system consisting of elements like upper-body and lower-body. The environment for the ballpoint pen includes papers on which it has to write, and an agent that uses the pen to write on the paper. The above example is given only at the physical structure level of abstraction; however, the system-environment view can exist at any other level of abstraction.

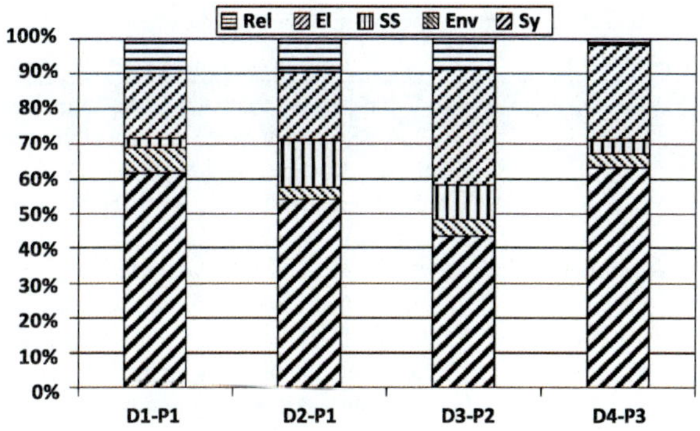

Fig. 7.1 Percentage of constructs of the system-environment view in observed design sessions

7.4 Validation of the System-Environment View

In order to validate the importance of incorporating a system-environment view in design theories, models and approaches, we analysed video protocols from a series of design sessions to check if the constructs of the view are used in designing. Four designing sessions are used for validating the importance of system-environment view. The intent was to check whether or not the constructs of the system-environment view proposed are naturally used, even if implicitly, in these design sessions. The video and audio protocols, their transcriptions, problem briefs, sketches of design session are taken from an earlier research [25], which was carried out well before this view was developed.

Each design session consisted of an individual designer solving a design problem under laboratory conditions. The designers were trained and instructed to discuss-and-think-aloud. These design sessions were video and audio recorded and each session was assisted by a researcher for any clarification during the session. Four designers [D1–D4] of varying background and experience were each given one problem brief from among the three product design problems (P1–P3), see Tables 7.1 and 7.2.

The transcriptions of these design sessions are analysed by coding the transcriptions using the proposed constructs, which together represent the system-environment view. The utterances of the designers from the transcription are used as instances of the above constructs (see Table 7.3).

Figure 7.1 shows the frequency of the instances of each construct in each of the four designing sessions analysed. All the constructs of the proposed system-environment view are observed to have been used in natural designing.

Table 7.3 Constructs of system-environment view and instances from transcriptions

Sy-En view		Instances from designing sessions
System	Sy	D: "I have basic chair would be I am making one conventional chair which can help in sit, write and eat, that is somewhere near a dining table" [Episode: Designer develops a chair as a solution for all the given requirements (i.e. sit, write, eat, and easy to transfer). Therefore chair is a System]
Environment	Env	D: "I have basic chair would be I am making one conventional chair which can help in sit, write and eat that is somewhere near a dining table" [Episode: Designer develops a chair (System) as a solution for the given requirements under specified environmental conditions such as presence of a table for writing and eating.]
Subsystem	SS	D: "so it (System) again becomes something like a suitcase and plus this retractable lid which will have to be carried in the rectangular frame kind of thing" [Episode: Suitcase and retractable lid forms a subsystem as both together will be put in a rectangular frame together forming a system]
Elements	El	D: "Velcro strip that can rest on these supports" [Episode: Velcro Strip is an Element developed]
Relationships	Rel	D: inner diameter handles is 25 mm which exactly matches with the outer dia of the tube [Episode: Here designer defines an organ level relationship "equality constraint" between inner dia of handles and outer dia of tubes

7.5 Illustration of System-Environment Co-Evolution

An example of the System-Environment co-evolution is taken from the session in which D1 solved P1. From the problem brief, D1 found various problems associated with existing furniture (i.e., the Existing Systems), two of which are: 'space for furniture', 'unfolding and packing' as shown in Fig. 7.2.

The problems identified were then evaluated and modified by D1, and requirements such as the following were generated: 'furniture will be foldable to save space' and 'fixed furniture (already)' (with the following sub points—'modular to save space' and 'It can be fixed to slots'), as shown in Fig. 7.2.

From all the requirements generated, D1 made a list of requirements to be fulfilled by the system. One of these—'foldable and space saving'—is shown in Fig. 7.3.

With the above list, D1 sketched an idea of a piece of furniture (Sy), see Fig. 7.4. According to this idea, when the user needs to use the furniture (Sy) as a bed or a storage space, it would be horizontal, but when not in use, (Sy) could be folded against the wall (Env) to save floor-space (Env). To fold the furniture (Sy) against the wall (Env), D1 defined the distance between the furniture (Sy) and the wall (Env); he also defined the relationship between the furniture (Sy) and the floor (Env) as a 'hinge joint', as shown in Fig. 7.5.

This example illustrates that, as the design process progressed, both the system and its environment evolved, simultaneously and through mutual influence.

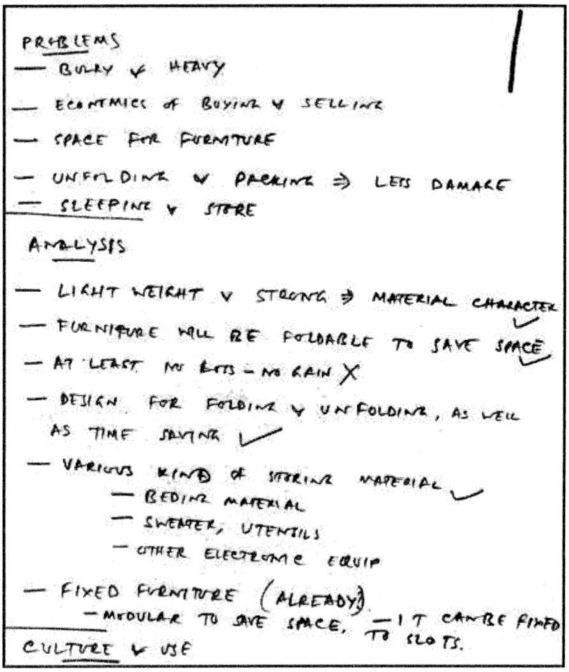

Fig. 7.2 Problems identified by D1

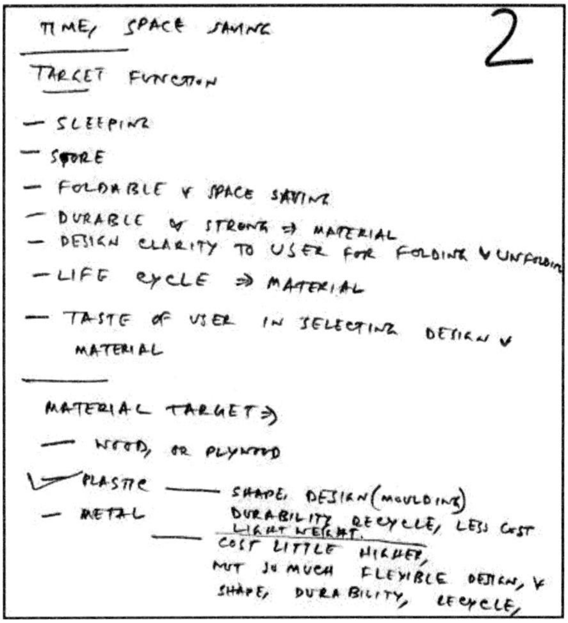

Fig. 7.3 Requirements identified by D1

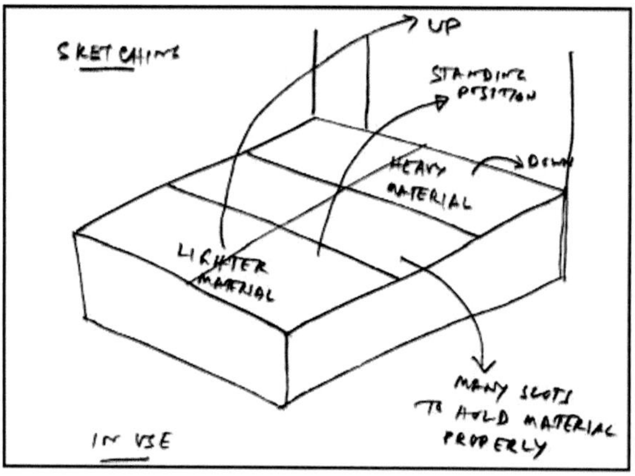

Fig. 7.4 One design idea generated by D1

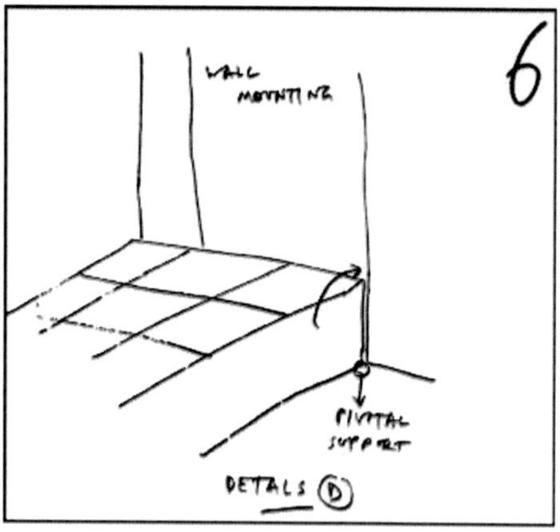

Fig. 7.5 Detailed description of idea in Fig. 7.4

7.6 Conclusions

The major findings in this work are as follows. Literature stresses the importance of the system-environment view, which includes environment as a construct, in designing. Current design models, however, do not consider environment as an explicit, evolvable construct in designing. Explicit representation of this view is necessary, e.g. for describing system-environment co-evolution. Based on literature, a new, system-environment view is proposed. Empirical studies show that designers consider both system and its environment as evolvable constructs, and change them both as necessary during designing. Further, this work shows the presence of all the constructs of system-environment view in designing, and their co-evolution.

References

1. Asimow M (1962) Introduction to design. Prentice-Hall Inc, Englewood Cliffs
2. Deng YM, Tor SB, Britton GA (2000) Abstracting and exploring functional design information for conceptual mechanical product design, engineering with computers 16, Springer-Verlag London Limited, pp 36–52
3. Hall AD (1962) A methodology for systems engineering, 6th edn. Van Nostrand, Princeton
4. Hubka V, Eder W (1988) Theory of technical systems. A total concept theory for engineering design. Springer, Berlin
5. http://en.wikipedia.org/wiki/Pen
6. Groover PM (2006) Automation, production systems and computer integrated manufacturing, 2nd edn. Prentice-Hall Inc, New Jersey
7. Cross N (2000) Engineering design methods–strategies for product design. Wiley, England
8. Visser W (1991) Evocation and elaboration of solutions: different types of problem solving actions In: Proceedings of 3rd COGNITIVA symposium on at the crossroads of artificial intelligence, cognitive science, and neuroscience, Madrid, Spain, pp 161–168
9. Ulrich KT, Eppinger SD (2003) Product design and development, 3rd edn. Tata McGraw Hill, New Delhi
10. French M (1999) Conceptual design for engineers. Springer, Berlin
11. Chakrabarti A, Johnson A, Kiriyama T (1997) An approach to automated synthesis of solution principles for microsensor designs. In: Proceedings of international conference on engineering design (ICED97), Helsinki, Finland, pp 125–128
12. Srinivasan V, Chakrabarti A (2010) An integrated model of designing. In: Ashok KG, Andrés Gómez de Silva Garza (eds) Special issue on knowledge based design. ASME J Comput Inf Sci Eng (JCISE) 10(3):031013
13. Lossack R (2002) Design processes and the context for the support of design synthesis. In: Chakrabarti A (ed) Engineering design synthesis—understanding, approaches and tools. Springer, London, pp 213–227
14. Pahl G, Beitz W (1996) Engineering design: a systematic approach, 1st edn. Spriger, London
15. Hubka V, Eder W (1996) Design science: introduction to the needs, scope and organization of engineering design knowledge, Springer-Verlag London Limited, Web Translation by W. Ernst Eder and Filippo A. Salustri
16. Hansen C, Andreasen M (2002) Two approaches to synthesis based on the domain theory. In: Chakrabarti A (ed) Engineering design synthesis—understanding, approaches and tools. Springer, London, pp 93–108
17. INCOSE (2004) Systems engineering handbook—a "What To" Guide for all se Practitioners, INCOSE-TP-2003-016-02, Version 2a, Released by—Technical Board, International Council on Systems Engineering (INCOSE)
18. Howard T, Culley S, Dekoninck E (2009) The integration of systems levels and design activities to position creativity support tools, In: Research into design: supporting multiple facets of product development, ICoRD'09, Bangalore, India, pp 43–50
19. Blessing L (1994) A process-based approach to computer-supported engineering design, Ph D thesis, University of Twente, The Netherlands
20. Prabhakr S, Goel A (1998) Functional modeling for enabling adaptive design of devices for new environments. Artif Intell Eng 12:417–444
21. Zeng Y (2011) Environment-based design (EBD) methodology, ASME 2011 international design engineering technical conferences & computers and information in engineering conference, Washington, DETC 2011-48263
22. Zeng Y (2004) Environment-based formulation of design problem. Trans SDPS: J Integr Des Process Sci 8(4):45–63
23. Bhatta S, Goel A (2002) Design patterns and creative design. In: Chakrabarti A (ed) Engineering design synthesis—understanding, approaches and tools. Springer, London, pp 271–284

24. Chakrabarti A, Sarkar P, Leelavathamma B, Nataraju B (2005) A functional representation for aiding biomimetic and artificial inspiration of new ideas. Artif Intell Eng Des Anal Manuf (AI EDAM) 19(2):113–132
25. Kota S (2010) An interactive support for developing environmentally friendly product lifecycles, Ph D thesis, Indian Institute of Science

Part II
Creative and Inventive Design (TRIZ)

Chapter 8
Managing Design Constraints in Synthesis Reasoning

S. C.-Y. Lu and A. Liu

Abstract Constraints in the context of design synthesis represent various bounds on acceptable technical systems. In early design stages, constraints must be carefully managed in order to keep synthesis on the right track. This paper presents a new constraints management model which is developed based on a domain-independent synthesis reasoning framework. Within the proposed model, design constraints are classified into four types: internal input constraint, external input constraint, internal system constraint, and external system constraint. For each type of design constraints, the model prescribes a respective management strategy.

8.1 Introduction

In general, design synthesis can be seen as a reasoning process from an intangible intent (e.g., what) to some more tangible instantiations (e.g., how) [1]. This process is especially difficult at early design stages (i.e., functional design and conceptual design phases) when both design intent and design constraints are still intangible and subjective. Design intent identifies the goal for synthesis reasoning, whereas design constraint establishes a bound within which synthesis reasoning is carried out. Since the design intent and design constraint play different roles in synthesis reasoning, they should be explicitly distinguished. Otherwise, synthesis could be mistakenly diverged from goal-driven design to constraint-driven design.

S. C.-Y. Lu (✉) · A. Liu
University of Southern California, Los Angeles, CA 90089, USA
e-mail: sclu@usc.edu

A. Liu
e-mail: angliu@usc.edu

A. Chakrabarti (ed.), *CIRP Design 2012*, DOI: 10.1007/978-1-4471-4507-3_8,

Design constraint plays an important role within design synthesis in specific to creative alternative creation. In the alternative creation stage of design synthesis, "novelty" and "appropriateness" of solution alternatives ideated are two important factors that must be considered [2]. The importance of constraints in such creative tasks has already been extensively studied by cognitive psychology [3, 4]. Despite such importance, few efforts have been devoted to studying ways to manage design constraints in the context of synthesis for early-stage creative design. Most existing constraint management models heavily rely on the traditional problem-solving methods [5]; hence they often ignore the impact of constraints on problem-framing which is particularly important in creative design synthesis. Hence, these existing models cannot be directly used to address those early-stage design constraints which are all intangible, subjective and dynamic.

This paper presents a new constraint management model to support design synthesis. Various design constraints are firstly classified into several kinds, then for each kind, the model prescribes a unique management strategy. Constraints, although important, are only partial description of the design synthesis process. The management of constraints can only be effective when incorporated into the whole picture of design synthesis. For this reason, our constraint management model is developed based on an existing domain-independent synthesis reasoning framework.

The rest of the paper is organized as following. Section 8.2 summarizes the distinguishing characteristics of design constraints, and explains the reasons why a new constraint management model is needed. Section 8.3 formally presents the proposed model. Section 8.4 concludes this paper with outlooks for future works.

8.2 Constraints in Design Synthesis

8.2.1 Characteristics of Constraints in Design Synthesis

At early design stages, design constraints are mostly intangible and hard to quantify. Constraint can be seen an element factor that restricts a system from achieving a potentially higher goal [6]. Such definition suggests that the specification and valuation of design constraints rely on two variables: a specific goal and a tangible system. In the context of design, the former refers to the initial design intent (i.e., an input of design), whereas the latter means the final technical system (i.e., an output of design). At early design stages, the initial design intent needs to be further specified/structured and the final technical system is yet "to-be" created via synthesis. As a result, by definition, the design constraints cannot be expressed very explicitly and precisely at early stages. For instance, at the conceptual design phase, designers are unable to precisely estimate the actual product deliver time which is an important constraint within the entire product development cycle. The lack of quantitative information at early stages significantly increases the management difficulty of design constraints that have major impacts on later design decisions and final design outcomes.

All types of design decisions comprise subjective and objective parts. The former is more evident during the early design stages (e.g., functional design and conceptual design phases); whereas the latter is apparent toward the end (e.g., technical design phase) [7]. Design constraints are no exception. In the past, design constraints are often treated as purely objective, whereas their subjectivities are often neglected. This is largely because most of traditional approaches focus on analysis activity that occurs in later stages, as opposed to the synthesis activity that exists in early stages. Similar to those objectively defined constraints (e.g., geometric shape and physical laws) which play important roles in analysis during technical design phase, the subjectively constructed constraints are equivalently important to synthesis during the functional and conceptual design phase. Nevertheless, the impact of latter to design is far from fully explored. Additionally, as today's product development task becomes increasingly complex, diverse expertise from different stakeholders are often required in order to collaboratively carry out design synthesis. When multiple stakeholders each has individual preferences are required to jointly decide the boundary conditions (e.g., the bounds, limits, etc,) of design synthesis, it becomes even more difficult to capture and manage the subjectivity of design constraints.

Design constraints are dynamic at early design stages. This is particularly evident for the type of subjective constraints that are resulted from designer's previous decisions. As a particular design decision is changed, so do the associated constraints. Even for some external constraints (e.g., budget, schedule, etc.) which are imposed to designers by outside parties, their specific value of limitation may remain negotiable until the technical systems are finalized. In addition, as new objectives, components, information, and knowledge are increasingly added to the technical system via synthesis, new constraints will constantly arise and grow.

Constraints are intangible, subjective and dynamic at early design stages. Hence, they are often confused with the functional requirements (i.e., FRs). FRs are the real targets of design, whereas design constraints are only the bounds to acceptable solutions which must satisfy the desired FRs. Unlike FRs which should be stated and maintained independent of each other, design constraints do not have to satisfy such an independence axiom. In addition, it is often unnecessary to specify the tolerance for the constraints, whereas the FRs normally have a design range associated with them [8]. In terms of the mutual relationship between FR and design constraints, it becomes more efficient to select FRs when the design synthesis is appropriately constrained [8]. In any case, a true creative design synthesis should be more target-driven than constraint-driven.

8.2.2 Need for a New Constraint Management Model to Support Design Synthesis

It is important to note that design synthesis, in both sprite and process, is not equivalent to constraint satisfaction. There have been many early studies to formulate design task as a constraint satisfaction problem (CSP), and then to adopt the existing constraint-based systems (CBS) to manage design constraints [5]. Despite few successful implementations on simple design task, it has been proven to be very difficult to directly apply CBS to support design synthesis. This is largely because the natures of constraints (i.e., intangible, subjective, and dynamic) at early design stages are very different from the implementation prerequisites (i.e., tangible, objective, and static) of typical CBS. Constraint satisfaction is a part of design synthesis; however, design synthesis cannot be simplified as a pure constraint satisfaction process.

Recently there have been many attempts to apply preference aggregation principles from social choice to support design decisions [9]. Specifically, multiple individual preferences are properly combined in order to rank-order alternatives from most to least desirable. Such thinking, although useful in certain type of design decisions, cannot be directly adopted to manage constraints for design synthesis. Unlike design objectives which can be compared by their relative importance and design solutions that can be measured by their absolute performance, all types of design constraints are equally important in terms of preventing non-functional or unacceptable solutions. Hence, there is no strong need for an absolute ranking of all constraints. Nevertheless, the diverse preferences towards constraints are still important in design; they should be comprehensively collected and systemically combined to ensure that the "fence" of design space is seamless and flawless. Alternatively, we can say that the preference towards certain design constraint should be treated as a sort of "veto power" from individual designers to the final group design decision. As a matter of fact, the existence of constraints in design is one major difference between engineering design and social choice [9].

There are also many efforts devoted to develop certain constraint-based automatic reasoning and logic programming to support design decisions [10, 11]. These methods, although are effective in some specific domains (e.g., geometry design and digital circuit verification), do not meet the general applicability requirement of the new constraint management model for design synthesis. The "subjectivity" of design constraints must be carefully controlled in order to maintain such domain-independent requirement. Specifically, the new model must provide the clear definition, criteria, and classification of design constraints in order to objectively sort different kinds of constraints and distinguish the design target from design constraints. As well, for each type of constraints, the new model must prescribe an appropriate strategy to address it objectively.

8.3 A Constraint Management Model for Design Synthesis

8.3.1 Classification of Design Constraints

Various constraints must be considered in design synthesis. In general, these constraints can be classified into two types: "input constraints" which apply to the overall design task, and "system constraints" that apply to specific design decisions [8]. The input constraints are closely associated with the given design task (i.e., the assignment); hence they are designer-independent and must be satisfied by all proposed solutions. Whereas the system constraints are resulted from designers' previous decisions, therefore they are always designer-dependent and specific to certain design decisions. For examples, in industries, corporate strategy, market competition, government regulation, budget, and schedule, which are imposed to designers associated with the initial design task, can all be categorized as input constraints. Behaviors of certain device, capability of particular manufacturing machines and some domain-dependent physical laws, which relate to the realization of specific design objective, should be classified as system constraints. In short, the major difference between input constraint and system constraint lies in the original source. If the constraint comes from a general design assignment, it is defined as an input constraint; whereas if the constraint results from designers' specific decisions, it is defined as a system constraint.

Meanwhile, constraints can be either internal or external of the technical system being designed [6]. Design begins with an initial intent (i.e., goal), and this intent grows to become a sophisticated technical system by purposefully synthesizing relevant resources, information and constraints. During such synthesis process, the gradual evolution of the technical system is constrained by both internal and external forces. The internal constraint is a part of the technical system; hence, it limits the evolvement of the technical system only from inside. The internal constraint is evident when design targets demand more than the current technical system can deliver. For instance, if the chosen machine cannot successfully produce the required component, then it becomes the internal constraint of the manufacturing process. In contrast with internal constraints, external constraints are not part of the technical system; as a result, it bounds the expansion (rather than evolution) of technical system from the outside. The external constraint appears when the technical system tries to function more than it currently capable of or jump out of the scope of a given design task.

Meanwhile, constraints for design synthesis can come from both social and brute realities. Social reality knowledge is those stakeholder-dependent agreements resulted from social interactions; whereas the brute reality knowledge is those stakeholder-independent natural laws derived from domain physics. For example, the constraints from social reality can include the preferred strategies and business objectives of the company in terms of the product outcomes as well as the existing market competitions identified through benchmarking. The constraints derived from brute reality (which must be treated as non-negotiable "hard"

Table 8.1 Classification of constraints in design synthesis

Type of constraint	Input constraint		System constraints	
	Internal	External	Internal	External
Level of abstraction	General	Specific	Mostly general	Specific
Level of flexibility	Mostly Rigid	Rigid	Flexible	Rigid
Level of variability	Dynamic	Static	Dynamic	Static
Lever of subjectivity	Objective	Objective	Subjective	Mostly Objective
Practical example	Initial CN	Competitions	Behaviors of DP	Physical laws

constraints in synthesis) include, for examples, available production facilities, budget limits, and particular physical knowledge of the application domains, such as physics, materials, etc.

Based on the above discussions, design constraints can be classified into four types: internal input constraint, external input constraint, internal system constraint, and external system constraints. Table 8.1 summarizes their different characteristics. Specifically, internal input constraint defines the constraint which must be part of the technical system but are not chosen by designers themselves. External input constraint represents the constraint that is not contained in the technical system but is part of the design task description or particular assignment. Internal system constraint refers to the constraint which is chosen by the designers to be part of the technical system. External system constraint describes the constraint that is resulted from designer's previous decisions but not part of the final technical system. As Table 8.1 indicates, among all types of constraints, internal system constraints are most difficult to manage due to their special characteristics at early stages of design.

To create a new technical system, the design synthesis process starts with the functional design phase when the designers rationally choose a set of functional requirements (FR) to fully satisfy the given customer needs (CN). The internal input constraints are determined by the initial CN, whereas the external input constraints (e.g., budget and schedule) are imposed to the designers as part of the given design task. At this stage, due to the specific choice of FR, certain external system constraints (e.g., market competition, corporate strategy, government regulations, etc.) are incorporated in the decision making process (Fig. 8.1). Next, the designers proceed to the conceptual design phase to select certain design parameters (i.e., DPs) that can satisfy FRs. At this stage, many internal system constraints will emerge to limit the realization and further decomposition of FRs. For instance, if the behaviors of a particular DP cannot successfully satisfy the respective FR, this DP will become the internal constraint of the evolving technical system. Similarly, according to the Axiomatic Design Theory, the DPs at higher level of the DP hierarchy may become constraints of FR at the next level of the FR hierarchy [7]. The last stage is the technical design phase, when the process variables (PVs), which can satisfy the DPs, must be determined to manufacture the technical system. In this stage, more external system constraints must be

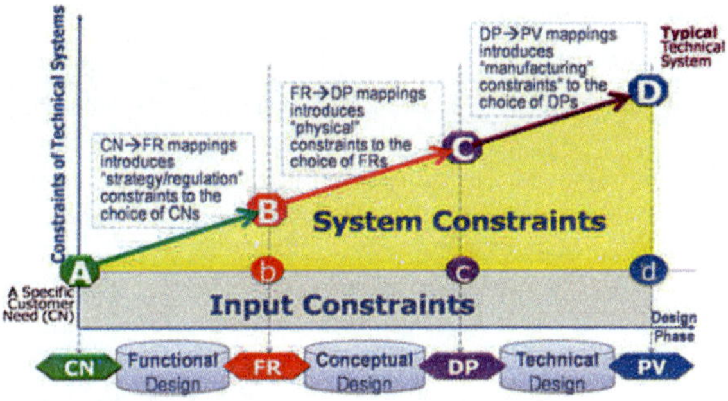

Fig. 8.1 Input/system constraints in different design phases

considered (Fig. 8.1). For instance, certain physical laws and available manufacturing capability, which are external of the technical system, will become constraints of production of the chosen DP.

8.3.2 Strategies of Resolving Design Constraints

For each type of design constraint, we prescribe a unique strategy to address it in design synthesis. This is necessary, because different kinds of constraints appear in different design phases and play diverse roles in synthesis. Hence, they cannot be resolved using one universal strategy. For the internal input constraints (e.g., customer demands), since they are derived from the initial design intent (i.e., to satisfy CNs), they cannot be simply removed or ignored. Because such constraints are often started as intangible and lack of details, the best strategy is to define more specifications to avoid any potential violation in the upcoming decisions. For the external input constraints (e.g., budget and schedule), because they are mostly imposed on designers by outside parties (e.g., management), the best strategy is to carry out collaborative engineering negotiation [12].

For the internal system constraints (e.g., behaviours of certain device), since they are resulted from designers' own decisions, they are subject to change. Note that even if certain internal system constraints can be removed by changing the previous decision, new internal system constraints will always arise due to the same decision change. Because the internal system constraint is mostly subjective, flexible and dynamic, whether to remove it or comply with it often relies on designers' preference aggregation result. For the external system constrains (e.g., physical laws and manufacturing capacity), since they are introduced to designers due to certain system requirements (e.g., security and quality), they should be treated as hard constraints and cannot by violated in any circumstance. Therefore, the best strategy is to add extra buffers (e.g., safety factors) to the technical system to ensure that the external system constraint is never starved [6]. Table 8.2 summarizes the specific strategy for each type of constraints.

Table 8.2 Strategies to resolve different types of constraints

Type of constraint		Type of strategy
Input constraint	Internal of system	Define more specifications to avoid violation
	External of system	Engineering negotiation
System constraint	Internal of system	Preference aggregation
	External of system	Add extra buffers to avoid violation

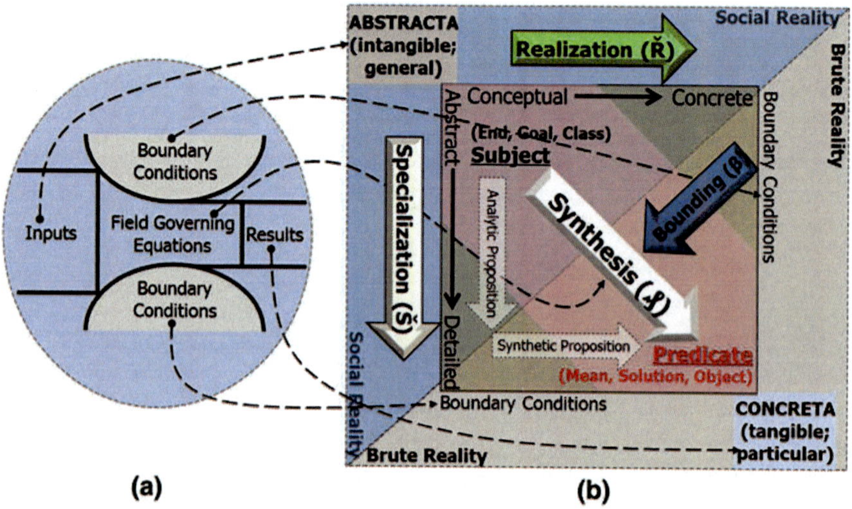

Fig. 8.2 Design synthesis as a progressive constraining process. **a** An initial-boundary-valued problem. **b** A structured synthesis reasoning framework

8.3.3 Design Constraints in Synthesis Reasoning

When viewed as a generic reasoning activity, synthesis is intrinsically an ill-posted (or under-constrained) problem. This suggests that synthesis reasoning can only produce a specific result under a proper set of constraints. Conceptually speaking, synthesis reasoning can be modelled as solving an initial-and-boundary-valued problem [1]. Figure 8.2 shows how an initial-and-boundary-valued system model (Fig. 8.2a) is conceptually mapped to an existing synthesis reasoning framework (Fig. 8.2b). The squared region in Fig. 8.2b represents a bounded (or constrained) "synthesis reasoning field" within which synthesis reasoning constantly transforms relatively conceptual/abstract design intent to more concrete/detailed design instantiation by following certain governing equations.

In this conceptual framework for synthesis reasoning, constraints are framed as the boundary conditions that take advantage of various bounding information to limit the creation and consideration of possible options. In Fig. 8.2b, the four sides of the squared area describe the internal and external input constraints, the two

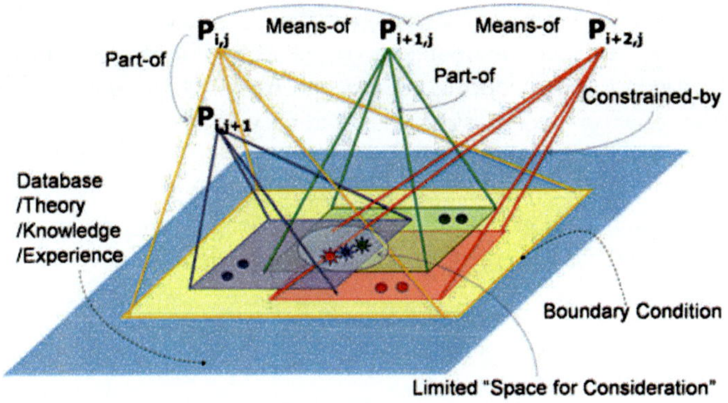

Fig. 8.3 Design synthesis as a progressive constraining process

edges of the diagonal band represent the external system constraints, and the Bounding (β) operation indicates the reasoning "forces" from constantly emerging internal system constraints. Note that in synthesis reasoning, the Bounding (β) operation establishes a "constrained-by" dependency relationship.

Using such a framework, design synthesis can be regarded as a progressive refinement process. Figure 8.3 below provides an alternative perspective to look at various design constraints in design synthesis. Generally speaking, synthesis is, given an abstract $P_{i,\ j}$, to arrive at a concrete $P_{i\ +\ 1,\ j\ +\ 1}$ that "is-a" tangible "thing". Rather than randomly mapping from $P_{i,\ j}$ to $P_{i\ +\ 1,\ j\ +\ 1}$, design synthesis must follow a systemic manner to firstly establish a "space for consideration", then to ideate possible solutions within this limited space. In this process, $P_{i,\ j}$ as the given intent, serves to define the input constraints; whereas $P_{i\ +\ m,\ j\ +\ n}$ (i.e., the representation of previous decisions made) functions to progressively construct the system constraints via "constrained-by" dependency relationship. Note that, the proposed synthesis reasoning process is based on the Axiomatic Design, and its application scope is limited to the creative engineering design at early stages.

8.4 Conclusion

In design synthesis, to impose certain design constraints to a technical system is an important decision which will affect further evolvement and final outcome of the technical system. Such decision cannot be made randomly, but rather it must be consistent with the initial design goal and all previously decisions. This task is particularly challenging at early design stages where constraints are still intangible, subjective and dynamic. This paper presents the initial development of a new constraint management model for synthesis reasoning. Within the proposed model, design constraints are firstly classified into four types: internal input constraint, external input constraint, internal system constraint, and external system constraint. For each type, the model prescribes a unique management strategy. Finally, we further incorporate the above studies into a domain-independent synthesis reasoning framework in order to systemically build boundary conditions (i.e., constraints) for design synthesis. Future works of this research include developing a tracking method to quantitatively trace the dynamic changes of design constraints and an adductive reasoning based diagnosis method to identify the early violation of constraints. A series of design experiments will also be conducted to illustrate the performance of this new model.

References

1. Lu SC-Y, Liu A (2011) A Synthesis Decision Framework for Early-stage Innovative Design. Proceeding of 2011 CIRP Design
2. Moreau CP, Dahl DW (2005) Designing the solution: the impact of constraints on consumers' creativity. J Consum Res 32(1):13–22
3. Costello F, Mark K (2000) Efficient creativity: constraint-guided conceptual combination. Cogn Sci 24(2):299–349
4. Stokes PD (2001) Variability, constraints, and creativity: shedding light on claude Monet. Am Psychol 56(4):355–359
5. Jeon H (2000) Dynamic constraint management in collective design. Pro Quest Dissertations and Theses
6. Goldratt E (1999) Theory of constraints. North River Press, Croton-on-Hudson
7. Lu SC-Y, Liu A (2011) Subjectivity and objectivity in design decisions. CIRP 7 Ann Manuf Technol 60(1):161–164
8. Suh NP (1990) The principles of design. Oxford University Press, New York, p 401
9. Scott MJ, Antonsson EK (1999) Arrow's theorem and engineering design. Decis Making Res Eng Des 11(4):218–228
10. Shpitalni M, Lipson H (1997) Automatic reasoning for design under geometrical constraints. CIRP Ann: Manuf Technol 46(1):85–88
11. Jaffar J, Maher MJ (1994) Constraint logic programming: a survey. J Logic Program 19(20):503–581
12. Lottaz C, Smith IFC, Robert NY, Faltings BV (2000) Constraint-based support for negotiation in collaborative design. Artif Intell Eng 14:261–280

Chapter 9
Webcrawling for a Biological Strategy Corpus to Support Biologically-Inspired Design

D. Vandevenne, J. Caicedo, P.-A. Verhaegen, S. Dewulf and J. R. Duflou

Abstract In the context of a larger effort to develop a tool that supports ideation in the early stage of Biologically-Inspired Design, this paper describes how the first important research question is tackled: any scalable approach towards such a tool requires a large corpus of biological strategies. This corpus should contain as much of the world's knowledge about how organisms tackle problems as possible and it should be updated in an automated way. However, currently such a resource or system does not exist. This paper presents a scalable webcrawling approach that allows to continuously search the Internet for biological strategies and to keep its knowledge base up-to-date without manual interaction. The webcrawler solves this needle-in-a-haystack task by combining different classifiers to score the relevance of web documents to the envisaged corpus. It uses these scores to focus future crawling and to gain efficiency. In this way, it becomes possible to continuously harvest new biological strategy documents in a scalable way. Finally, the possible applications of this contribution are positioned in the different existing approaches for systematic BID.

D. Vandevenne (✉) · J. Caicedo · P.-A. Verhaegen · J. R. Duflou
Department of Mechanical Engineering, Centre for Industrial Management,
K.U.Leuven, Belgium
e-mail: dennis@cib.kuleuven.be

S. Dewulf
CREAX NV, Ieper, Belgium

A. Chakrabarti (ed.), *CIRP Design 2012*, DOI: 10.1007/978-1-4471-4507-3_9,

9.1 Introduction

Biologically-Inspired Design (BID), also known as Biomimicry, Biomimetics, Bio-Inspiration and Bionics, is the discipline where inspiration is taken from nature to solve problems humans encounter. BID is receiving increasingly more attention from research and industry because of the two main advantages the field is often associated with: sustainability and proven performance [1, 2]. Other noteworthy advantages of biomimetic products are their enhanced marketability caused by the entailed green image of such products, by the savings they often imply (e.g. less energy consumption) and by the association to the organism itself (e.g. to swim as fast as a shark with a biomimetic swimsuit). Furthermore, drawing inspiration from a largely unused biological knowledge domain entails a higher probability of identifying leapfrog innovations.

These high expectations of biomimetic products are currently not met with adequate methods and algorithms to enable designers to identify candidate biological strategies for biomimetic design. Most existing biomimetic ideas currently originate from spontaneous inspiration, e.g. the invention of Velcro. The inventor, George de Mestral, spontaneously observed the ability of the cockleburs to attach itself to the fur of his dog. This inspired him to study the phenomenon in detail and to develop the well-known innovation. Another way to integrate bio-inspiration into the innovation process is the employment of a multidisciplinary design team, an approach which, although expensive, provides no guarantee for success. Therefore, a knowledge-based, systematic BID process is envisaged to generate more biomimetic ideas in less time by reducing the element of chance in ideation. However, a large repository, containing the world's knowledge about biological strategies, that can support such a knowledge-based ideation system does not exist currently. Therefore, this paper presents a first iteration of a new webcrawling approach that automatically identifies biological strategies on the Internet and that has the ability to stay up-to-date without requiring human extensive involvement.

9.2 Related Research

To the best of our knowledge, no webcrawler has been reported on that collects biological strategy documents. However, webcrawling has been used before as a corpus building strategy, especially for applications where it is difficult or expensive to build a large document collection of a specific type. The typical application domain is linguistics, e.g. for building under-resourced languages corpora [3] or building a cross-language domain-specific dictionary [4]. Because of the immense size of the Internet, trying to crawl every page is never attempted for topic-specific corpus building. Instead, a more directed approach is taken: focused webcrawling. A focused or topical crawler searches for pages relevant to a predefined topic. It is a useful tool for collecting information on very specific

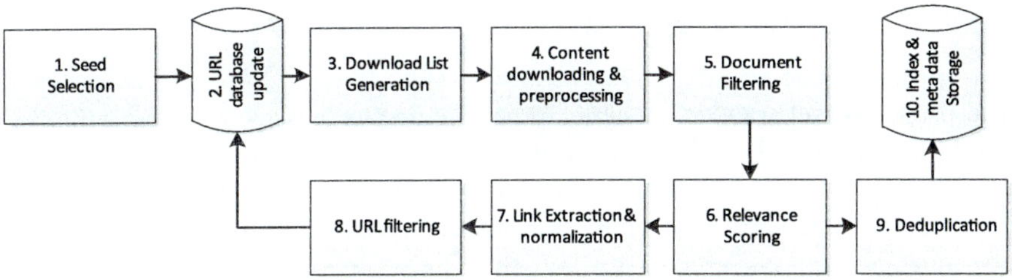

Fig. 9.1 Main components of the proposed focused webcrawler for the collection of biological strategy documents

topics in a cost-effective way. A focused crawler can be interpreted as a best-first crawler in which an evaluation mechanism measures the relevance of Web documents to a given topic. This evaluation mechanism can range from simple keyword matching to more complex artificial intelligence classification techniques. In the next section, the proposed focused biological strategy webcrawler and its components are detailed.

9.3 Webcrawler Outline

Figure 9.1 details the different components which together form the proposed focused webcrawler. Any general (unfocused) webcrawler, e.g. one that indexes a large number of pages, needs a way to select its seed URLs (1) which are injected into the URL database (2) when crawling starts. Next, a list of pages to download is generated (3) which are then downloaded and preprocessed (4). From the content of the downloaded web documents, a general webcrawler extracts and normalizes child links (7), and adds them to the URL database (2). These processes are iterated until a stopping criterion is reached. In each iteration, metadata about the visited documents is added to a database (10). Such an unfocused webcrawler visits all the URLs it encounters and is best typed as a breadth-first crawler where the download list generation process results in a first in–first out behavior (FIFO).

A focused crawler, like the one depicted in Fig. 9.1, modifies and adds components to the general scheme to increase crawling efficiency. In this way, it avoids having to crawl the whole Web to identify a relatively small number of relevant web pages. The proposed focused crawler performs a best-N-first URL selection strategy in the download list generation process (3). In this way, only a predefined number of most promising URLs from the URL database are fetched. In this database, the URLs are associated with scores which are assigned by an evaluation mechanism (6). Furthermore, the proposed focused webcrawler also integrates document filtering to retain only English documents and to remove biomedical documents (5), URL filtering to predict whether a new URL is likely to point to a biological document (8) and document deduplication (9).

9.3.1 Filtering Processes

The proposed webcrawler integrates three filtering processes: document filtering (5), URL-filtering (8) and document deduplication (9). Document filtering eliminates documents even before scoring them for relevance. First, all non-English documents are filtered by n-gram language analysis, where sequences of letters are modelled for each language and the closest model to a given document determines its language. Next, biomedical documents are filtered by the presence of common biomedical terms in the title of the downloaded documents. These common biomedical terms were identified by a term frequency analysis of 200 biomedical documents, such as: cancer, genome, health, disease, treatment, clinical, drug, etc.

Before new URLs are added to the URL database, a filtering step (8) is implemented to eliminate documents that are unlikely to discuss biology based on just their URL. Such pure URL-based classification has been reported to have potential by [5]. Feature selection is essential, where URLs first are split into tokens (e.g. AskNature) and then into words (e.g. Ask, Nature). These features are transformed into a feature vector to train and test a Support Vector Machine (SVM) classifier, which was found to perform best by [5]. The biology URL filter, developed for the webcrawler, has 93.9 % precision with 81.0 % recall. Other, more straightforward, URL filters eliminate URLs that point to unwanted file types like images, audio files or videos; or filter URLs with language codes that are unlikely to lead to English documents.

Duplicate document detection (9) is necessary because the same description of a particular biological strategy can exist on multiple pages with different URLs. For each document a hash value is calculated by a hash function. Such a hash function is able to transform document content to a much shorter hash value with two important properties: identical documents result in the same hash value and it is very unlikely that different documents result in the same hash value. In this way, exact duplicates are detected. Currently, only a rudimentary check is implemented to detect near duplicates by comparing page titles (see Future Work).

9.3.2 Scoring Processes

Relevance scoring, (6) in Fig. 9.1, is a process that estimates how likely it is that a downloaded document contains a biological strategy. It is an essential component of the proposed focused webcrawler because the scores it generates are not just used as evaluations of particular documents but they are also associated with the documents' child links as relevance predictions. This allows the crawling process to be focused.

Relevance scoring is performed by combining the following three components. For a detailed explanation of each component, please refer to [6].

- A supervised, content-based *biological strategy SVM classifier* is developed to assign a score that reflects the likeliness of a document being relevant to the corpus. The essential component of this classifier is feature selection, where only verbs and adverbs are retained from the documents. This rationale is supported by related research in design-by-analogy [7] and by a biomimetic design case, described by [8], where biological strategies are searched for, mainly by choosing verbs as search keywords. Furthermore, the selected term list is manually filtered for assumed relevance in the context of knowledge transfer between biology and engineering. For each term, the question is asked: could looking at how an organism does "the term", be interesting for design-by-analogy? This method, for example, selects terms like: protect, move, make, produce, distribute, etc. and eliminates terms like occur, result, report, propose, etc. In this way, the thousand most frequent, relevant verbs and adverbs are selected as features to train and test the SVM.
- A supervised, content-based *biology SVM classifier* is deployed to filter the potential non-biological downloaded documents. Downloaded documents are transformed into feature vectors by removing stop words, performing stemming [9] and applying term frequency–inverse document frequency transformation.
- A *mentions ratio filter* is used to filter documents which are not mentioning enough or which are mentioning too many organisms. Mention detection is performed by an open source species name identification system called LINNAEUS [10]. The database is expanded to include all scientific and common organism names of the NCBI taxonomy, spread over 26 biological ranks. Furthermore, the standard stop list is expanded to include organism names that are rare but frequently appear in texts in a non-relevant context (e.g. Alexandra, Nevada, Erica, laser, tapes, etc.) or organism names that are more likely to be mentioned in non-strategy documents because of their popularity as test animal (e.g. rats, mice, cattle, etc.). The mention ratio is defined as the ratio of organism mentions to the total word count.

The above three components are combined in the following way [6]. Documents that have a content biology score in [0.8,1.0] and a mentions ratio in [0.01,0.06] are assigned the score of the content-based strategy classifier; other documents are given score 0. For the target application, precision is much more important than recall as precision determines the focusing quality of the crawler. With a cut-off value of 0.80, the combined scoring process has 78.1 % precision.

9.4 Webcrawling Results

To test the proposed biological strategy crawler, a seed list of 29 URLs was composed: some obtained from journal directories on biological and life sciences, others from Web directories. The seed list is composed of academic seeds because such resources are expected to demonstrate a higher average quality and detail

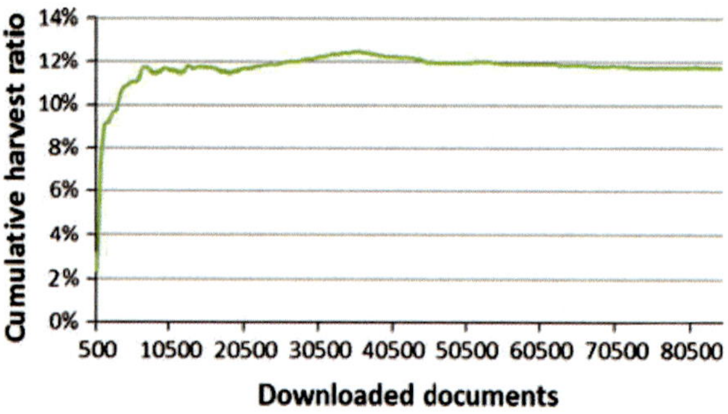

Fig. 9.2 The cumulative harvest rate

compared to, for instance, blog posts. The complete list and seeding strategy can be consulted in [6]. The number of downloaded URLs per iteration, cf. the N in best-N-first crawling, is set to 256, as [11] found that this value gives a good balance between exploration and exploitation.

To evaluate crawling efficiency, the harvest rate, the rate at which relevant documents are found [12], is used. To measure harvest rate, ideally one would expect manual inspection of the downloaded document collection. However, this task is subjective and very costly for a large corpus, hence the scores of the combined evaluation method, as described above, are used to measure the harvest rate. This use of the combined evaluation method is testing the assumption that highly relevant pages tend to refer to highly relevant pages. The substitution of expert opinion with the combined evaluation method implies that the resulting harvest rate should be interpreted with a precision of 78 % on the resulting biological strategy corpus.

A focused crawl of 110000 documents has been performed. After duplicate removal, and taking into account that about one in five links are dead, this resulted in 84309 processed documents of which 9890 are classified as biological strategies. As can be seen from Fig. 9.2, at first a couple of thousands of downloaded documents are necessary to reach a steady cumulative harvest rate. From then on, harvest rate is and remains steady at about 12 %. Until present, the largest experiment conducted went up to 350,000 documents with no drop in harvest rate.

A focused crawler should achieve a higher harvest rate than an unfocused crawler to justify additional focusing effort. To verify this, the crawler was also tested with two other implementations: once with only the biology SVN classifier to focus the crawl and once with a modified version of the Online Page Importance Computation (OPIC) algorithm [13], hence unfocused. The cumulative harvest rate of both versions never exceeded 1 %, clearly justifying the extra computation required by the proposed focused crawler.

Tests are performed on a quad-core (2.83 GHz) with 16 GB RAM and a broadband Internet connection. No technical limitations are met that restricted, or are expected to impede, the scalability of the webcrawler to more extended runs. This is the consequence of the best-N-first URL selection strategy, which causes each iteration to deal with the same number of input URLs. Therefore, mainly depending on the number of broken links and the impact of active politeness policies, processes 4–8 take about the same time each iteration. Processes 2 and 3, respectively updating the URL database and generating the download list, increase very slowly with the size of the URL database, but they do not impede large crawls significantly. In general, the main bottleneck for the proposed crawler is politeness, which is necessary to avoid causing Denial of Service (DoS) attacks or excessive server bandwidth costs and is considered as good practice for webcrawling. Politeness prevents a host to be consulted more than once every 5 s.

9.5 Discussion and Applications

Over the last decade, the awareness of the advantages of BID in the design community has increased significantly, which has resulted in a research community aiming at developing methods and tools to support ideation in BID. A detailed overview of the different methods and tools is provided in Ref. [14], positioning the major contributions in four sequential steps that characterize a BID process: problem formulation, search, analysis and transfer. Although the presented webcrawling approach, with a constant cumulative harvest rate of about 12 %, is designed to support a scalable ideation tool for BID, an automatically generated and updated biological strategy corpus has the potential to support all ideation methods and tools for systematic Biologically-Inspired Design, as explained below.

In the case of a manual search one tries to summarize the problem by selection of just a few search words which subsequently are used to query a database, e.g. biological library databases or the Internet. Besides the cumber of the inherent terminology gap between the engineering and biological domains, such resources contain mostly non-relevant documents which makes the task at hand a needle-in-a-haystack search problem. By using the biological strategy corpus generated by the outlined webcrawler, a much more efficient search becomes possible which enables the designer to evaluate more candidate ideas in the same time.

The three methodologies below focus on the first two steps of the BID process: problem formulation and search, and can benefit from a large biological strategy corpus in the following ways. AskNature [15] is a tool based on a functional hierarchical taxonomy, called the biomimicry taxonomy, which structures a database of currently 1385 biological strategies. The large number of strategies, generated by the webcrawler, could be positioned in this taxonomy to drastically increase the size of their biologically strategy database. BioTRIZ, a TRIZ-based [16] methodology proposed by [17], requires engineering problems as well as biological solutions to be abstracted to the contradiction in design parameters that characterizes the systems in both domains and the inventive principles that apply for the resolution of the conflicts are identified for reuse later. The authors of [17] currently integrate 2500 conflicts, originating from 500 biological phenomena, which is less than a fraction of the data behind classical TRIZ. The biological corpus, generated by the proposed focused webcrawler, could be used to expand the BioTRIZ knowledge base. A third contribution, proposed by [18], attempts to bridge the terminology gap between the engineering and biological domain by means of a systematic, semi-automatic search method that requires the design problem to be expressed in functional keywords and then generates biological meaningful bridge verbs and text passages containing them [18, 19]. Although it is explained by [14] that this method does not scale for large corpora, the general introductory biological textbook "Life, the science of Biology" [20], currently used as input, can be replaced by a relatively small selection of biological strategies generated by the webcrawler.

In contradiction to the previous three methodologies, the following three focus on the last two steps of the BID process: analysis and transfer. They require the instantiation of detailed models for each biological and engineering system resulting from the search phase. This time-consuming and expensive manual task encompasses a detailed analysis of the systems in both domains and expresses them to a common abstraction level in order to facilitate knowledge transfer. To the best of our knowledge, such a methodology has currently been reported for functional basis models by [21], for Structure-Behavior-Function (SBF) models by [22] and for SAPPhIRE models of causality by [23]. As explained by [14], the creation of such model meta data for very large biological strategy repositories, like the one generated by the proposed webcrawler, is not a task that scales well; as a single model instantiation typically requires hours of work and experts in both domains. Therefore, a large biological strategy corpus cannot be plugged-into these methods and systems directly. It is however possible to consider automated, or more likely semi-automated, instantiation of such models in the future.

The problem of scalability, which all of the above methods entail, is further detailed and addressed in [14]. The authors propose a scalable preprocessing step based on a high-level conceptual representation of the biological and engineering domains. This methodology is able to directly integrate a very large biological strategy corpus in the search phase of the BID process and proposes a small number of relevant biological strategies as input to the analysis and transfer phases. In this way, the three methodologies based on transfer models, described above, also become scalable.

9.6 Conclusion

The paper presents a novel approach for identifying large amounts of biological strategies on the Web using a focused webcrawler. This contribution addresses an important bottleneck for systematic BID, based on a large knowledge repository. The webcrawler is a software application that surfs the Internet for biological strategies without human intervention and that is able to automatically update its database to integrate current and future knowledge. The presented webcrawler has a steady cumulative harvest rate of 12 % which is high considering that identifying biological strategies on the Internet is a needle-in-a-haystack problem. The first iteration of the webcrawler has already resulted in promising results and a number of ideas for future work. Although the webcrawling approach is designed to support scalable systematic BID tools, its applicability to all types of existing contributions in the domain is detailed.

9.7 Future Work

This paper reports on a new research line, using webcrawling for building a corpus of biological strategies. The proposed crawler represents a first stable iteration, which allows demonstrating the potential of the approach. However, there are quite a number of ideas for improvement which can be addressed in future research, e.g.:

- Alternative seeding strategies, than the one explained above, can be researched.
- Crawl focusing, as described above, is mainly content-based. However, other, link-based, approaches estimate the likelihood of a link being relevant by analyzing the graph-like structure formed by its parent links. Furthermore, a combination of both approaches is possible for further performance improvements [24].
- More elaborate near-duplicate detection strategies can be researched to clean the resulting corpus.
- The proposed crawler is static, i.e. its evaluation mechanisms do not update their models during the crawl. An adaptive crawler would use the new information found during a crawl to improve its evaluation mechanisms, e.g. [25].

References

1. Bar-Cohen Y (2006) Biomimetics: biologically inspired technologies. CRC/Taylor & Francis. ISBN: 0849331633
2. Benyus JM (1997) Biomimicry: innovation inspired by nature. Harper Perennial. ISBN:0060533226
3. Scannell K (2007) The crubadan project: corpus building for under-resourced languages. Building and exploring web corpora. In: Proceedings of the 3rd web as corpus workshop, incorporating Cleaneval, p 5
4. Talvensaari T, Pirkola A, Järvelin K, Juhola M, Laurikkala J (2008) Focused web crawling in the acquisition of comparable corpora. Inf Retrieval 11(5):427–445
5. Baykan E, Henzinger M, Marian L, Weber I (2009) Purely URL-based topic classification. In: Proceedings of the 18th international conference on world wide web, New York, pp 1109–1110
6. Caicedo J (2011) Web crawling for the acquisition of a biological corpus, master thesis for master of industrial management. K.U.Leuven, Belgium

7. Verhaegen P, D'hondt J, Vandevenne D, Dewulf S, Duflou J (2011) Identifying candidates for design-by-analogy. Comput Ind 62(4):446–459
8. Lenau T, Dentel A, Ingvarsdóttir Þ, Guðlaugsson T (2010) Engineering design of an adaptive leg prothesis using biological principles. Design 2010, Dubrovnik—Croatia, 17–20 May 2010
9. Porter MF (1997) An algorithm for suffix stripping. Readings in information retrieval. Morgan Kaufmann Publishers Inc., San Francisco, pp 313–316
10. Gerner M, Nenadic G, Bergman C (2010) LINNAEUS: a species name identification system for biomedical literature. BMC Bioinformatics 11(1):85
11. Menczer F, Pant G, Srinivasan P, Ruiz ME (2001) evaluating topic driven web crawlers. In: Proceedings of the 24th annual international ACM SIGIR conference on research and development in information retrieval, pp 241–249
12. Chakrabarti S, Van Den Berg M, Dom B (1999) Focused crawling: a new approach to topic-specific web resource discovery. Comput Netw 31(1999):1623–1640
13. Abiteboul S, Preda M, Cobena G (2003) Adaptive on-line page importance computation. In: Proceedings of the 12th international conference on world wide web, pp 280–290
14. Vandevenne D, Verhaegen P-A, Dewulf S, Duflou JR (2011) A scalable approach for the integration of large knowledge repositories in the biologically-inspired design process. ICED11, Denmark, pp 15–18
15. Ask nature, the biomimicry design portal. Available from http://www.asknature.org. Accessed 30 Sep 2011
16. Altshuller GS (1984) Creativity as an exact science: the theory of the solution of inventive problems. Gordon and Breach Science Publishers. ISBN: 0677212305
17. Vincent JFV, Bogatyreva OA, Bogatyrev NR, Bowyer A, Pahl AK (2006) Biomimetics: its practice and theory. J R Soc Interface 3(9):471–482
18. Chiu I, Shu LH (2007) Biomimetic design through natural language analysis to facilitate cross-domain information retrieval. Artif Intell Eng Des Anal Manuf 21(1):45–59
19. Chiu I, Shu LH (2005) Bridging cross-domain terminology for biomimetic design. In: Proceedings of ASME 2005 international design engineering technical conferences and computers and information in engineering conference. Paper No. DETC2005/DTM-84908, Long Beach, CA, pp 24–28
20. Purves WK, Sadava D, Orians GH, Heller HC (2001) Life: the science of biology. W.H.Freeman & Co Ltd. ISBN: 0716726297
21. Nagel JKS, Nagel BI, Stone RB, McAdams DA (2010) Function-based, biologically inspired concept generation. Artif Intell Eng Des Anal Manuf 24(4):521–535
22. Vattam S, Wiltgen B, Helms M, Goel A, Yen J (2010) DANE: fostering creativity in and through biologically inspired design. In: First international conference on design creativity, Kobe, Japan
23. Chakrabarti A, Sarkar P, Leelavathamma B, Nataraju BS (2005) A functional representation for aiding biomimetic and artificial inspiration of new ideas. Artif Intell Eng Des Anal Manuf 19(2):113–132
24. Wang C, Guan Z, Chen C, Bu J, Wang J, Lin H (2009) On-line topical importance estimation: an effective focused crawling algorithm combining link and content analysis. Journal of Zhejiang University SCIENCE A 10(8):1114–1124
25. Chakrabarti S, Punera K, Subramanyam M (2002) Accelerated focused crawling through online relevance feedback. In: Proceedings of the 11th international conference on world wide web, pp 148–159

Chapter 10
Assessing the Performance of Computerized Tools for Inventive Design: Insights From Unsatisfactory Outcomes

N. Becattini, Y. Borgianni, G. Cascini and F. Rotini

Abstract Computers actually support, almost automatically, routine tasks such as those related to the optimization in design. Besides, the scientific community shows a growing interest in developing computer systems to aid non-routine tasks as a key to enhance individuals' creativity and innovation potential. In such a context, several attempts have been made to create tools based on the TRIZ logic to support inventive problem solving; some of them have been commercialized since decades, but still there is no established paradigm and all of them suffer from several limitations. So far the analysis of those limitations has been focused on the structure and on the nominal features of the software tools, while no in-depth and systematic investigation has been made to identify the reasons behind the partial failure of the existing systems. This paper proposes a set of general criteria to perform the evaluation of computerized tools supporting inventive design and reports an exemplary application, through protocol analysis, to the dialogue-based computerized algorithm for problem analysis, published by the authors in the past.

N. Becattini · G. Cascini (✉)
Politecnico di Milano, Dipartimento di Meccanica, Via La Masa 1,
20156 MI, Italy
e-mail: gaetano.cascini@polimi.it

N. Becattini
e-mail: niccolo.becattini@kaemart.it

Y. Borgianni · F. Rotini
Università degli Studi di Firenze, Dipartimento di Meccanica e Tecnologie Ind.li,
Via di Santa Marta 3, 50139 FI, Italy
e-mail: yuri.borgianni@unifi.it

F. Rotini
e-mail: federico.rotini@unifi.it

A. Chakrabarti (ed.), *CIRP Design 2012*, DOI: 10.1007/978-1-4471-4507-3_10,

10.1 Introduction

Computer support to the product development process is widespread in the industry since the last decades. However, current artificial intelligence resources allow to automate just routine activities, such as those involved in optimization problems, i.e. when computers are used to choose the most appropriate value of a pre-defined set of variables, but no significant qualitative modifications are expected. Typically, those tasks emerge in the last phases of the design process, when the range of possible choices is limited to the details of the system.

On the other hand, in the last years, the scientific literature is collecting a growing number of contributions about the introduction and the development of computerized systems for supporting the early stages of product development cycle, namely the design stages where it is required to solve inventive problems and where creativity plays a paramount importance [1]. Yet, cognitive processes involved in those design stages, e.g. ideas association, analogies, concept blending, are still activities completely in charge of human beings and computers role is barely aimed at supporting and fostering creativity either through a more efficient visualization of the mental model under study, or by guiding the reasoning path according to standardized strategies [2].

According to Funke and Frensch [3], problem solving is the most complex intellectual activity, because it deals with a number of tough characteristics as the presence of a large set of design variables mutually tangled, the need to satisfy multiple goals (politely) and the lack of clarity. With reference to this subject, Simon [4] suggested the distinction between ill and well-structured problems in design, according to a specific set of characteristics. An emerging branch of study is constituted by Computer-Aided Innovation (CAI). CAI Systems, often based on TRIZ theory, aim at supporting problem setting by guiding the designer to the formalization of a problem in terms of contradictions. TRIZ potential contribution to enhance industrial innovation has been largely acknowledged, as for example in [5]. Besides, all the existing software applications show several lacks in terms of real usability and usefulness.

In order to face this limitation, the authors have firstly dedicated proper efforts to the identification of the requirements a computer-aided inventive problem solving tool should satisfy; then they have developed a system implemented in a dialogue-based framework, namely OPEN-IT, which supports the problem setting phase by structuring the information according to TRIZ logic and fosters a learning-by-doing process (e.g. by teaching how to recognize the relevant aspects of a problematic situation) [6]. Two different groups of testers, holding a degree in Mechanical Engineering and with a scarce exposition to TRIZ theory, have demonstrated that the algorithm gives a significant contribution in the analysis of an inventive problem as presented in [6] and [7]. However, such experimental activity also highlighted that some testers still encounter difficulties during the definition of the problem characteristics, potentially leading to unsuccessful analyses.

An immediate conclusion arises from this experience: either it is hard to properly identify the requirements of a computerized tool for supporting inventive design, or it is harder than expected defining an algorithm capable to satisfy those requirements. None of the papers in the TRIZ or in the CAI literature provide relevant directions to address this dichotomy; thus, it is a relevant matter of investigation the definition of some general criteria and practices to assess the functionality and the usability of TRIZ-based computerized aid to inventive design.

Several attempts have been proposed in scientific literature about methods for determining why human–computer interaction does not produce the expected results (e.g.: there is plenty of papers about self-efficacy studies, as well reviewed for gender differences in [8]). Hewett et al. in [9] presented a method for evaluating the performances of a computer tool supporting creativity. Nevertheless, nowadays, a standardized procedure that provides precise measurements about failures of computer-aided innovation systems in producing good results is still missing.

This paper proposes a set of general criteria to study the unsuccessful analyses of technical problems carried out by users of a TRIZ-based tool for problem solving, with the aim of identifying the sources of inefficiency and, consequently, the directions for improvement. A first exemplary application of these criteria is done through an in-depth investigation of the failures emerged during the tests carried out by engineers and industrial designers that used the above-mentioned OPEN-IT computer based system. The aim is to identify the main criticalities of the software application, so that it is possible to clarify if new requirements emerge and if those formally fulfilled have yet to be satisfied. Protocol Analysis [10] has been taken into account as a well established approach for inspecting design cognitive processes and characterizing mental paths and behaviours.

A brief description of the characteristics and of the capabilities of the algorithm is shown in Sect. 10.2, together with contributions from literature that point out the attention to the role and requirements of computer applications within creative tasks. In addition, brief references about protocol analysis in design phases are mentioned. Section 10.3 clarifies the criteria for assessing the performance of a computerized tool for inventive design and Sect. 10.4 presents a detailed examination of the outcomes achieved through their application to the OPEN-IT experimental results and discusses about the main directions of development to be undertaken in order to improve the system. Eventually, Sect. 10.5 summarizes the original contribution of the paper and briefly discusses the evidences emerged in the analysis.

10.2 Computer-Aided Problem Solving: Lessons Learned from Past Experiences

As mentioned above, in order to improve the characteristics of the existing computerized systems for problem solving in design tasks, the authors proposed a computerizable algorithm for problem analysis, implemented in the OPEN-IT dialogue-based framework, whose characteristics are here briefly presented.

Insights from Scientific Literature Computer-aided systems for Problem Solving have to embed different characteristics [6]; some of those requirements were already discussed by other scholars. Lubart [2] recognize that the coaching of designers, acting as an expert system that guides the user throughout cognitive processes, is one of the utmost roles a computer can play. Hewett [11], from a different perspective, claimed that the problem analysis should be carried out by taking into account different facets of the problem, enlarging the range of investigation. Aurisicchio et al. [12] pointed out that the information gathering is a time consuming activity in design phases, concluding that it is necessary to ease the research of relevant contents from knowledge sources.

TRIZ-related requirements Consistently with TRIZ [13] and OTSM-TRIZ [14], a successful problem solving activity, capable to produce breakthroughs, is characterized by different aspects. The user should be supported throughout an abstraction process of problem features, as to focus just on the characteristics the solution should have until the convergence towards a unique and formalized description. Moreover, this abstraction activity should foster the user in defining technical barriers (in TRIZ terms contradictions) that prevent a direct implementation of typical solutions. At last, it is strictly required that such a system does not need the user to hold a long education period to become effective, in order to improve its usability in contexts where scarce resources for training courses are available.

OPEN-IT—Algorithm for Problem Analysis In order to embed all the above-mentioned characteristics into a computer aided-system for problem solving, the authors built a dialogue-based algorithm for the analysis of technical problems, whose latest updates have been published in [7]. A full description of the algorithm is out of the scope of the present paper. In brief, the computerized procedure is composed by more than 200 nodes organized in eight logical blocks. The nodes represent an articulate set of questions, choices or written messages exploiting a common terminology, rather than TRIZ jargon. Several nodes are aimed at checking the correctness of previous user answers, so that they can be employed in contextualizing the text of the following questions. The ultimate objective of the procedure is identifying the most critical TRIZ contradiction behind a given problem.

As briefly mentioned in the introduction, all the TRIZ-based systems supporting inventive design suffer from poor efficacy [15], especially for those individuals who are not experienced in using abstract models. Despite the intention to go beyond the limits of current commercial systems, some failures have been recorded also by the authors in the testing campaign of the OPEN-IT framework.

Generally speaking, the existence of unsuccessful results may highlight that some requirements are just partially satisfied, either in terms of functionality or in terms of usability; alternatively, new requirements still need to be elicited. Thus, a relevant objective is the definition of a set of criteria capable to shed light on the sources of limited performance. In this paper, the proposed criteria are applied in combination with a protocol analysis approach [10], due to its suitability in examining designers' behaviour along the design process [16].

10.3 Criteria for Examining the Results of a Computer-Aided Problem Solving Activity

A designer carries out his design activity by focusing on different problem features, such as the performances to be achieved, the drawbacks to avoid or the consumption of resources. In order to overcome problems, the problem definition strategy can be carried out at a more or less abstract level, thinking about structures and embodiments, exploitable physical principles, as well as requirements and goals. Traditionally, a design protocol analysis is carried out by processing each design step performed by the designer or the design team; in this case, since the study is dedicated to computer-aided tools supporting inventive design activities, the protocol analysis is focused on the interactions between the user and the software system. The below proposed criteria are aimed at classifying the steps of a design activity, regardless the Computer-Aided tool adopted.

Criteria for Strategy Assessment Computer-Aided systems for problem solving may leave a complete freedom or, on the other hand, force the user into a predefined set of steps or instruments to cope with. Therefore, these criteria have to take into account both the extreme situations, so to encompass all possible cases. Then, regardless of designers' choice or input request by the computer, the steps can be classified according to the following set of six criteria:

- *Functional Requirement:* Human–computer interactions (HCI) related to the elicitation of the objectives to be achieved by a given technical system.
- *Behavioural Variable:* HCI focusing on the mechanisms (physical, chemical, geometrical,...) that allow a certain phenomenon to take place.
- *Structural characteristic:* HCI taking into account specific design variables that allow to leverage a physical principle.
- *Choice:* HCI concerning decisions which are made without a particular reference to the strategy and the path to solve a problem.
- *Communication:* HCI dedicated just to transfer the designer information about the progressing process.
- *Check:* HCI through which the computer asks the designer about the correctness of previous steps.

Criteria for Assessing the Focalization on Problem Features As seen before, a designer may focus on different aspects of a problem. The below defined criteria are aimed at classifying them.

- *General features:* HCI focused on features and characteristics that are not directly related to the problem, but generally refer to the technical system.
- *Removal of drawbacks:* HCI characterized by undesired consequences emerging during the functioning of the technical system.
- *Presence of conflicts:* HCI aimed at individuating the elements of the problem that prevent the elimination of drawbacks.
- *Improvement of performance:* HCI that take into account a better achievement of performances for which the technical system has been designed.
- *Requirements for system functioning:* HCI coping with the means that allow the technical system to properly work.
- *Broadening spectrum of investigation:* HCI addressed at helping the designer to avoid fixation by exploring alternatives from a wider perspective.

Criteria for Assessing the Kind of Errors A traditional design activity, e.g. by using trial-and-error, is characterized by mistakes, or useless solution attempts, that reduce the efficiency of the problem solving process. The criteria defined hereafter are aimed at making a distinction between those mistakes:

- *Content:* Mistakes due to misunderstandings about how the system works, wrong interpretations of the mechanisms causing the undesired phenomena, poor investigation of the problem due to neglected elements or effects.
- *Form:* This class is a residual of the first one. Such mistakes are typically characterized by wrong insertions due to language issues, disregard of the directions to follow, as recommended by the computer system or dictated by the principles of the employed design methodology, etc.

The above definitions are clarified by the examples reported in the following Section. The overall set of criteria allows the exploration of designers' reasoning path according to different perspectives; however, it is required to specify which is the investigation sequence to follow, so that the analysis can be repeated in different contexts. The authors suggest to:

1. Record all the steps of a problem solving activity by means of an appropriate method for protocol analysis (think-aloud or conversational, concurrent or intro/retrospective, combinations…);
2. Determine the characteristics of each step, both in terms of strategy and focus;
3. Evaluate the correctness of the steps, following their original sequence, maximally through objective criteria (e.g. for form mistakes) or by experts' assessment when required;
4. Determine, for each incorrect step, whether the error regards the content or the form.

The above-presented criteria can successfully describe the steps of a problem solving activity carried out with the support of a computer tool, as confirmed by several tests carried out on different software applications. In detail, the questionnaire of Innovation Workbench (www.ideationtriz.com) easily allows to record the designer's activity and all the provided answers can be classified with reference to the suggested criteria. They can be also used to classify the steps carried out with Invention Machine's Tech Optimizer (www.inventionmachine.com), as well as South beach Modeller (www.southbeachinc.com). The recording of those design steps requires the additional employment of software for logging keystrokes to accomplish the above step 1.

10.4 Insights from Unsuccessful Problem Analyses and Discussion

The authors have carried out a high number of tests with their problem solving algorithm, owning an intrinsic capability to record design steps. Therefore it is possible to examine a significant sample of problem solving session logs, by using the proposed criteria and investigation procedure.

A group of graduates, composed by both males and females holding a MS or a PhD in Mechanical Engineering or in Industrial Design, has been asked to face two technical problems [7] emerged in real industrial contexts by means of the OPEN-IT platform presented in Sect. 10.2. Testers' competencies on systematic problem solving were almost completely absent, since just an individual of the sample claimed to have been submitted to 20 training hours in TRIZ. Each of them holds at least a First Certificate in English or higher. At the end of the testing session, 24 analyses were selected for the present study, because they showed at least one error in any node of the questioning procedure and did not result in particularly valuable outcomes within the scope of solving the encountered problem.

A list of exemplary errors related to several criteria among those defined in Sect. 10.3, is reported below (words in brackets refer to terms previously introduced by the designer):

- *Functional Requirement* Question: "Which technical function is carried out by the <calendaring system> in order to <apply a film on surface>? Use the infinitive form of the verb without "to" (i.e. keep ink, dry the clothes, deliver a box...)"; Answer: "air bubble".
- *Behavioural Variable* Question: "Which is the undesired effect that arises in the system as a consequence of getting the satisfactory level of the <cleanliness of the frying pan>? Use a noun without the article or a verb in the -ing form (e.g. high noise, overheating,...)"; A: "avoid rivets, introducing a new film".
- *Removal of drawbacks* Question: "Which is the undesired effect that arises in the system? Use a noun without the article (e.g. noise, bone breaking, vibrations, obstructed view...)"; A: "limited effect". (*Note* the user does not specify what effect he is talking about).

Table 10.1 Steps where users have provided a wrong answer both in terms of content and form. The percentage of wrong steps has been calculated as the ratio between the number of mistakes for the specific feature and the overall number of errors

Feature under analysis	% of wrong steps
General features	58.72
Removal of drawbacks	10.47
Presence of conflicts	8.72
Improvement of performances	8.72
Requirements for system functioning	0.00
Broadening the spectrum of investigation	13.37

- *Presence of conflicts* Question: "Do any bad consequences come out if you <increase> the <roughness of the external surface> of the <driving roller>?"; A: "No". (*Note* the user does not realize that there is an undesired consequence).
- *Content* Question: "Define the instant or the initial condition in which the <rivets> start to <fix the mutual position>"; A: "dirt".
- *Form* Question: "Define the instant or the initial condition in which the <mechanical joint> starts/start to <clamping the handle>"; A: "assembly process". (*Note* the user is referring to the whole time interval, instead of the initial instant as requested).

The following examination counts the revealed punctual mistakes, regardless the quality of the final outcomes of the questioning procedure. Its results will be presented in an aggregate form, so to highlight whether the system globally satisfies the requirements it has been designed for. As shown in Table 10.1, the main evidence refers to the highest percentage of mistakes occurring along the steps concerning the "General Features" of a technical system. Furthermore, Table 10.2 shows that the most severe difficulties arise when the designer is asked to specify aspects of the technical system that are strictly related to mechanisms and the physical principles that determine the presence of both desired and undesired effects.

In order to gather with greater accuracy information about the sort of encountered problems, the authors collected detailed insights about the marginal distribution of this kind of errors, characterizing them also in terms of "content" and "form" as shown in Table 10.3. In the Table, a further cluster of "mistakes" is illustrated, which is relevant for the specific CAI application adopted or for any dialogue-based system. The Table includes the counting of those queries for which the designer gave no answer.

A noticeable presence of mistakes related to the form may highlight that the main observed limitations could be due to the user interface of the computer-aided system, rather than on the poor knowledge of the designer about the physics of the specific issue under investigation. Nevertheless, the results of Table 10.3 do not confirm such hypothesis, although further investigations are needed to obtain more robust indications. An in-depth analysis of the results shows that the algorithm still lacks in the capability of abstracting problem features. Indeed, according to the

Table 10.2 Summary of incorrectly answered steps. Communication and Check mistakes have been taken into account because they present errors caused by previous wrong steps

Focus of the analysis	% of wrong steps
Functional requirement	28.68
Behavioural variable	46.51
Structural characteristic	15.50
Choice	1.55
Communication	6.50
Check	1.55

Table 10.3 Correlations between strategy and kind of errors. Misalignments in sums are due to the rounding of percentages

	Content (%)	Form (%)	Not assigned (%)	(%)
Functional	23.26	5.43	0.00	28.68
Behavioural	36.43	10.08	0.00	46.51
Structural	14.73	0.78	0.00	15.50
Choice	1.55	0.00	0.00	1.55
Communication	0.00	0.00	6.20	6.50
Check	0.00	0.00	1.55	1.55
	75.97	16.28	7.75	100.00

data illustrated in Table 10.2, the most critical aspects to be investigated are related to functional features and behavioural variables, while just a minor percentage of errors concern the structural characteristics of a technical system.

Moreover, it is worth to reflect upon the uneven distribution of feature related errors, since diverging conclusions may emerge according to different interpretations. On the one hand, designers could have encountered troubles since the beginning of the procedure, where "General Features" related questions are asked; on the other hand, the same testers could have paid less attention to those aspects that they do not consider essential for the description of the problem.

Therefore, the above considerations show how this preliminary criteria-based analysis is capable to reveal the main essence of flaws in computer-aided systems and allows to plan appropriate strategies to overcome the arisen shortcomings. A detailed examination of individuals' behaviour may result as an important element to distinguish the facets that are still ambiguous, or, as well, to determine the extent of the factors that generate problems along the analysis of technical systems.

10.5 Conclusions

This work briefly summarizes the requirements of a Computer-Aided Problem Solving System. The authors propose an original metric to evaluate the behaviour of designers using this sort of tools, so that it is possible to analyze the steps of the design process according to definite criteria. The purpose is to start to examine why specific human–computer interactions produce failures, allowing to highlight whether the requirements already addressed by the literature are still far to be met and, if needed, to elicit new ones.

An application of this kind of protocol analysis has been conducted on 24 tests using a computer-aided system for problem analysis developed by the authors. The examination of the results in an aggregate fashion allows to make preliminary considerations about the degree of achievement of the above-mentioned requirements. The obtained results show the directions of development to be prioritized in order to improve the framework of the computer-aided system. Specifically, a relevant feature to be addressed concerns its capability to support the user in abstracting the problem and allowing him/her to focus with more attention on facets that may appear as marginally related to the problem, but that can hide potential direction for its solution.

Further insights about the analysis of the results obtained by the individuals could represent an interesting point to be discussed, especially considering the branch of protocol analysis in computer-aided design that is among the purposes of future investigations by the authors.

Eventually, this general approach can be also easily replicated on different computer-aided systems for problem solving that use a different way to structure the designers' knowledge. Related results can constitute a starting point to share a common vision on what should be done for obtaining a more mature and reliable computerized means for supporting the creative stages of the design process.

References

1. Hüsig S, Kohn S (2011) Open CAI 2.0—computer aided innovation in the era of open innovation and web 2.0. Comput Ind 62(4):407–413
2. Lubart T (2005) How can computers be partners in the creative process: classification and commentary on the special issue. Int. J Human–Comput Stud 63:365–369
3. Funke J, Frensch PA (2007) Complex problem solving: the European perspective. In: Jonassen DH (ed) Learning to solve complex scientific problems. Lawrence Erilbaum, New York, pp 25–47
4. Simon HA (1973) The structure of ill-structured problems. Artif Intell 4:181–201
5. Tomiyama T, Gu P, Jin Y et al (2009) Design methodologies: industrial and educational applications. CIRP Ann—Manuf Technol 58(2):543–565
6. Becattini N, Borgianni Y, Cascini G, Rotini F (2012) Model and algorithm for computer-aided inventive problem analysis. Comput Aided Des 44(10):961–986
7. Becattini N, Borgianni Y, Cascini G, Rotini F (2011) A TRIZ-based CAI framework to guide engineering students towards a broad-spectrum investigation of inventive technical problems. Int J Eng Edu (in press)
8. Burnett MM, Beckwith L, Wiedenbeck S et al (2011) Gender pluralism in problem-solving software. Interact Comput 23(5):450–460
9. Hewett T, Czerwinski M, Terry M et al. (2005) Creativity support tool evaluation methods and metrics. Report of NSF workshop on creativity support tools (available at http://www.cs.umd.edu/hcil/CST/report.html), 10–24

10. Ericsson KA, Simon HA (1993) Protocol analysis: verbal reports as data (rev. ed.). The MIT Press, Cambridge
11. Hewett T (2005) Informing the design of computer-based environments to support creativity. Int J Human–Comput Stud 63:383–409
12. Aurisicchio M, Bracewell R, Wallace K (2010) Characterising the information requests of engineering designers. J Eng Des 21(6):707–730
13. Altshuller GS (1984) Creativity as an exact science. Gordon and Breach Science Publisher, New York
14. Cavallucci D, Khomenko N (2007) From TRIZ to OTSM-TRIZ: addressing complexity challenges in inventive design. Int J Product Dev 4:4–21
15. Cavallucci D (2011) A research agenda for computing developments associated with innovation pipelines. Comput Ind 4:377–383
16. Jiang H, Yen CC, (2009) Protocol analysis in design research: a review. Paper presented at the design | rigor and relevance, international association of societies of design research (IASDR) conference Seoul, Korea

Chapter 11
Comparing a Graph-Grammar Approach to Genetic Algorithms for Computational Synthesis of PV Arrays

Corinna Königseder, Kristina Shea and Matthew I. Campbell

Abstract The research presented in this paper compares two different computational approaches to topology synthesis: graph grammars (GG) and a genetic algorithm (GA). Both are applied to the domain of solar panel networks, which are configured to maximize power under a variety of sunlight conditions. Networks mostly connected in series create high voltages but are susceptible to shading. On the other hand, highly parallel topologies tend to create too low voltages. Thus, there is an opportunity to optimize photovoltaic (PV) array topologies given sunlight conditions. The paper describes methods to model the PV topology optimization problem using both a synthesis and an optimization approach. The focus is on computationally representing the design problem as well as the problem specific knowledge. Implementations of two different grammar rule sets in the GG are compared to four different encodings of the design problem in the GA. Results are shown for both approaches and a discussion of their advantages and drawbacks is given.

C. Königseder · K. Shea (✉)
Technische Universität München, Virtual Product Development Group,
Boltzmannstr. 15, 85748 Garching, Germany
e-mail: shea@pe.mw.tum.de

C. Königseder
e-mail: koenigseder@pe.mw.tum.de

M. I. Campbell
Department of Mechanical Engineering, University of Texas at Austin,
Austin, Texas, USA
e-mail: mc1@mail.utexas.edu

A. Chakrabarti (ed.), *CIRP Design 2012*, DOI: 10.1007/978-1-4471-4507-3_11,

11.1 Introduction

Energy generation from renewable resources is one way to approach the increasing demand for energy. Photovoltaic (PV) modules are one of the most commonly known renewable energy sources. Usually PV modules are connected in series to create PV arrays and obtain the desired voltage level [1]. Much effort has been put into the development of intelligent control strategies to maximize the amount of power a PV array generates [2]. Most of them include Maximum Power Point Tracking (MPPT), which aims to maximize the power output by controlling the load of the PV array. Shading of a single PV module however, highly affects the amount of power a PV array can produce. Especially under partial shadow conditions it has been proven that not only MPPT but also appropriate topologies can optimize the overall power generation [1].

In this paper a PV array topology optimization problem is used as an application for illustrating different approaches in modeling topology optimization problems. Modeling refers to defining or determining an approximate representation of a real world system in order to analyze it. Determining a topology or configuration is one of the most difficult tasks in the design process and is the basis for computational design synthesis (CDS). CDS supports the creation of new, innovative and complex products. It is often applied to assist the human designer in generating and evaluating many alternatives and leverages the computer's ability to search the space for preferred and optimal solutions [3]. The CDS process can be divided into four major steps that are presented in [3]: representation, generation, evaluation and guidance.

All of these four essential steps are regarded in the PV array topology optimization example and the main focus of this paper is on the representation. The aim is to compare different modeling approaches to this topology optimization problem. One approach is based on graph grammars (GG) and synthesizes the PV array network using GG rules. It will be referred to as the "GG approach". In the other approach, referred to as the "GA approach", possible connections between single PV modules are defined as a bitstring and a genetic algorithm (GA) is used to search for the best design to connect them. Both, GG synthesis and GA optimization are generic methods that lead to their wide range of applicability.

This paper is organized as follows. In Sect. 11.2 background knowledge on both PV modules and CDS is given. The PV array topology optimization problem is presented in Sect. 11.3. Two representations of the problem and the implementation of the constraints in the GG (Sect. 11.4) and GA (Sect. 11.5) approach are given. The approaches are compared and results are shown (Sect. 11.6) and discussed (Sect. 11.7). Section 11.8 concludes the paper with an outlook.

11.2 Background

PV Cells PV cells form the basis for most PV power systems. In general, a PV array consists of several interconnected PV modules where each consists of several interconnected PV cells [4].

The connection of modules to create PV arrays is important because of its high potential to improve the overall power generation especially under shaded conditions [1], [4]. Partial shading not only influences the shaded PV modules, but the whole array. The amount of influence is based on the topology in which the modules are arranged. In general, the shading of a single module results in a lower current production that affects all modules within the shaded module's series. In 2007, at least 19 distinct methods were compared in literature for tracking the maximum power point (MPP) of PV systems [2]. Thus, the final tracking of the MPP is assumed to be possible. The research focus here is on creating topologies that allow for the highest MPP under non-optimal conditions such as partial shading. For the research described in this paper no lower boundary for voltage is given.

Computational Design Synthesis (CDS) Two different CDS methods are compared in this paper regarding their ease to model a real world PV array system for topology optimization. In this Section, these methods are characterized in general and the state-of-art for each is reviewed.

In mechanical engineering, *graph grammars* (*GG*) have been and still are developed by several researchers. In the four step approach for CDS, the specification of graph rules is executed in the representation step. In this phase the problem specific knowledge is formulated using graph rules and an initial graph for starting the synthesis process is specified. In the generation, the applicable rules are identified. Either the computer or the human designer chooses which one(s) to apply. Through the recursive execution of the generation step, new designs can be synthesized. To influence this generation process and to achieve better designs, the evaluation and guidance step are used. In the evaluation step, the generated candidates are evaluated while the guidance step includes strategies for preserving and eliminating candidates, changing search direction, and terminating the process.

Several design tasks have been supported with graph grammars e.g. the synthesis of mechanisms and epicyclic gear trains [5], the automated synthesis of gear boxes [6] and sheet metal design [7]. Besides assisting synthesis, graph grammars allow the support of the whole design process, e.g. flexibly providing knowledge that was previously stored in paper-based design catalogues [8] and reusing concepts from extracted design knowledge [9]. One of the key enablers to establish the use of graph grammars in engineering is the ability of a graphical interface to define and interpret grammars. GG interpreters, such as booggie [10] or GraphSynth [11], allow for the design of grammar rules and ease their application. For a review of recent approaches see Chakrabarti et al. [12].

Genetic algorithms (GA) stem from evolutionary strategies that have been introduced in the 1950s. All of these strategies were based on the evolution of a population of candidate solutions to a given problem. Soon evolutionary strategies were used for solving engineering optimization problems [13]. One deficit of GAs is that the designer has to put great effort into modeling the design problem such that the design variables are represented on a genotype-level. This usually leads to transformations between a string representation and one that can be evaluated.

11.3 Problem Formulation

The PV array topology optimization problem is used as an application to compare GGs and GAs to represent and design a real world system.

The objective of the topology optimization problem is to find a topology producing maximum power for a given shading condition. Topologies are defined by connections between neighboring modules specified by their compass directions. This means module A can have a neighbor B to its north, east, south or west. As no analytical formula for PV array power generation using arbitrary topologies and shading exists, a simulation model is used for evaluation. To keep the problem size at a reasonable level and to allow for a generic simulation model, the following restrictions are made:

1. the model is restricted to 16 PV modules on a 4×4 module grid
2. only neighboring modules can be connected
3. bypassing modules is not allowed which includes:
    a. a connection between a PV module and the positive terminal of the grid is only allowed if there is no other incoming connection to the PV array
    b. a connection between a PV module and the negative terminal of the grid is only allowed if there is no other outgoing connection from the PV module
4. no connections to modules from outside are allowed
5. cycles between two modules (coexisting connections from module A to B and from module B to A) are not allowed

Restrictions 1–3 limit the creation of infeasible, unnecessary complex or known suboptimal solutions, restrictions 4 and 5 are mostly based on the electrical solver to keep the simulation converging in a timely manner (i.e. not "crashing").

The size of the solution space for this problem is 3^{24} ($\approx 28 \times 10^{10}$) as there are 3 options for each connection between two modules A and B (no connection, connection from A to B, connection from B to A).

11.4 GG Approaches

In the generative grammar approach, a GG is used for the representation and generation of the PV array topology. It comprises rules and an initial graph to represent the real world model and generate designs. The evaluation is left to a simulation model but two GG implementations are treated in this Section. The first

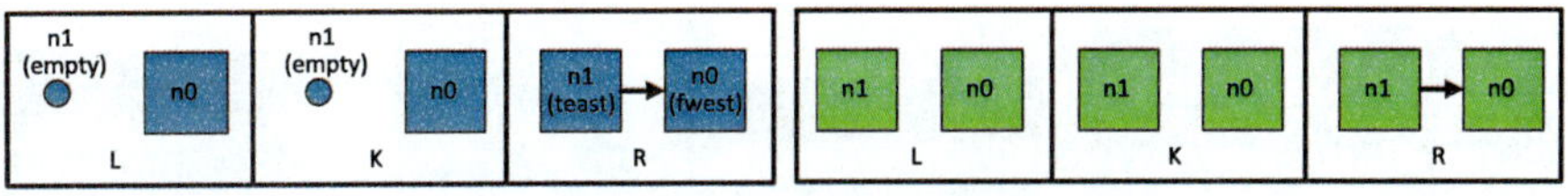

Fig. 11.1 Rule in "modules" (*left*) and "connections" (*right*) approach

one concentrates on the nodes (representing PV modules) while the second one focuses on the arcs (representing connections between existing modules) in the graph representation.

GG "modules" Approach In the first GG approach the focus is on developing the PV array adding one module after another. Two rule sets are implemented, the first one to generate new topologies for the PV array, the second to preprocess the generated model to enable simulation. An initial seed graph is created which defines 16 (four by four) possible positions for PV modules. Rule set one consists of five rules. Four rules search for an empty position for a PV module in the neighborhood of an existing one, add a new module in respective direction and connect the new module to the existing one. The fifth rule adds a new module to an empty position on the grid but does not connect this module to neighboring ones. An example rule application that adds a new module on the west side of an existing one is shown in Fig. 11.1 (left). The rule is defined by its left hand side (L), right hand side (R) and context (K). Placeholders are represented as small circles and PV modules as squares. Labels on PV modules on the right hand side (teast, fwest) and the arrow of the arc indicate the direction of the connection.

Connections to the positive and negative terminal of the PV array are only implicitly defined in these rules and are determined by the rule of rule set two according to constraint 3 in Sect. 11.3. This rule also transfers the graph into a representation that can be used as input for the simulation model in the evaluation step.

GG "connections" Approach In the second GG approach the focus is on developing the topology of the PV array by connecting already existing PV modules. Similar to the previous approach, two rule sets are implemented to generate, respectively preprocess the topology.

The initial seed graph is a field with four by four PV modules and arbitrary connections of these 16 PV modules. Rule set one consists of two rules. Rule one connects two neighboring modules; rule two removes the connection between two modules. Connections to the positive and negative terminal of the PV array are calculated by the rule of the second rule set as in the "modules" approach. An example rule application that adds a new connection between modules is shown in Fig. 11.1(right). Rule set two is similar to the one in the "modules" approach.

11.5 GA Approaches

As a comparison to the GG approach, a GA is used. The topology of the physical system is represented as a bitstring. The generation and guidance step of the CDS

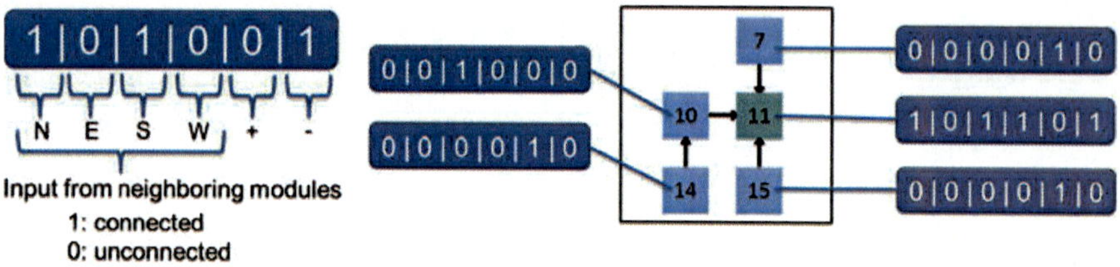

Fig. 11.2 Definition of the bitstring in "modules" approach for one module (*left*) and examples for bitstring representations (*right*)

process are done by the algorithm. The evaluation is carried out by running the simulation model. Two different bitstring encodings are presented in the following. Similar to Sect. 11.4 these are referred to as "modules" and "connections" approaches.

GA "modules" Approach for the PV array in this paper's example, each of the 16 modules can be defined by six switches. This results in 96 bits to represent the topology of the PV array analogously to the "modules" approach in the GG approach. These bits indicate whether the module has incoming connections from north, east, south or west or a connection to the positive or negative terminal of the entire array, whereby "1" means that a connection exists, "0" that it does not (see Fig. 11.2). Similar to the GG approach, the connections to the positive and negative terminal can also be determined. This reduces the bitstring from 96 to 64 bits.

There are several ways to implement the constraints in the GA approach which depend on the representation (96 or 64 bits) of the system. The three following methods are presented to model constraints (1) as explicit constraints, (2) as explicit and implicit constraints, or (3) using a penalty function. Explicit constraints are given as inequality constraints to the GA and are to be solved before the objective function is calculated, i.e. triggering the simulation to be run. The formulation of the 64-bit genome implicitly captures some of the constraints and is hereafter referred to as implicit constraints.

Requirements 1, 2 and 4 of Sect. 11.3 are implemented similarly for all three methods in the GA bitstring representation and the simulation model.

Method (a) which handles all constraints as *explicit constraints* uses the 96-*bit* bitstring representation. Restrictions 3 and 5 are defined by 56 inequality constraints.

Method (b) uses *explicit and implicit constraints* and a bitstring representation that is reduced to 64 *bits*. Compared to method (a) the amount of explicit constraints is reduced from 56 to 24 inequality equations.

Method (c) uses the 64 *bit* representation introduced above and a *penalty function*, which rejects infeasible solutions. In method (c), infeasible designs are penalized and the run of the simulation for these topologies is avoided but invalid designs are not removed from the population. With this method the constraints of the GA can be understood as a validity check implemented in the objective function.

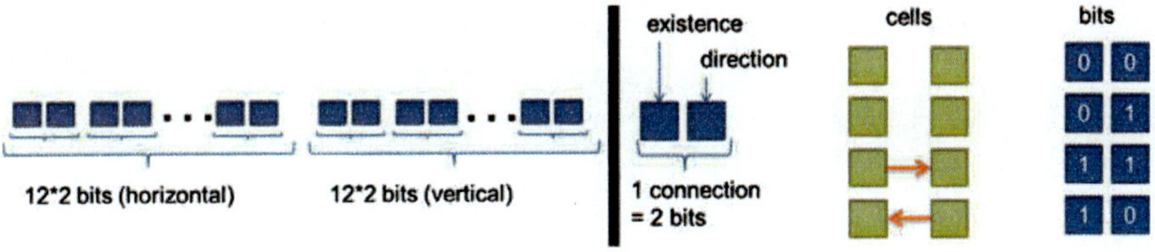

Fig. 11.3 Definition of the bitstring in "connections" approach in general (*left*) and examples for encoding of arcs in horizontal direction (*right*)

GA "connections" Approach The second GA approach is implemented similar to the "connections" approach in the GG. The bitstring represents the connections between the modules in a different way compared to the "modules" approach. This results in 12 horizontal and 12 vertical connections between the 16 modules on the PV array grid. Based on these 24 connections a 48 bit bitstring representation is used in which each of these connections is represented by two bits. The first bit indicates the existence of a connection and the second gives the direction of the connection (see Fig. 11.3). However, this biases the representation to parallel connected designs. The determination of the connections to the positive and negative terminal is carried out in the objective function. With this encoding no explicit constraints or penalty functions are necessary, as all restrictions are implemented in the bitstring representation or the mapping.

11.6 Experiments and Results

The GG and the GA approaches are compared through several experiments.

The simulation model for the PV array is modeled in Matlab R2010b using Simulink 7 and the SimScape and SimElectronic toolboxes. To enable the creation of arbitrary topologies, all possible connections are defined and equipped with switches. Depending on how these switches are set, the topology of the PV array changes. For the representation, generation and guidance step in the GG approach, the GG synthesis software GraphSynth [11] is used. Using the rule in rule set two in the GG approaches, the simulation model is loaded in Simulink and a simulation run is performed with the topology created. The objective function value, in this case the maximum power of the generated PV array, is returned to the search process. In the GA approach, the GA of the Matlab optimization toolbox is used. The mapping of the bitstring to the inputs for the simulation model as well as the simulation itself is done in the objective function of the GA. The GA is not adapted to the specific problem as it is not the goal of this paper's research to tune a specific method but to show how different problem representations can influence the results. *For the GGs* 20 test runs are executed wherein in each run 1,000 designs are created by randomly applying the rules of rule set one. Each design is evaluated and the best candidate's topology is stored. *For the GAs* 20 test runs are

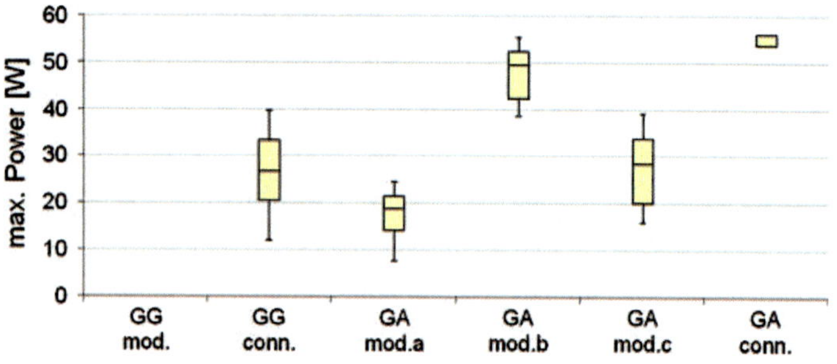

	Method	Constraints	(a)	(b) [W]	(c)	(d) [h:mm]
GG	modules	none	yes	56.2	guaranteed	00:49
GG	connections	none	no	27.2	guaranteed	00:52
GA	modules a	explicit	no	17.8	not guaranteed	03:48
GA	modules b	explicit and implicit	yes	48.3	not guaranteed	03:49
GA	modules c	penalty function	no	27.2	not guaranteed	00:43
GA	connections	none	yes	54.7	guaranteed	00:55

Fig. 11.4 Results of 20 experiments with 1,000 runs (boxplot (*top*): lines define minimum and maximum and box defines lower quartile, median and upper quartile)

executed and in each run 1,000 designs are created using the Matlab GA default settings with a population size of 20 and a maximum of 50 generations. For all runs the best individual is stored.

The results of 20 experiments with 1,000 simulations each are shown in Fig. 11.4. No shading pattern is used, which means all modules are fully illuminated. This is done because in this case the optimum design is known. Therefore, and due to the huge search space (compare Sect. 11.3) no experiments with an exhaustive search method are conducted to compare the results to. The optimal design for such a configuration regarding power solely is a configuration with all modules connected in parallel. For all approaches the following results are presented: (1) attainment of the optimal design (yes/no), (2) average value for the maximum power of the best design found in each optimization run, (3) feasibility of generated designs, (4) run-time.

For the *GG "modules" approach* all 20 runs creating random designs produce feasible results and reach the optimal design. The average over all designs created with this approach however is much lower (P_{mean} = 27.0 W). With the *GG "connections" approach* only feasible designs are created and the run-times are similar to the "modules" approach.

In the *GA* "modules" *approach* the optimal design is attained only once. Using method (a) and (b) simulation interruptions occur caused by non-solvable linear algebra errors. These interruptions represent infeasible designs not solvable by the simulation tool. They are not taken into consideration for the results shown in Fig. 11.4 but are discussed in Sect. 11.8. Run-time is higher than for the GG approach and method (c) due to constraint solving. Method (c) runs without interruptions but does not create feasible designs in each experiment. To allow comparison, only feasible designs are considered as results. The *GA "connections" approach* shows the best results among the GA approaches but again simulation interruptions occur during the runs.

11.7 Discussion

The GG and GA approach differ in how the physical system is modeled. The GG approach is more intuitive and allows an easier encoding of constraints but it is hard to check whether the rules capture the whole feasible space. In the GA approach the whole design space is captured but the difficulty is to encode the system as a bitstring and to sufficiently restrict the design space to feasible solutions.

All presented approaches allow modeling the PV topology optimization problem however, only the GG approaches generates feasible designs for all experiments. Referring to the results for the maximum power averaged over all experiments, the GG "modules" approach shows the best results and is the only approach to find the global optimum in each run. This is a result of the random generation of designs and the chosen shading pattern since it prefers a solution with parallel modules. The probability to select the necessary rule to add a module in parallel is higher than for other rules, since it can be applied on any empty position. Further research can include different shading patterns and algorithms to guide the generation step.

The GG "connections" approach shows considerably worse results compared to the GG "modules" approach and the GA "connections" approach. This is caused by the nonexistent guidance and the infinite search tree for this approach as the application of one rule can undo a previous rule application (e.g. add/remove connection). It shows the direct coupling between the representation and search space.

Among the GA constraint methods ("modules" approach), method (b) performs best. Method (a) performs worse because of its higher number of explicit constraints that have to be solved and that do not contribute to optimizing the design but keeping the solution in the feasible space. In method (c) numerous designs are rejected because the constraints are not met, so this approach resembles more a naive generate and test loop than an optimization. The GA "connections" approach shows best results among the GA approaches and has the most efficient encoding.

Comparing the "modules" and "connections" approaches, the GG performs best for the "modules" approach while the GA performs better in the "connections" approach. This is because each representation fits more naturally with the representation in the search algorithm in each case. Which representation to choose therefore is not only dependent on the problem to be solved but also on the used CDS approach.

11.8 Conclusion

In summary, for the given problem, the GG approach allows an intuitive representation of the feasible space of solutions through graph rules to both generate valid alternatives and map a design model to a simulation model. Using a GA, the implementation of problem specific knowledge shows the most promising results when realized without constraints. The issue of optimizing the topology for various shading conditions is not yet addressed but is equally possible with the CDS approaches and simulation model described above. Including shading will increase the problem complexity and is a possible extension along with constraining the lower boundary for voltage given by current step-up converters.

Acknowledgments The authors would like to acknowledge the Technische Universität München—Institute for Advanced Study, funded by the German Excellence Initiative.

References

1. Gao L, Dougal RA, Liu S, Iotova AP (2009) Parallel-connected solar PV system to address partial and rapidly fluctuating shadow conditions. IEEE Trans Industr Electron 56(5): 1548–1556
2. Esram T, Chapman PL (2007) Comparison of photovoltaic array maximum power point tracking techniques. IEEE Trans Energy Convers 22(2):439–449
3. Cagan J, Campbell MI, Finger S, Tomiyama T (2005) A framework for computational design synthesis: model and applications. J Comput Inf Sci Eng 5(3):171
4. Patel H, Agrawal V (2008) Maximum power point tracking scheme for PV systems operating under partially shaded conditions. IEEE Trans Industr Electron 55(4):1689–1698
5. Schmidt LC, Shetty H, Chase SC (2000) A graph grammar approach for structure synthesis of mechanisms. J Mech Des 122(4):371–376
6. Lin Y-s, Shea K, Johnson A, Coultate J, Pears J (2009) A method and software tool for automated gearbox synthesis. ASME, San Diego
7. Patel J, Campbell MI (2010) An approach to automate and optimize concept generation of sheet metal parts by topological and parametric decoupling. J Mech Des 132(5):051001
8. Helms B, Shea K, Hoisl F (2009) A framework for computational design synthesis based on graph-grammars and function-behavior-structure. ASME, San Diego
9. Kurtoglu T, Campbell MI (2006) A graph grammar based framework for automated concept generation. DESIGN
10. Helms B, Shea K (2010) Booggie—an object-oriented graph-grammar implementation for engineering design synthesis. In: Gero JS (ed) Fourth international conference on design computing and cognition dcc'10. Springer, Stuttgart
11. Campbell MI (2011) The official graph synth site. University of Austin, Texas [cited 2012 03.01.2012]. Available from: http://www.graphsynth.com
12. Chakrabarti A, Shea K, Stone R, Cagan J, Campbell M, Hernandez NV et al (2011) Computer-based design synthesis research: an overview. J Comput Inf Sci Eng 11(2):021003
13. Mitchell M (1996) An introduction to genetic algorithms. In: Press M (ed) MIT Press

Chapter 12
Toward an Automatic Extraction of IDM Concepts from Patents

Achille Souili and Denis Cavallucci

Abstract This paper presents a new approach for knowledge extraction from patent document for the use of design engineers. It is based on the Inventive Design Method (IDM) which derives from TRIZ, the theory of inventive problem solving whose goal is to theorize on the act of inventing partly on the bases of patent's observation. The purpose of this paper is to demonstrate that it is possible to automatically extract IDM concepts from patent document, using generic linguistic markers. This paper is the continuation of a previous work and mainly focuses on the retrieval of parameters which characterize the main components of contradictions related to an artifact evolution bottlenecks.

12.1 Introduction

A patent is a form of intellectual property that consists of a set of exclusive rights granted to an inventor or their assignee. The World Intellectual Property organization advances that 90 to 95 % of all the world's inventions are found in patented documents. Additionally, the European Patent Office also disclosed that "patents reveal solutions to technical problems, and they represent an inexhaustible source of information: more than 80 % of man's technical knowledge is described in patent literature" [1]. Thereby, patents constitute a first choice media, when

A. Souili (✉) · D. Cavallucci
LGECO/INSA Strasbourg, 24 Boulevard de la Victoire,
67084 Strasbourg Cedex, France
e-mail: wendemmi.souili@etu.unistra.fr

D. Cavallucci
e-mail: denis.cavallucci@insa-strasbourg.fr

A. Chakrabarti (ed.), *CIRP Design 2012*, DOI: 10.1007/978-1-4471-4507-3_12,

speaking of technological information. Thus, the observation of an artifact throughout the years enables the understanding of its evolution mechanism. In this context, and besides few contributions on patent mining, patents are underexploited in the process of designing new products; and there is a field of research in building relevant methodologies using patent mining, aiming at better assisting R&D activities.

Developed to assist engineers in their invention process, TRIZ and its derived method, Inventive Design, consist of logic, data; and research rather than intuition; and turns to be a relevant solution for these needs [2, 3].

TRIZ relies on the revealing of laws of technical systems evolution; and therefore enables an oriented design activity that anticipates future artefacts according to these laws. It includes three fundamental principles:

- Some problems and solutions are recurring in industry and science. A predictive solution to these problems can be found by categorizing the contradictions, which will be defined later.
- Patterns of technical evolutions may also be recurrent across many industries.
- Creative innovation, representing these technical evolutions, generally emerges outside the field where they were developed.

As for IDM, it was created to clarify the concepts involved in TRIZ [2]; and therefore solve its limits. These concepts were later developed by OTSM [4]. In fact, the comprehension of classical TRIZ which lacks formalized ontology is a complex task; and it is difficult to construct a computational design model upon these actual techniques and concepts.

In this context, patent documents can be considered as a reliable sign of human innovative activity; and IDM as well as TRIZ its underlying theory are based on the assumption that artifacts evolve according to a number of objective laws. The transition from a generation of an artifact to another is symbolized in a patent in a non-explicit way through the overcoming of a contradiction without any compromise. Thus, assuming that some information can be contained in patents, the latter are a vital knowledge source for an IDM survey. Problems, partial solutions and parameter concepts behind problems, which are most of the time characterized by one or more features of the artifact that are deteriorated or unsatisfactory, are worth noting here.

Unlike other traditional problem solving methods, TRIZ helps to break psychological inertia, often declared as vital in an inventive process, through the identification of contradictions. The research of the contradiction or the underlying inventive problem is currently manually made by experts. Besides, it is a time-consuming task and the results obtained are certainly not exhaustive.

This paper reports on an ongoing research to automate the retrieval of IDM concepts for the purpose to populate its ontology. Namely, it deals with the extraction of parameters [5] which are related to the concepts of contradiction within TRIZ. We pose the hypothesis that parameters may be localized through the use of generic linguistic markers and Natural Language Processing (NLP) tools. In this perspective, a literature review of contributions made so far on patent mining for TRIZ is introduced in Sect. 12.2. Section 12.3 is devoted to the presentation of IDM ontology. We later expose our method in detail in Sect. 12.4. As for Sect. 12.6, it is dedicated to conclusion and perspectives, after a discussion on the proposed method in Sect. 12.5.

12.2 Literature Review

This section presents major contributions made so far on patent mining in general and specifically on patent mining for TRIZ. With the development of the internet and the availability of patent database, several Information Retrieval (IR) systems, with some related to TRIZ, have intended to automate knowledge retrieval from patent documents. Such IR systems, addressing patents, typically use hybrid methods by associating statistics and linguistics. Feldman et al. [6] propose a document explorer which implements a text mining at the term level. A list of candidate words is produced to be keywords after a basic linguistic preprocessing. Such an approach is promising for processing unstructured part of patent document. In addition to this, Ghoula et al. [7] present a processing chain achieving an automatic semantic patents annotation through a structural ontology and domain ontology; biology as regarding their case.

However, the above mentioned works do not give access to the invention process. They do not consider artifact improvements. Most approaches considered thus far are limited in their domain of application and do not address all the patent sections.

With the development of TRIZ, many authors have intended to automate its use since its efficiency was obvious. For the purpose of using the inventive the inventive principles to solve problems in different domains, several researches were taken on the basis of NLP techniques such as SAO-based approaches. Closer to our work than the approaches mentioned above, SAO based approaches suit better for the unstructured patent section mining. SAO-based patent analysis which controls the syntactical structure of subject (Noun Phrase), Action (Verb Phrase) and object (Noun Phrase) explicitly represents relationships between the components of a patent. It is intrinsically connected to the concept of function understood differently by different authors. Savranski [8] calls it "the action changing the feature of any object" whereas for [9, 10] functions performed by or on components are represented by Action which constitutes with Subject the component of a system. Cascini et al. [9, 10] particularly advocate the use of functional analysis to identify a problem and generate innovative solutions. To solve a problem, this one is broken into its component functions which are later divided in sub functions until the function level for solving the problem is reached. Expressed as Subject-Actions-Object triads [10], functional analysis proves to be relevant for the representation of knowledge related to the patents key findings and the inventor's domain of expertise.

In addition, Moehrle et al. [11] focus on visualizing the similarities between patents, by means of Multi-Dimensional Scaling (MDS). The authors propose to adapt multidimensional scaling to SAO structures, for the purpose of mapping technological convergence between two companies. Besides, Yoon et al. [12] present a method to identify TRIZ trends automatically. Their approach, based on the use of adjectives and verbs, assumes that property refers to a specific characteristic of a system; and is usually described using adjectives whereas function indicates an action that alters the features of an object; and is usually described with verbs. Therefore, they propose to use the binary relations of "*verbs* + *nouns*" or "*adjectives* + *nouns*" to define specific trends and trend phases through semantic sentence similarity measuring. Furthermore, Dewulf describes properties as the attributes of a product. "What a product is or has". "They are mainly expressed in adjectives and is related to physical parameters" [13].

Still in TRIZ-based patent mining, an approach complementary to ours is Duflou et al.'s approach to identify contradictions in IDM-TRIZ (Duflou et al. [14]). Besides, Coelho's method [15] proposes to match and categorize human factors ergonomic principles in industry, under the light of the 40 inventive TRIZ principles. Furthermore, Verhaegen et al. [16] propose a method to automate the process of positioning the product on TRIZ trend in a radar plot, by analyzing and categorizing patents in known trend phases. Some researchers have also used patent citations to estimate the level of inventiveness of a given technology [17].

Our method is based on the use of generic markers of change to determine the evolution of the artifact. These changes may concern the deterioration or the improvement of one or more components of the artifact. Such characteristics may be defined through parameters which are features of the system. The purpose of such a determination is the retrieval of parameters, which are like problems and partial solutions, IDM-TRIZ concepts.

12.3 Our Method

This paper exposes on an ongoing project to automatically retrieve IDM related knowledge since this is manually done so far by experts. It explores the use of generic linguistic markers which may indicate concepts relevant to IDM ontology. Our approach tries to follow IDM's way of extracting relevant parameters when analyzing patent in the initial situation phase and indexing parameters that will allow defining the problem. Thus, parameters are of the utmost importance when dealing with inventive design studies.

12.3.1 IDM and Its Ontology

As stated above, IDM is an extension of TRIZ and deals with artifact evolution. Developed to address wider and more complex problematic situations, it assumes that any object created by human being is the result of evolution guided by objective laws. In patent documents, artifact evolves according to problems and partial solutions. Considering that the creation of an invention is its failure to change according to the previously mentioned laws, IDM advocates the clear formulation of contradictions to provide the inventor with a technique and knowledge base, which will enable him to generate design solutions that can be called inventive.

Ontology may be defined as the standard representation of a field or domain of the important categories of objects or concepts which exists in the field or domain, showing the relation between them. On the contrary of usual static and specific ontologies, IDM ontology is generic; and therefore intends to be applicable to all patents irrespective of their fields. It is dynamic i.e. it describes the impact of changes on each other. The approach tries to follow IDM expert's one when analyzing patents in the initial phase and indexing parameters that will allow him to find the problem. IDM basic concepts are problems, partial solutions and contradictions which include elements, parameters and values. However, this study will only focus on parameters, specifically parameters whose values change during the improvement provided by the patent. There are of two types: Evaluation parameters (EP), useful for evaluating the results of a design choice and Action parameters (AP) on which one can act. Thus, depending on the information to retrieve, one must know how patent document expresses them. Parameter concept is according to IDM ontology *object nouns* or *object noun phrases*. However, they do not always appear in this form.

12.3.2 An Automated Method to Retrieve IDM Parameters

To automate the retrieval of IDM parameters this paper proposes to use generic linguistic markers, likely to be found in all patents without any distinction of area, since IDM ontology is generic. Patent is a specific document which has a specific nomenclature and lexicon. Thus, regularities may be used to extract relevant knowledge. The present method is the result of research taken in the Laboratory for Design Engineering (LGéCo). The first step of this study is the collection of the linguistic markers.

12.3.3 Lexicon Database

Lexical information is essential for efficient NLP tasks. For the purpose of a fine tuning, a broad-coverage lexicon must be build. The patent domain is very spread out as it includes a great many technical fields, ranging from Chemistry to Engineering. As a result, collecting a broad-coverage lexicon of the language used in patent texts is a complex task. Thus, for collecting lexical data, a variety of resources, including patent corpora, computer readable lexicons and thesauri are used. It consists in the observation of how patent documents express parameters, in other word by determining regularities between the information and the morpho-syntactic structures of the patent document. A heterogeneous corpus of 100 patents was used to collect the candidate markers; then a second corpus of 87 patent documents was built to evaluate and validate the candidate markers and adapt the algorithm to retrieve the parameters. The patents constituting the first corpus of 100 were selected at random. As for the second corpus, it was build and already used by IDM experts in a manual retrieval.

IDM concepts are generic, independent of the domain of application; therefore the searched markers are generic too. These are adverbs of evaluation and adjectives, verbs and nouns introducing a concept change, called for the purpose of this study: "verbs and nouns of change". We identified from our study a list of: 60 parameters marker verbs, 273 maker verbs, 137 marker adverbs and 473 marker adjectives.

Parameters marker verbs

Verbs inducing a change of state are those typically used in patent documents. Our study reveals that they are the most productive in terms of parameters and values. Examples of verbs are: *increase*, *decrease*, *release*, *damage*, *raise*, *change*, *maximize*, *augment*, *minimize*, *diminish*, *change*, *differentiate*, *etc*.

Furthermore, the frequent use of modal verbs is also to be noted. Grammatically speaking, modal verbs generally express a certainty or a necessity. Precisely speaking modal verbs used in patent document are mainly verbs like *can*, *could*, *may*, *might*, *will*, *shall*, *would*, *should or must*. Such verbs are, in most cases, followed by verbs (active or passive) or adjectives. However, relevant information is introduced by those followed by the verb "to be".

Parameters marker nouns

Nouns are used in patent document in a complex way. For the purpose of giving the maximum of information in a sentence, patent agents use complex compound nouns. However, we only include non-compositional lexical entries in the lexicon in order to minimize its size. In order to allow for the robust recognition of all compositional items by the parser, a grammar component is used in addition to the lexicon. Examples of nouns are: *deformity*, *density*, *diffusivity*, *dimensionality*, *discontinuity*, *ambiguity*, *durability*, *etc*.

Parameters marker adjectives

Adjectives likely to be indicators of parameters relevant to IDM are those used to express quality, degree, height, temperature, time, volume, shape, tension, speed, cost, etc. For example: *steady*, *dynamic*, *slow*, *fast*, *effective*, *erroneous*, *etc*. In addition, these adjectives are mostly used as adjective of comparison and may be located using signs of comparisons such as "-er" (simple adjectives) and "*more*, *less*, *worse*, *better*, *worse*" (complex adjectives). For the latter, the preposition "than", which follows the adjective of comparison, may be used to identify either parameters or elements. However, the nature of the concepts located after the said proposition is for the currently difficult to be determined.

Table 12.1 Examples of structures

Structure	Example
Adj[a] + SP[b]	(...) the casting *speed* was reduced
SP + SP	molten metal *surface height* differences
SP + NP[c]	fluctuation occurs on the *surface* of the molten steel
(...) + NC[d] + in/of/ on + (DET[e]) + (...) + P[f]	(...) percentage *change* in the *coil resistance*.
(...) + NC + in/of/on + (DET) + P + and/or + (DET) + P	(...), and the change in *pressure* and/or *flow*

[a] Adjective
[b] Super Parameter
[c] Noun phrase
[d] Noun of change
[e] Determinant
[f] Parameter

12.3.4 Parameters Retrieval

The IR methodology adopted to retrieve knowledge relevant to IDM is exposed in [18]. The conclusion of this study is that there are several ways to reach parameters. First, we have parameters which can be localized with the 39 general TRIZ parameters. For the purpose of this study, these parameters will be called super parameters (SP). In addition to these structures, parameters can be located through verbs. We have for example the case of verbs of change like "generate, improve, impair" exposed in [18]. Nouns of change (NC) can also introduce parameters. Thus, the following structures were determined in Table 12.1

However, it is worth noting that the above structure is far to be exhaustive. Furthermore, one can also notice the possibility to localize parameter using adverbs like *significantly*, *slightly*, *highly*, *effectively*, *substantially*, *expensively*, *inexpensively*, etc. For example: "The middle pellet 37 comprises a wide single perforation 40 of *slightly* low *diameter* to that of tube 20, and plays, in fact, a role of spacer separating the two other pellets". Nevertheless, the evaluation of this premise is still on and need to be furthered.

12.4 Case Example in Continuous Casting of Steel

Continuous casting of steels is a well known steel-making technology. It consists in a casting with an indefinite length and a given cross-section. The casting process requires the observance of consequent technological procedures, standard production conditions and advanced control equipment. Despite various feedback controls, the formation of defects cannot be fully avoided. Amongst others, the entrapment of inclusions, bubbles, slag, and other particles into solidified steel products is a critically-important quality concern [19]. These particles require

Table 12.2 Comparison between manual and automatic retrieval

Human retrieved parameters (HRP)	Computer retrieved parameters (CRP)
Nozzle clogging (%)	Nozzle clogging
Nozzle cost ( )	Nozzle cost
Transformation of alumina inclusions in solid	
Quality requirements	quality requirements
Spatial distribution of energy (Number/mm thick)	spatial distributions of energy
Critical tearing speed in the meniscus (m/s)	
Depth of deep horn (mm)	Depth of deep horn
horns density (nb/m)	horns density
Argon flow injected into the mold through the nozzle (l/min)	argon flow injection in the mold through the nozzle
Powder presence in ingot mould (Kg/steel T)	Presence of powder in ingot mould
	anything with the implementation

Perfect match Reformulated in HRP Not a valid parameter No match found in CRP

expensive inspection, surface grinding and rejection of steel. If undetected, large particles lower the fatigue life, while captured bubbles and inclusion clusters cause slivers, blisters, and other surface defects in rolled products. During continuous casting, particles may enter the mold with the steel flowing through the submerged nozzle. In addition, mold slag may be entrained from the top surface.

ArcelorMittal, number one steelmaker worldwide decided to partner with us regarding our research in providing us with their expertise on this particular subject and test our methodology of automatic parameters extraction. This exercise consisted in spending (thus measuring) the necessary quantity of human effort to fully describe the problem, then operating our methodology for automatic extraction and observing and discussing the obtained results in both situation.

Human extraction required us three full open days of work with five experts in the domain (two experts in continuous casting process, two experts in chemistry, one expert in metallurgy). Automatic extraction required us half a day of a methodologist and of a metallurgist (Table 12.2).

12.5 Discussions

For the purpose of evaluating the performance of the method, proposed, we made a comparison between automatic and manual retrieval of parameters from the ArcelorMittal corpus [18] using recall[1] and precision[2] measure (Table 12.3).

[1] Recall is the ability of an algorithm to present all relevant concepts.

[2] Precision is the ability of an algorithm to present only relevant concept.

Table 12.3 Recall and precision results

Recall (%)	Precision (%)
47, 35	82, 15

The above figure summarizes the result obtained parameters retrieved. Recall score is low while precision one is high. Thus, the approach hereby presented, whatever encouraging is still unsatisfactory. It is essential to improve different rates to further the study of patent documents and multiply algorithm to include the maximum of syntactic structures expressed in the patent document.

12.6 Conclusion and Perspectives

In this paper, we first reviewed the literature that exists in patent mining, specifically contribution made so far on patent mining related to TRIZ. Then, we proposed our method to automate the retrieval of IDM concept like parameters. As a reminder, parameters are variables of a system. This premise of the method is to retrieve IDM concepts using generic linguistic markers. The originality of the method is that it will provide a solution to assist IDM experts in their task. The evaluation of the result obtained reveals that although the method is encouraging, it is not fully satisfactory yet. Indeed, the evaluation performed shows a low recall rate with a relative high precision rate.

However, two aspects have to be underlined. First, at this stage of the research, we are not able to automatically determine whether the parameters retrieved are Evaluation Parameters or Action Parameters yet. Second, the linguistic resources and the algorithm may not be applied to non-covered patent domain like software industry. Thus, this research is only an encouraging start and it will be interesting to increase the accuracy of parameters consideration to better automate IDM core concepts retrieval.

References

1. Yeap T, Loo GH, Pang S (2003) Computational patent mapping: intelligent agents for nanotechnology. In: IEEE proceedings of international conference on MEMS, NANO and Smart systems, pp 274–178
2. Altshuller GS (1998) 40 Principles: TRIZ keys to technical innovation. (Lev Shulyak et Steven Rodman, Trans.). Technical Innovation Center, INC., Worcester, MA, 1998, p 141. (1st ed 1998), ISBN-10: 0964074036
3. Cavallucci D, Khomenko N (2007) From TRIZ to OTSM-TRIZ, addressing complexity challenges in inventive design. In: International Journal of Product Development 4(1−2):4–21
4. Khomenko N, Ashtiani M (2007) Classical TRIZ and OTSM as a scientific theoretical background for non-typical problem solving instruments. In: TRIZ Future Conference 2007, Frankfurt, Germany

5. Zanni-Merk C, Cavallucci D, Rousselot F (2010) Using patents to populate inventive design ontology. In: Proceedings of the TRIZ Future conference pp 52–62
6. Feldman R, Fresko M, Hirsh H, Aumann Y, Liphstat O, Schler Y, Rajman M (1998) Knowledge management: a text mining approach. In: Proceedings of the 2nd international conference on practical. Aspects of Knowledge Management, Basel
7. Ghoula N, Khelif K, Dieng-Kuntz R (2007) Supporting patent mining by using ontology-based semantic annotations. In: Proceedings of IEEE/WIC/ACM international conference on web intelligence, Silicon Valley, USA
8. Savransky S (2000) Engineering of creativity: introduction to Triz methodology of inventive problem solving. Boca Raton
9. Cascini G, Fantechi A, Spinicci E (2004) Natural language processing of patents and technical documentation. In: Lecture notes in computer science, vol 3163. pp 508–520
10. Cascini G, Russo D (2007) Computer-aided analysis of patents and search for TRIZ contradictions. Int J Product Development 4(1−2):52−67
11. Moehrle M, Geritz A (2004) Developing acquisition strategies based on patent maps. In: Proceedings of the 13th international conference on management of technology, R&D Management, Washington
12. Yoon J, Kim K (2011) An automated method for identifying TRIZ evolution trends from patents. Experts Systems with Applications 38(12):15540−15548
13. Dewulf S (2006) Directed variation: variation of properties for new or improved function product DNA, a base for 'connect and develop'. In: Proceeding of the TRIZ future conference 2006, Kortrijk, 9–11 October 2006
14. Duflou JR, Wim D (2006) On the complementarity of TRIZ and axiomatic design: from decoupling objective to contradiction identification. In: TRIZ Future Conference 2006
15. Coelho DA (2009) Matching TRIZ engineering parameters to human factors issues in manufacturing. WSEAS Trans Bus Econ 6:547–556
16. Verhaegen PA, D'Hondt J, Vertommen J, Dewulf S, Duflou JR (2009) Relating properties and functions from patents to TRIZ trends. CIRP J Manuf Sci Technol 1(3):126
17. Li Z, Tate D, Lane C, Adams C (2011) A framework for automatic TRIZ level of invention estimation of patents using natural language processing, knowledge-transfer and patent citation metrics. In: Computer-aided design 44(10):987–1010 doi: 10.1016
18. Souili A, Cavallucci D, Rousselot F, Zanni C (2011) Starting from patent to find inputs to the Problem Graph model of IDM-TRIZ. In: TRIZ Future Conference 2011, Dublin–Ireland
19. Fuhr F, Cicutti C, Walter G, Torga G (2003) Relationship between nozzle deposits and inclusion composition in the continuous casting of steels. In: Iron & steel society international technology conference and exposition, pp 165–176

Chapter 13
Virtual Reality Technologies for Creative Design

Julian Adenauer, Johann Habakuk Israel and Rainer Stark

Abstract Due to its immense visualization and interaction possibilities virtual reality (VR) is often regarded as the "ultimate" future technology for product development and engineering tasks. Although the majority of current industrial use cases for VR lie in fields of reviewing and validating, this article argues that VR has enormous potential to support creative design in the early conceptional phases of product development. To demonstrate this potential three sample VR systems are presented and matched to the "design principles for tools to support creative thinking" developed by Shneiderman et al. [1]. Based on this exemplary systems and related studies, it is argued that VR holds high potential to considerably support creative design and to improve the early phases of product development.

13.1 Introduction

For a growing share of products the user's experience and context of use are major factors for the success and acceptance of a product [2]. Furthermore current transformations in lifestyle and environmental conditions pose fundamental

J. Adenauer (✉)
GRK Prometei, ZMMS, TU Berlin Franklinstr. 28, Sekratarial 2-6,
10587 Berlin, Germany
e-mail: jad@zmms.tu-berlin.de

J. H. Israel · R. Stark
Fraunhofer IPK, Virtual Product Creation, Pascalstr. 8-9,
10587 Berlin, Germany
e-mail: johann.habakuk.israel@ipk.fraunhofer.de

R. Stark
e-mail: rainer.stark@ipk.fraunhofer.de

A. Chakrabarti (ed.), *CIRP Design 2012*, DOI: 10.1007/978-1-4471-4507-3_13,

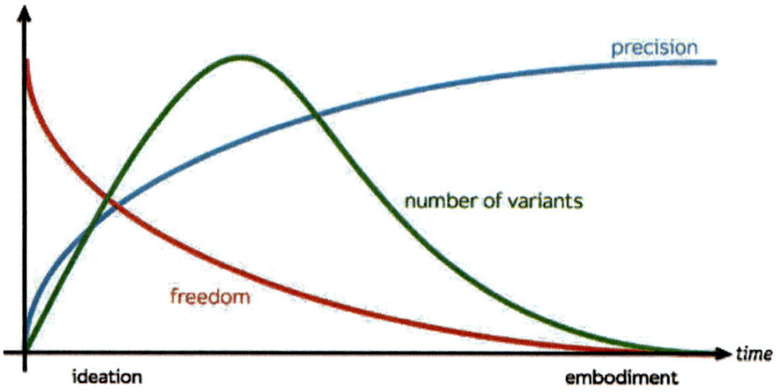

Fig. 13.1 Varying requirements towards creativity support along the product development process

changes to established product concepts e.g. in the fields of urban transportation, telecommunication, personal computing, and business and personal living environment. Thus the requirements to match external objectives and integrate product better into use contexts grow rapidly and pose enormous challenges. Engineers and designers have to come up with novel system solutions and innovative products to solve or mitigate urgent technological and social problems. Creative thinking—one major pre-requisite for new product design and innovation—has become a major constraint for the design process and should be supported by novel creativity support tools [3].

It is generally accepted that factors like properties of modeling media and tools as well as human-related factors, such as spatial perception, sensorimotor coordination and spatial reasoning, have a strong influence on the design problem solving process and its overall result [4–7]. While many of the criteria towards creative design can be addressed with existing tools, some requirements are not covered yet. In this article, novel creativity support tools based on virtual reality technology are described which fill existing gaps in the creativity support for product design; namely spatiality, ambiguity, embodiment and interaction prototyping.

During the design process, the requirements towards creativity support are varying, which is illustrated in Fig. 13.1. In the early phases of product design, creativity support tools should support creation of many alternatives, support exploration and provide sufficient information to select promising concepts. Models created in that early phases should furthermore have an ambiguous character and trigger further creative ideas [4]. In later stages of product development, the creative process includes working on details and partial solutions. Creativity support tools for the later phases should therefore support refinement and detail work.

This article focuses on creativity support for the early phase of the design process where many design decisions are not yet definite. Decisions made in this phase have great effect on both cost and quality.

In order to assess the potential of a technology to support creativity, the twelve "design principles for tools to support creative thinking" that have been defined by Shneiderman et al. [1] can be considered. In this paper three exemplary virtual reality systems are presented and matched against these principles in order to review the creativity support potential of the virtual reality technology.

13.2 Computerized Support for Design

The industrial design process nowadays is largely supported by computer tools. The use of PDM/PLM-systems and CAD/CAE-packages has become the standard in almost all product development industries. These tools have largely improved the efficiency of the development process, especially in the later phases, where precise modeling, complex simulations, and efficient organization are the key factors. The computerized support tools focus on integrity, accuracy, precision and efficiency rather than exploration and ideation.

In the early phases, however, computerized support is rare. Designers, therefore, predominantly make use of analogue physical tools in this phase. The first tools in the design process are usually pen and paper to draw the first ideas in form of simple sketches. While sketching stays an important medium during the entire design process [4] to explore the ideas further, a variety of prototyping tools is used in the later phases. Dependant on the preferences of the designer and the given design task, these range from paper prototypes, foam models to interactive physical prototypes.

Though, it has often been claimed that Virtual Reality has great potential to support the early phases of product development, current virtual reality tools fail to meet this promise [8]. They are more commonly used in later phases for the evaluation of product concepts in design review meetings or for simulating assembly/disassembly tasks. They are seldomly used for content creation, thus, the creative potential offered by this technology is hardly harnessed.

13.3 VR Tools for Creative Design

Virtual Reality technologies offer enhanced interaction capabilities compared to regular computer workspaces. Using 3D imaging and head tracking, virtual objects appear more realistic and the understanding of spatial relationships is enhanced. Furthermore, the possibility to spatially interact with virtual objects offers a great amount of freedom and novel capabilities of expression to the user. As argued before [9] the authors see the most promising features of immersive virtual environments in:

- *Three-dimensionality:* Immersive VR systems allow for stereoscopic visualization of objects in one-to-one scale within the working space of the user [10]. Seminal work in mental imagery shows that humans represent objects mentally in a three-dimensional manner [11]. A medium allowing for direct creation of external three-dimensional models should thus require less mental transformation than traditional two-dimensional media. As product development always results in three-dimensional objects, the intensified usage of immersive environments is likely to help understanding any designed object more directly and provide additional information regarding central product properties.
- *Interactivity:* Fast and fluent idea development is a pre-requisite for creative and innovative product development [5, 12]. Assumed that proper interaction-techniques are employed, VR is a highly interactive technology which provides high-bandwidth information exchange between user and system. Furthermore, users can manipulate virtual product models and review them interactively from various perspectives [10].
- *Digitality:* In contrast to paper-and-pencil sketches or physical modeling, digital media allow for a direct integration of developed product models into the overall toolset of virtual product creation [13, 14]. Also, as creativity can be stimulated by external cues [15], digital modeling environments may employ digitally created artificial stimuli in order to stimulate the development process, e.g. by providing abstraction techniques or by visualizing various use scenarios.

In order to evaluate which potential to support creativity virtual reality has in practice, general principles can be applied. A list of design principles for tools that support creative thinking has been suggested by Shneiderman et al. [1]:

1. *Support exploration.* Being able to try out different alternatives is an important requirement for creative work.
2. *Low threshold, high ceiling, and wide walls.* Tools should be easy to learn for novices but should also be powerful enough to support more sophisticated projects. Furthermore, a wide range of exploration should be possible ("wide walls").
3. *Support many paths and many styles.* Since working styles are very different among people, tools should not prescribe one right way of performing creative work but be open to different approaches.
4. *Support collaboration.* Since most creative work in product development is done in teams, tools should support the collaboration of group members.
5. *Support open interchange.* The result of the work with the tool should be reusable in other applications. This is especially important since the development process will almost certainly make use of a variety of different tools.
6. *Make it as simple as possible—and maybe even simpler.* A reduced number of features can improve user's experience.

7. *Choose black boxes carefully.* 'Magic' black boxes can significantly improve a tool's simplicity but also reduces the possibilities of the user to influence and adapt it. Thus a balanced trade-off between simplicity and adaptability of the tool has to be found.
8. *Invent things that you would want to use yourself.*
9. *Balance user suggestions with observation and participatory processes.* Including users in the design process increases the chances that the tool will be accepted by a wider user community.
10. *Iterate, iterate—then iterate again.* Just as iteration is crucial to design tasks it is as important for the development of creative support tools.
11. *Design for designers.* Create tools that enable others to invent and create.
12. *Evaluate your tools.* Even though measuring the benefit of using creative support tools is not trivial, it should be made sure that tools rather support but hinder creative work in evaluation studies.

To explore the capabilities of VR technology to support creative product design, three exemplary tools have been reviewed based on these principles: Immersive Sketching, Virtual Clay Modeling and Interaction Prototyping. These tools exist as research prototypes and differ from standard VR tools used in industry since their primary focus is on creating and manipulating digital models rather than viewing and validating.

13.3.1 Immersive Sketching

Sketching is a primary method for creative design in product development. As per Buxton [4] sketches have various properties needed in creative design processes, namely are they quick to make, inexpensive, disposable, plentiful, they have a clear vocabulary, minimal detail, and they are ambiguous and suggest as well as explore solutions rather than confirm them.

Based on the large body of work which provides empirical support for the utility of conventional sketching in creative design (e.g. [7, 16]), a number of prototypical systems [17–20] for immersive sketching in 3d virtual environments and corresponding empirical studies [6, 9, 21–23] emerged. Immersive sketching allows users to create free-hand sketches in 3d space just at the tips of their fingers or sketching tools. A 3d modeling system developed at Fraunhofer IPK provides only three different kinds of physical sketching tools (matching Shneiderman's above mentioned criterion *make it as simple as possible*) but allows for free-hand creation of various 3d shapes (matching the *support many paths and many styles* criterion): a pen for drawing 3d strokes, a free-hand tool which allows for extruding Bezier-curves, and a snapping tool that allows for moving and extruding any arbitrary 3d sketch. The system was developed in a user centered design process which involved the intended users in the requirement analysis and in frequent evaluation studies [21, 24, 25] (matching Shneiderman's *iterate, evaluate*

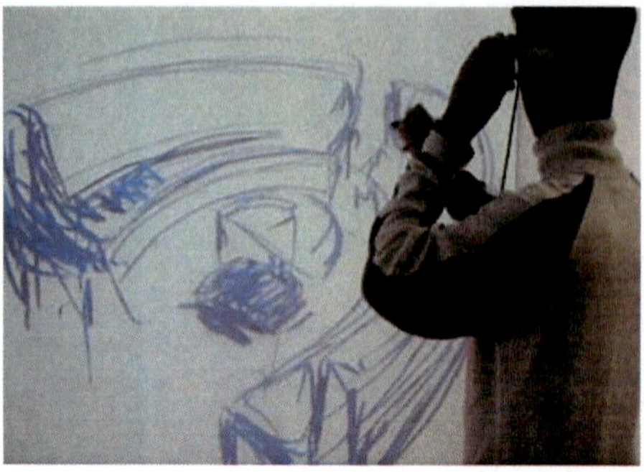

Fig. 13.2 Immersive sketching [26]

and *design for designers* criterion). Finally, the 3d sketches created with the systems can be exported to be used in CAD systems, matching the *support open interchange* criterion.

Using the immersive sketching system a comparative study on 2d and 3d sketching was conducted among 24 furniture designers and interior architects in order to investigate—among others—the utility of immersive sketching for creative design (Fig. 13.2) [21, 26].

With respect to Shneiderman's criteria the following findings could be made. Designers actively explored the interaction space and their own sketches, they moved significantly faster in the 3d than in the 2d condition (37 %), they also actively sketched significantly longer in the 3d than in the 2d condition (33 %) and sketched more details (50 %). Considered together it can be assumed that the designers actively explored their inner concepts and models of the intended product by externalizing it in the immersive environment and developing it further (criterion *support exploration).*

Another study on the learnability of 3d sketching revealed that users show a steep learning curve in their abilities to sketch primitive 3d objects in space, matching the *low threshold, high ceiling, and wide walls* criterion [24]. Taking this criterion literally it could also be shown that 3d sketches span a five times bigger volume than 2d sketches [21, 26]. Immersive environments provide more physical space to externalize mental images than conventional media, and this space is actually being used by the designers.

The utility of immersive sketching for creative design is also supported by user statements expressed in the studies. The main unique qualities mentioned were [21, 27]:

- Spatiality: "to work with the space", to create spatial models and to perceive their spatial impact,
- One-to-one proportionality: to draw models in a one-to-one scale, investigating the model with respect to the own body and its proportions,
- Associations: to "take existing objects into the virtual space and work with them",

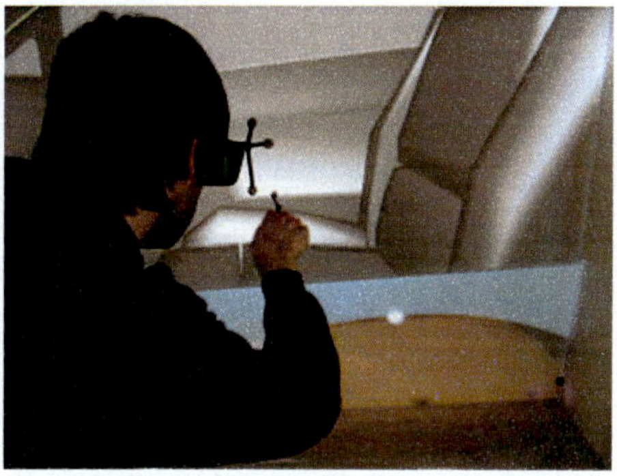

Fig. 13.3 Virtual clay modeling in an immersive CAVE environment [32]

- Formability: to manually warp virtual models and sketches which in turn allows for a gradual developing and testing of ideas until they are "mature",
- Stimulation: the hedonic stimulation of the designer.

Considering the empirical results, immersive free-hand sketching seems to support important elements of early ideation phases in product development processes (Fig. 13.1), because it supports the free exploration of product models and concepts, and the creation of product variants.

13.3.2 Virtual Clay Modeling

Virtual Clay Modeling (VCM, Fig. 13.3) allows for generating design models in analogy to conventional clay modeling. Using VCM designers can generate arbitrarily shaped models by simply adding or removing virtual "material". This "Cut & Paste" can be achieved with graphically represented spline-based modeling tools which the designer directly manipulates using tracked physical handles (tangible user interfaces). The graphical (virtual) tools can take the forms of plane rakes or templates of true sweeps. VCM systems are usually realized based on elementary Boolean operations which are performed on virtual clay models represented as compressed voxel models with a high degree of resolution.

The VCM modeling system of the Fraunhofer IPK [28] allows designers to create their own library of shaping-tools which can be applied to the virtual clay, matching Shneiderman's *support many paths and many styles* criterion. Some of the interaction tools are bi-manual tools which refer to the traditional process (matching the *design for designers* criterion), even though no haptic feedback is provided. Virtual Clay Modeling finds its application primarily in the refinement of initial design concepts. Thus, VCM does not generate new variants but generates more detail and adds more information to the product model (Fig. 13.1). While immersive sketching allows to create shapes and lines, VCM gives the user the impression of working with virtual material, allowing for gradually evolving

Fig. 13.4 Hybrid tailgate prototype

the product model and iteratively changing design details, which matches Shneiderman's *support exploration* criterion.

13.3.3 Interaction Prototype Designer

While Immersive Sketching and Virtual Clay Modeling are focused on the design of the product's shape, Interaction Prototyping Designer (IPD) is a method for designing and evaluating the product's interactive properties in an early stage of its development.

The Interaction Prototype Designer (IPD) developed at Fraunhofer IPK is part of the Smart Hybrid Prototyping (SHP) approach [29] that aims at establishing tools for centering the product development process around virtual prototypes. These prototypes are enhanced with simulation capabilities and physical elements to *hybrid prototypes*. One example for this approach is the hybrid tailgate (Fig. 13.4) that combines VR-visualization, advanced simulation of the mass-spring-damper system (pre-simulated) and a haptic device. SHP combines the advantages of digital tools with realistic and interactive experience of the product.

While the SHP approach in general is aiming at an integrated and continuous support throughout all phases of the product development process, the IPD is specifically targeting the early conceptual phases. Thus, accurate simulation is less important than the support of fast iterations and a continuous workflow (matching the *support exploration* criterion). To achieve this, the IPD combines several open source tools, matching Shneiderman's *support open interchange* criterion.

The IPD lets developers not just design the physical shape of a product but also allows to define sensors and actuators. Dependant on the values of these virtual sensors the product's interactive behavior can be defined (e.g. a sound starts or a light flashes when a virtual sensor indicates a specific distance to a virtual object). This is done using visual programming for basic tasks and Python scripts for more complex ones. Beside these editing tools also a game engine is integrated that allows real-time rendering and basic simulation of physical behavior.

To interact and experience the prototype, the IPD can be coupled to immersive displays that are driven by a single computer (e.g. powerwall) or a PC clusters (e.g. CAVE system).

Beside these virtual components of the prototype, physical objects can also be integrated to enhance the realism and make the interaction with the virtual prototype more natural. This is done by an abstraction layer provided by the TUI framework [30]. This makes it easy to integrate real-time tracking data, sensor inputs and physical actuator outputs into the hybrid prototype. By integrating widely used prototyping platforms such as Arduino or Phidgets [31], users can apply their knowledge using these platforms (meeting the *support many paths and many styles* and the *low threshold, high ceiling, and wide walls* criteria). Furthermore, the large online communities dedicated to these tools offer deep knowledge and support and are a constant source of new and improved tool-capabilities.

Since the modeling environment and the simulation are integrated into the same tool, the developer can move fast between creating, evaluating and changing the prototype—allowing fast iteration steps which are essential for early concept refinement. This complies with Shneiderman's *support exploration* criterion.

13.4 Conclusion

This article argues for further explorations of virtual reality technologies for creativity support in conceptual product development. Three VR-based tools were presented and reviewed in relation to the "design principles for tools to support creative thinking" by Shneiderman et al. [1]. Immersive sketching shows its potentials it the rapid creation of design variants in 1-to-1 scale, Virtual Clay Modelling has its strength in elaborating on a product's shape, and Interaction Prototyping allows for experiencing interactive product features in early phases of the design process. It could be shown that a significant number of principles was addressed by the presented tools supporting the assumption that VR in general offers unique possibilities for creative design and can assist the emergence of creative product solutions. While traditional tools like CAD and PDM remain central for detailed engineering, VR-based tools might experience a more widespread use in early conceptual design phases.

Acknowledgments Support for this work has come from the Fraunhofer Institute for Production Systems and Design Technology Berlin, the BMBF project AVILUSplus (01IM08002) and the DFG research group prometei (project number 1013/2).

References

1. Shneiderman B, Fischer G, Czerwinski M et al (2006) Creativity support tools: report from a U.S. national science foundation sponsored workshop. Int J Hum Comput Interact 20(2):61–77
2. Hummels C, Djajadiningrat T, Overbecke K (2001) Knowing, doing and feeling: communication with your digital product. Tagungsband Interdisziplinäres Kolleg Kognitions- und Neurowissenschaften, pp 289–308
3. Shneiderman B (2007) Creativity support tools: accelerating discovery and innovation. Commun ACM 50(12):20–32
4. Buxton B (2007) Sketching user experiences: getting the design right and the right design. Morgan Kaufmann Publishers, Los Altos
5. Hacker W (2007) Innovationsmanagement der frühen Phasen der Produktentwicklung: Stärkung des "intellektuellen Kapitals."XII. Int. Produktionstechnisches Kolloquium PTK2007, Fraunhofer IPK, 121:I–VIII
6. Pache MW (2005) Sketching for conceptual design. Empirical results and future tools. Technische Universität München, München
7. Tversky B (2003) Sketching for design and design of sketches. In: (ed) Human behaviour in design, Springer Berlin
8. Dangelmaier M (2009) Concept Experience. Virtuelle Realität in frühen Phasen der Produktentstehung. 8. Paderborner Workshop Augmented & Virtual Reality in der Produktentstehung, HNI, pp 321–332
9. Stark R, Israel JH, Wöhler T (2010) Towards hybrid modelling environments—merging desktop-CAD and virtual reality-technologies. CIRP Ann Manuf Technol 2010(59):179–182
10. Bowman DA (2005) 3D user interfaces. Addison-Wesley, Boston
11. Kosslyn SM (1995) Mental Imagery. In: Kosslyn SM, Osherson DN (eds) An invitation to cognitive science: visual cognition and action. MIT Press, Cambridge, pp 267–296
12. Krause F-L, Jansen H, Kind C, Rothenburg U (2007) Virtual product development as an engine for innovation. In: Krause F-L (ed) The future of product development. Proceedings of the 17th CIRP design conference. Springer, Berlin, pp 703–713
13. Krause F-L, Lüddemann J (1996) Virtual clay modelling. 5th IFIP WG5.2 Workshop on geometrical modeling in Computer-Aided Design, International Federation for Information Processing IFIP, pp 162–175
14. Krause F-L, Kimura F, Kjelberg T, Lu SC-Y (1993) Product modelling. In: CIRP annals 1993, manufacturing technology. Hallwag Publishers, Bern, pp 695–706
15. Amabile TM (1983) The social psychology of creativity. Springer, New York
16. Sachse P, Leinert S, Ulich, E (2002) Skizzen und Modelle: Wieso Hilfsmittel des Denkens und Handelns beim Konstruieren? In: Hacker W (ed) Denken in der Produktentwicklung. Psychologische Unterstützung in den frühen Phasen. vdf Hochschulverlag AG an der ETH Zürich, Rainer Hampp Verlag, München, pp 63–81
17. Fiorentino M, de Amicis R, Stork A, Monno G (2002) Surface design in virtual reality as industrial application. DESIGN Conference, pp 477–482
18. Haselberger F, Bues M, Schuetz T (2007) An immersive CAD testbed
19. Kara LB, Shimada K (2008) Supporting early styling design of automobiles using sketch-based 3D shape construction. Comput Aided Des Appl 5(6):867–876
20. Stork A (2000) Effiziente 3D-Interaktions- und Visualisierungstechniken für benutzerzentrierte Modellierungssysteme. Fachbereich Informatik Dr.-Ing
21. Israel JH, Wiese E, Mateescu M, Stark R (2009) Investigating three-dimensional sketching for early conceptual design—results from expert discussions and user studies. Comput Graph 33(4):462–473
22. Keefe DF, Zeleznik RC, Laidlaw DH (2007) Drawing on air: input techniques for controlled 3D line illustration. IEEE Trans Visual Comput Graphics 13(5):1067–1080

23. Mäkelä W, Reunanen M, Takala T (2002) Possibilities and limitations of immersive freehand expression: a case study with professional artists. ACM multimedia conference MM'04, ACM press, pp 504–507
24. Wiese E, Israel JH, Meyer A et al (2010) Investigating the learnability of immersive free-hand sketching. ACM SIGGRAPH/Eurographics symposium on sketch-based interfaces and modeling SBIM'10, ACM SIGGRAPH and the eurographics association, pp 135–142
25. Wiese E, Israel JH, Zöllner C, Pohlmeyer AE, Stark R (2009) The potential of immersive 3D-sketching environments for design problem-solving. 13th International conference on human-computer interaction HCI 2009, pp 485–489
26. Israel JH (2010) Hybride Interaktionstechniken des immersiven Skizzierens in frühen Phasen der Produktentwicklung. Institut für Werkzeugmaschinen und Fabrikbetrieb IWF. PhD
27. Israel JH, Zacharias E (2007) Möglichkciten und Grenzen des Skizzierens im Dreidimensionalen Raum aus Designersicht. 7. Berliner Werkstatt Mensch-Maschine-Systeme, VDI-Verlag, pp 465–470
28. Biahmou Tchebetchou AR, Krause F-L (2005) Methoden für das Industriedesign in Virtueller Realität. Institut für Werkzeugmaschinen und Fabrikbetrieb IWF. PhD
29. Stark R, Beckmann-Dobrev B, Schulze E-E, Adenauer J, Israel JH (2009) Smart Hybrid Prototyping zur multimodalen Erlebbarkeit virtueller Prototypen innerhalb der Produktentstehung. Der Mensch im Mittelpunkt technischer Systeme 8. Berliner Werkstatt Mensch-Maschine-Systeme (ZMMS Spektrum Band 22), VDI Verlag, Düsseldorf
30. Israel JH, Belaifa O, Gispen A, Stark R (2011) An object-centric interaction framework for tangible interfaces in virtual environments. Proceedings of the 5th international conference on tangible, embedded, and embodied interaction, ACM, pp 325–332
31. Greenberg S, Fitchett C (2001) Phidgets: easy development of physical interfaces through physical widgets. Proceedings of the 14th annual ACM symposium on user interface software and technology, ACM, pp 209–218
32. Röper D Entwicklung voxelbasierter Modellierungsmethoden für immersives Design im Kontext

Part III
Enabling Technologies and Tools

Chapter 14
Design of CAM-Interfaces for Two Robots Based Incremental Sheet Metal Forming

H. Meier, J. Zhu, B. Buff and C. Magnus

Abstract Robot based incremental sheet metal forming (*Roboforming*) is an innovative die less forming process, which is developed by the Chair of Production Systems at Ruhr-University Bochum. Suitable for rapid prototyping and manufacture of small batch sizes with low costs, this forming process is based on flexible shaping through the synchronous movement of two tools hold by two industrial robots. Since there is no existing solution to quickly and accurately generate two synchronized tool paths according to the will of users, the research of this method should first be focused on the tool path generation. In this paper, different forming strategies and their algorithms of generating synchronized path points are explained at first. Based on these established algorithms, three approaches to create CAM-interfaces for the generation of robot programs are introduced thereafter. Depending on different users, they are individually based on postprocessors, Application Program Interface (API) and a stand-alone program. Each introduced approach for *Roboforming* is validated by forming experiments and they also give a good reference to the third-party development of a CAD/CAM-system.

14.1 Introduction

Developing new individual products within a short time and with low costs is often the aim of industrial research. Robot based incremental sheet metal forming, which is also called *Roboforming*, is a new rapid prototyping method to form sheet metals

H. Meier · J. Zhu (✉) · B. Buff · C. Magnus
Lehrstuhl für Produktionssysteme (LPS), Ruhr-Universität Bochum,
Universitätsstraße 150, D-44780 Bochum, Germany
e-mail: zhu@lps.rub.dc

A. Chakrabarti (ed.), *CIRP Design 2012*, DOI: 10.1007/978-1-4471-4507-3_14,

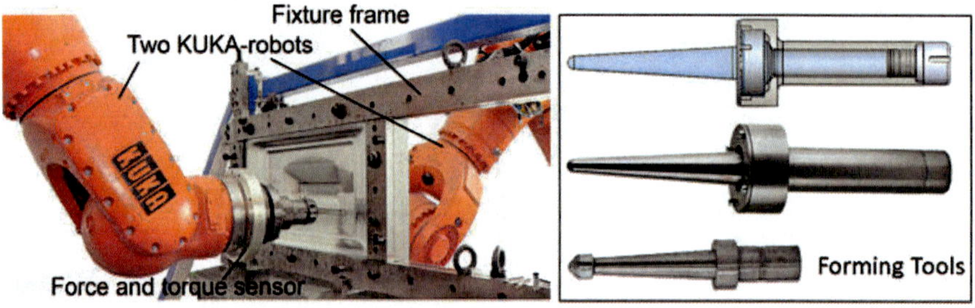

Fig. 14.1 Experimental set-up

by means of two universal forming tools driven by two industrial robots (Fig. 14.1). The sheets are firmly clamped on a fixture frame and both robots are connected to a robot system realizing the synchronization. Similar to other incremental forming methods [1] [2], the final shapes of workpieces are kinematically generated by the movement of the forming tool along the lateral direction and its gradual infeed in depth direction.

In many incremental forming methods, workpiece-dependent dies are often used for complex workpieces with convex and concave structures in order to maintain good geometric accuracy [3]. In *Roboforming* the supporting tool driven by the slave robot can support the part on its backside according to individual surface structures. Furthermore, the use of an industrial robot system enables large-sized structures to be formed with lower equipment costs compared to a CNC-machine with a large working area. Therefore, low costs, fast production speed and high manufacture flexibility are realized in this method by entirely abandoning dies and geometry specific forming tools. However, there is no existing CAM-solution to quickly and accurately generate two synchronized tool paths according to the will of users. The first challenge is to develop a qualified CAM-interface planning the tool path based on different forming strategies.

14.2 Forming Strategies and Algorithms

Concerning incremental sheet metal forming by using a CNC-machine, two different methods are commonly used. The first one is called Single Point Incremental Forming (SPIF) [4], which is a relatively simple variant (Fig. 14.2a). Because of the unformed areas between the clamp frame and the corners of the geometry, the forming accuracy is to a large extent influenced by the low structure stiffness. Therefore, back plates with geometry specific outlines are added to the backside of the sheet for a stiffer support. Correspondingly in *Roboforming* a supporting tool is synchronized with the forming tool moving on the boundary of the part to substitute the backing plates (Fig. 14.2c). This forming strategy is called Duplex Incremental Forming with Periphery supporting tool (DPIF-P). For its path generation, the path of the forming tool is first generated and then a specified movement trajectory of the

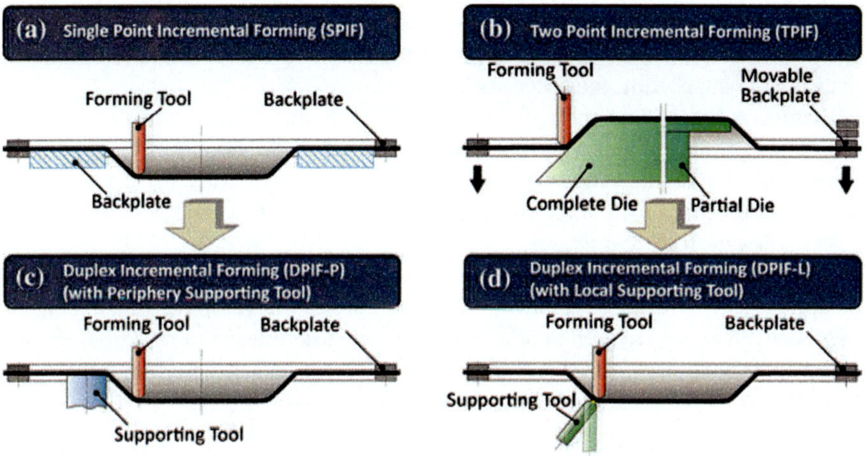

Fig. 14.2 Two derived forming strategies in *Roboforming*. **a** Single point incremental forming (SPIF). **b** Two point incremental forming (TPIF). **c** Duplex incremental forming (DPIF-P) (with periphery supporting tool). **d** Duplex incremental forming (DPIF-L) (with local supporting tool)

supporting tool has to be defined. According to each position of the forming tool, the precise position of the slave tool is determined on this trajectory at last [5].

The second forming variant is called Two Point Incremental Forming (TPIF), where either a partial or a complete die with geometry specific structures is used on the backside of workpieces (Fig. 14.2b). Although better geometric accuracy can be achieved, a longer product delivery time and higher material costs are unavoidable. To solve these problems, *Roboforming* uses a locally supported forming strategy (DPIF-L), where the supporting tool moves at the opposite side of the sheet directly against the forming tool, generating a forming gap between both tools (Fig. 14.2d). Especially for complex geometries, through the exchange of both tools' forming and supporting functions, convex and concave structures can also be formed.

For its path generation, the points on the geometrical surface facing the forming tool are firstly generated based on CAD-models. According to both tools' radius and the sheet thickness, the forming points are shifted along the surface normal in both directions to obtain positions of the tool centre points (TCP) [5]. Moreover, the orientations due to the surface normal can also be given to each tool.

14.3 Interfaces

Postprocessors in a CAM-system are always provided by the software seller or by the machine manufacturer. After tool path planning in the CAM-system, proper programs with special statements and commands for the individual CNC-machines used are generated by these postprocessors. So the tools loaded on the machines can move to the right positions planned in the CAM-system. Unfortunately, most powerful CAM-systems are only designed for CNC-machines and even if there is a postprocessor provided for industrial robots, it is still incapable to generate the path for the supporting tool used in *Roboforming*.

Based on the API of the software FAMOS robotic®, which was designed for offline-programming and simulation of industrial robots, an add-in was specially developed to generate synchronized tool paths for *Roboforming* [6]. However, its functions and the calculation speed are not competitive with other well-known CAM-software for NC-systems.

Modern CAM-systems often provide users the possibilities to develop their own postprocessors or even the realization of a third-party development hereupon. So these conveniences are used to develop the special user interfaces for *Roboforming*. According to different users and the complexities of path generation, three interfaces based on different methods have been developed.

14.3.1 Postprocessor

The CAD/CAM-system Solid Works/CAM Works is used here for the functional expansion. The first approach is to create user-defined postprocessors, which can not only generate robot programs, but also realize the position calculation for the TCP.

By using the 5-axis milling block in CAM Works, pattern, step size, point density and other parameters describing the tool path on the geometrical surface can be controlled. Tools with a ball nose are used for the path generation and all path points determine the positions of the tool tip. The tool axis is controlled to be always vertical to the surface, in order that its orientation described by the 4th and 5th axis can be used to deduce the surface normal vector.

After the preliminary path generation in the CAM-system, some important processing parameters are defined, such as tool and machine type, positions and orientations of the tool tip, etc. In the programming language of postprocessors, these parameters can be called through the system variables. Moreover, new variables and functions can also be added to a central library file used for the postprocessors. The left side of Fig. 14.3 shows the internal compositions of the postprocessor. There are different main section categories and each main section has subsections for a better classification. All sections used are modified to meet the offline-programming language of the used robots.

During the program generation, the section "Start of Tape" is firstly called (Fig. 14.4), in which all initial parameters of the robot are defined, such as movement speed, move accuracy, sensor initialization, etc. According to the type of tool movement one relevant section is called and this section searches for the matched function for its new variables in the library file. Due to the system variables and the function, the values of these new variables are sent back to the section and at the end, a proper command for the robot system is outputted by this section. For example, during the forming process the section "5 Axis Line Move" is mostly called (Fig. 14.3), through which the robot drives the tool on the surfaces

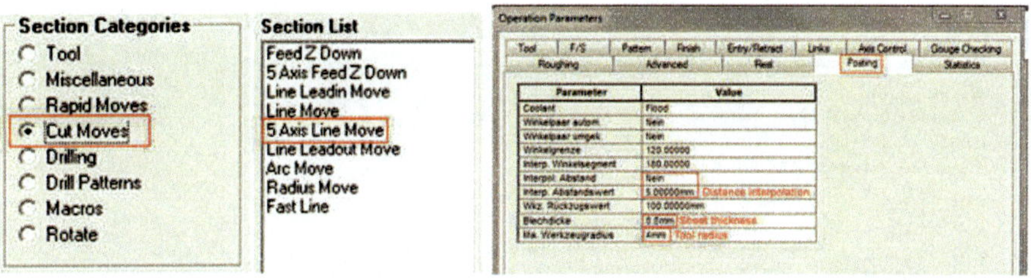

Fig. 14.3 Sections in the programming structure and new defined variables in the user interface

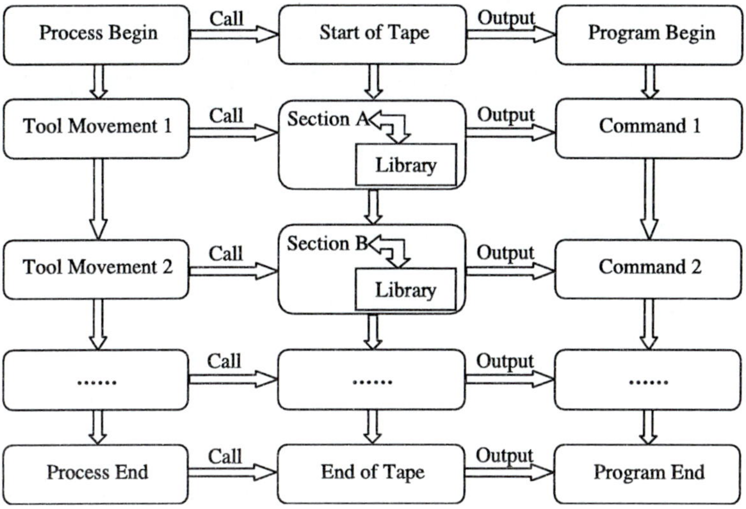

Fig. 14.4 Internal operating structure of the postprocessor

to be formed. New variables like the positions and orientations of the TCP, which are initialized and exactly defined in the library file, are inserted in this section.

To get the final postprocessor used in the CAM-system, the library file and the programmed main file containing all sections are compiled by a post-generator provided by the CAM-system. Most variables are defined in the library file and once compiled, they cannot be changed any more. But some variables like the tool radius and the sheet thickness, which always need to be modified, can be changed later in the tab "Posting" in CAM Works (right side of Fig. 14.3), where the defined variables therein will be called by the postprocessor.

In the forming strategy with local support (Fig. 14.2d), two postprocessors respectively for the forming and supporting tool are created to generate the program. After the tool planning in CAM Works, each postprocessor is chosen in turn (left side of Fig. 14.5) for the output of programs used in the robot system (right side of Fig. 14.5).

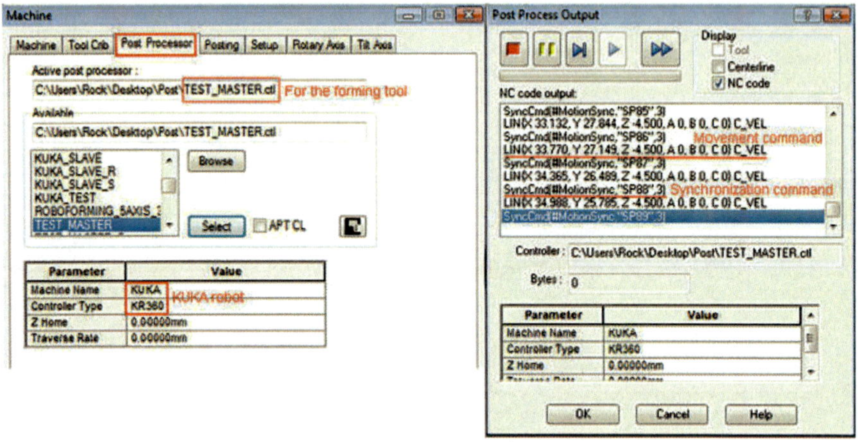

Fig. 14.5 Choosing the right postprocessor and printing the robot program

14.3.2 Application Program Interface

The first approach introduced in Sect. 3.1 is for junior users, who need simplicity, convenience and quickness. But there is not much freedom provided for users, because except several changeable variables (right side of Fig. 14.3) everything else is packed in the predefined postprocessor, which cannot be changed later. Moreover the postprocessor cannot realize more complex functions, because a rather simple and limited programming language provided by the CAM-system must be used. Consequently it is impossible to generate tool paths for the other forming strategy with a peripheral support (Fig. 14.2c), where a relatively more complicated algorithm is needed to determine positions of the supporting tool on the sheet edges.

Therefore the second approach should give advanced users more options to generate final robot programs and this is realized through a new add-in for the CAD/CAM-system created by the API. Any program written with.NET Framework can be connected to Solid Works/CAM Works and with the use of C#, new written programs can call the existing features in the CAD/CAM-system or even realize new functions based on these features.

After the start of Solid Works, users can open the add-in name *IBU Works* directly from the main menu. The add-in consists of different components, which are responsible for different forming strategies. Figure 14.6 shows a window for the various options, which should be modified before the program generation. For example, in the tab used for the setup of working conditions, the tool orientations can be set always vertical to the sheet surface or the orientation can also be restricted to a value domain. Sheet thickness, tool radius, safety planes for both tools, parameters for the force-control support, shifting angle of the supporting tool, variable tool movement speed and other parameters should be defined at first to realize different forming strategies.

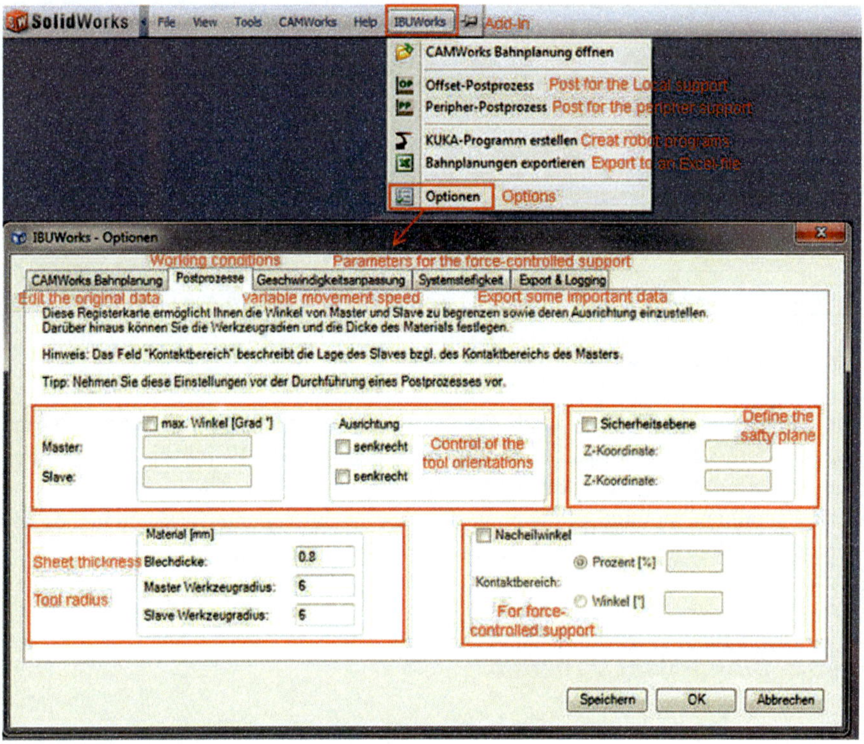

Fig. 14.6 User-interface programmed with API

Figure 14.7 shows the internal programming structure of this add-in, which is distributed due to different tasks. The first namespace has a name *Controller* and it owns three classes having the core functions. The class named *Coordinates Controller* is composed of all coordinates, including the original ones from the CAM-system and the calculated ones for the robot system. Additionally, *Coordinates Controller* is also responsible for the loading and editing of path points, the calculation of suitable movement speed according to the point distance as well as the coordination of the post process.

The class named *KUKA Source Controller* has the function of generating adequate robot programs including commands for line movement, movement speed and other parameters for the robot system. If another kind of robot is used for *Roboforming*, only the change in this class is necessary for a quick adjustment.

All adjustable parameters in the option window (Fig. 14.6) are preserved in the class *Options Controller*, which has also the responsibility of sending saved parameters to the post process. The namespace *Forms*, which determines interfaces of the add-in shown in Fig. 14.6, is separated from other namespaces. Its sub-classes listed in Fig. 14.7 are responsible for the layouts of all the 7 user windows.

The namespace *Postprozessor* consists of the algorithms for the two forming strategies DPIF-P and DPIF-L, which are contained in the two classes. The rest

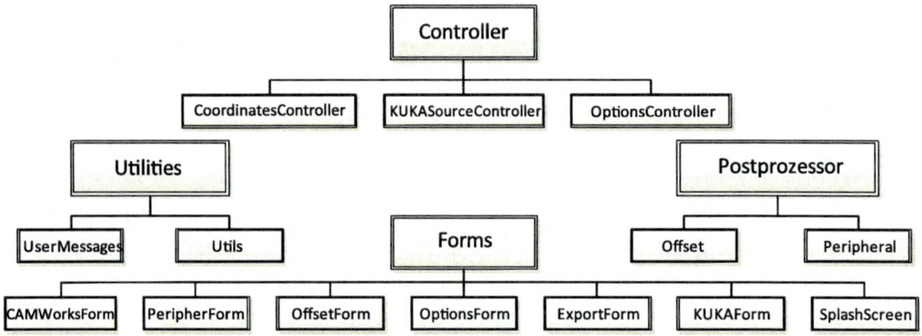

Fig. 14.7 Internal structure of the program

commands and functions, which are frequently used by the different parts, are contained in the namespace *Utilities*. The add-in *IBU Works* comprises large amounts of notifications, which give users the instructions and hints, like what kind of data should be inputted, string or number, etc. They are centrally saved in the class named *User Messages* and are frequently called by the other classes.

14.3.3 Stand-alone Program

If the computing speed, some more complex algorithms and even the graphic drawing of tool paths are considered, a stand-alone program for senior users is set up by the use of Matlab GUI. After path planning, original data from the CAM-system describing the points on geometrical surfaces is inputted into this stand-alone program. With the right setting of some necessary parameters like tool radius and sheet thickness, both tools' paths can quickly be generated according to different forming strategies.

Additionally, the graphic display area of the program can show every detail of the calculated path, such as distances of path points, distances between both TCPs, the whole path length, the 3D- or 2D-view of paths. Before the output of robot programs, the tool path can be verified from all aspects. Figure 14.8 shows a diagram with 3 paths, where the red path represents the processing points on the front forming surface, while the blue and green ones respectively stand for the paths of the forming tool and the supporting tool. To avoid a probable collision the path of the supporting tool must be ensured to stay above the other two paths, which means the offset of the processing points is basically in the right direction (Fig. 14.8).

Moreover, this program can also generate proper files as inputs for two different simulation models. One of them is a MBS-model (Multi Body System) simulating the status of both robots during the forming process and the other one is a FEM-model for a detailed deformation analysis of sheets. It means that the program serves as a pre-processor for the simulations too. Before the transfer of final programs to the robot system, all tool paths can be tested by the simulation.

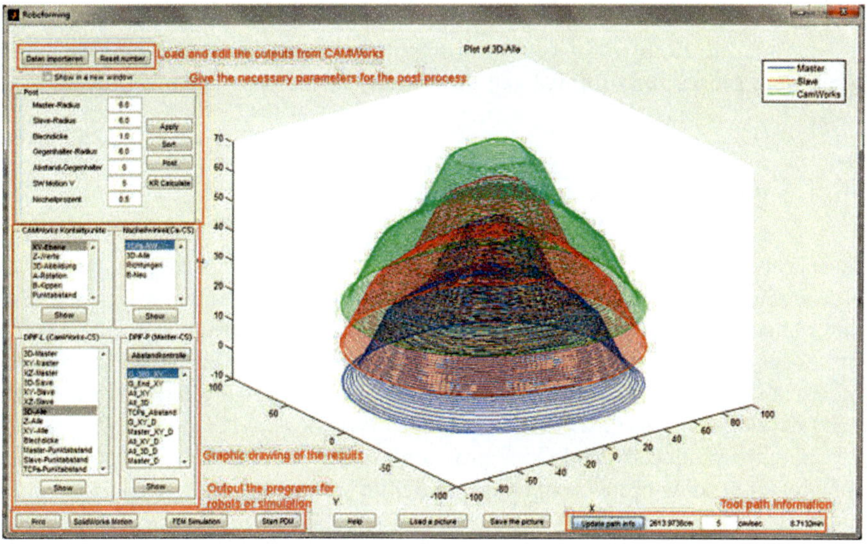

Fig. 14.8 User-interface established by Matlab GUI

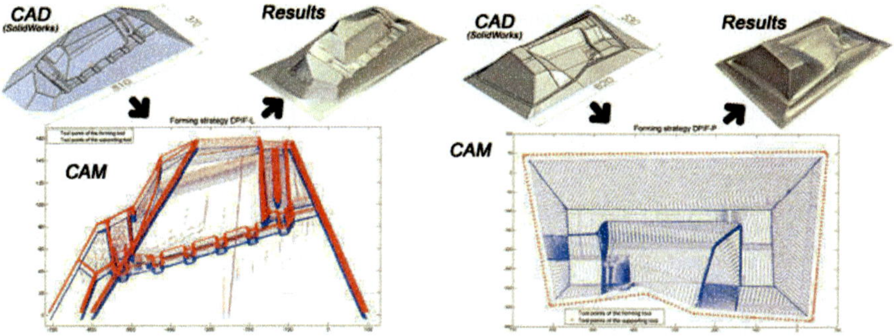

Fig. 14.9 Path planning with the designed interface and the corresponding forming results

For a central saving and management of all data including CAD-models, CAM data and files for simulation and experiment, this program is also connected to a PDM-system. According to different geometries and forming strategies, all data can be classified automatically in the PDM-system.

14.4 Experiments

By the use of these plugins, the path generation for *Roboforming* can quickly be completed. Figure 14.9 shows the process of achieving formed workpieces directly from CAD-models according to the forming strategies DPIF-L and DPIF-P. The blue and red points are respectively the path points of the forming and supporting tool generated in the designed CAM-interface, which is the stand-alone program last introduced. Both workpieces with a large geometric size and complex structures were successfully formed and the validity of the program was proved.

14.5 Conclusion and Future Work

In this paper three CAM-interfaces established with different methods for *Roboforming* are introduced. Based on postprocessors, API and a stand-alone program respectively, they are designed for various users whose requirements for software convenience and function are different. At the end, two forming experiments have proved the correctness of the path generation. These approaches and methods give a good reference to the third-party development of a CAD/CAM-system.

The last introduced stand-alone program made by Matlab GUI has already the function of sending proper programs to different simulation environments, but a lot more could be done later. For example, if the FEM environment sends proved simulation results like geometrical deviation back to the interface, the tool path could be corrected. There is another model describing the robot stiffness, and if processing forces and tool paths are simultaneously given to this model, the path deviation caused by the robot elasticity can be simulated and a corrected tool path can be generated under the consideration of robot elasticity [7]. All these models should be integrated in the designed CAM-interface, so that an improved path generation considering many actual influence factors during the forming process can be realized.

References

1. Jeswiet J (2001) Incremental single point forming with a tool post. In: Proceedings of the 9th international conference on sheet metal, Leuven, Belgium
2. Hirt G, Ames J, Bambach M (2006) Basic investigation into the characteristics of dies and support tools used in CNC-incremental sheet forming. In: Proceedings of the 25th IDDRG international deep drawing research group conference, Porto
3. Hirt G, Kopp R, Hofmann O, Franzke M (2007) Implementing a high accuracy multi-mesh method for incremental bulk metal forming. In: CIRP annals—manufacturing technology, vol 56. Elsevier, pp 313–316
4. Ham M, Jeswiet J (2007) Single point incremental forming limits using a box-behnken design of experiment. In: Key engineering materials, vol 344. Trans Tech Publ, pp 629–636
5. Kreimeier D, Zhu J, Smukala V et al (2011) CAM-solution for two robots based incremental sheet metal forming. Key Eng Mater 473:889–896
6. Meier H, Brüninghaus J, Buff B et al (2009) Tool path generation for free form surfaces in robot-based incremental sheet metal forming. In: Key engineering materials, sheet metal 2009 conference, vol 410. Birmingham, pp 143–150
7. Meier H, Laurischkat R, Zhu J (2011) A model based approach to increase the part accuracy in robot based incremental sheet metal forming. In: AIP conference proceedings, vol 1315, pp 1407–1412

Chapter 15
Development of Virtual Prototypes Based on Visuo/Tactile Interaction for the Preliminary Evaluation of Consumer Products Usage

Monica Bordegoni, Francesco Ferrise and Umberto Cugini

Abstract The paper describes an application based on low cost Virtual Reality technologies whose aim is to help designers in testing some functional aspects of their product concepts without the need of building physical mockups. In particular the application allows the designers to test aspects related to the usage of products. A case study that enables designer to verify the ergonomics aspects of the interactive Virtual Prototype (iVP) of a commercial refrigerator has been implemented. The iVP is based on a three-dimensional scale representation of the product and the use of a low cost user interface as input/output device. The paper describes the implementation of the iVP and the results of some preliminary user study that has been performed in order to validate the visuo/haptic interaction approach.

15.1 Introduction

In recent years, consumer products, e.g. domestic appliances and consumer electronics, that usually people buy for their own use are characterized by a strong emphasis on their interactive nature. In this sector more than others, a systematic

M. Bordegoni (✉) · F. Ferrise (✉) · U. Cugini (✉)
Dipartimento di Meccanica, Politecnico di Milano, Via La Masa 1,
20156 Milan, Italy
e-mail: monica.bordegoni@polimi.it

F. Ferrise
e-mail: francesco.ferrise@polimi.it

U. Cugini
e-mail: umberto.cugini@polimi.it

A. Chakrabarti (ed.), *CIRP Design 2012*, DOI: 10.1007/978-1-4471-4507-3_15,

validation of the design ideas, including the interaction design with potential customers, is crucial and strategic, as the definition and any modification of the product concept can affect customers satisfaction, in addition to its functional performances and manufacturing process. Actually, their success with customers does not only depend on their technical quality, functionalities, performances, robustness and so on, but also on the users appreciation of their aesthetic features, as well as of their ergonomic and usability features.

The evaluation of the concepts of new products is today made late in the product development process, after the product has been designed in detail (detail design, embodiment design) [1]. This aims at identifying the most appropriate and satisfactory product concept among those developed according to the product requirements defined a priori. Then, the product concept is eventually modified in line with the input collected in the evaluation phase, the design is released and the production starts [2].

Actually, various shortcomings of a product are often detected once the product is put on the market and potential customers try it at selling points or exhibitions, or when users use a product after purchasing. Therefore, methods for the continuous evaluation of the product features are crucial in order to refine its specifications and improve the quality of subsequent versions. In this way it is possible to plan evolutionary upgrades and incremental modifications of the products based on the evolving customers' feedbacks.

The evaluation of the aspects related to the usage of a specific product can be performed by observing and interviewing users while using the product. Actually, this practice allows us to record and notify the designers about problems and issues, but has the disadvantage that it is not possible to quickly propose new solutions and alternatives. Conversely, the use of virtual prototyping for performing similar evaluations could be more flexible, and could help in acquiring more useful information about variants and potential subsequent versions of that product.

Making final users use an interactive Virtual Prototype (iVP) of a product would allow us to get information about ergonomic and usability issues. The items of information that can be collected can be subjective, and can be captured by means of questionnaires compiled by users after using the product. Otherwise they can be objective, and so acquired by means of the analysis of data captured through measurement systems. Figure 15.1 shows how we can collect information about the evaluation of the use of consumer products performed in a virtual environment.

Actually, in order to use a virtual prototype for measuring the effectiveness of the proposed interaction with the product, we need to measure the effectiveness of the simulation of the interaction, which is implemented by the iVP.

This paper describes the implementation and the use of the interactive virtual prototype of an existing consumer product, a refrigerator. The interaction with the virtual prototype is based on a visuo/tactile user interface, which has been implemented using low cost Virtual Reality technologies. The aim of the research

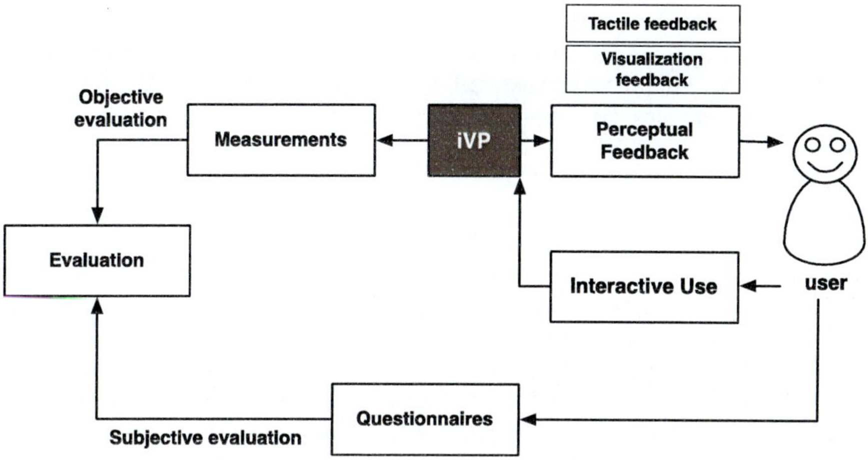

Fig. 15.1 How information on user's usage of the iVP is collected and used for the evaluation of human interaction

is to validate the effectiveness of the proposed interaction modality, so that it can be used in the future for testing specific usage aspects of product concepts.

15.2 Related Works

The evaluation of the design of products is typically made using physical prototypes, that are handcrafted or rapid prototyped [3]. Human factors disciplines have been criticized for not simulating and modelling human interactions with products at an early stage, relying instead on physical prototypes [4]. Thus, the design must have matured sufficiently for a physical prototype to be made, or tests have to be performed on operating products. The lack of simulation and modelling in human factors has been attributed to the inherent difficulty of predicting human interaction with products.

Early evaluations of human factors often rely on human-manikin tools such as Jack (http://www.plm.automation.siemens.com/), which are limited to mainly physical evaluations of anthropometric suitability [5]. Other traditional human factors approaches consider the tasks which humans must conduct, leading to the development of a branch of methods collectively named task analysis [6]. Other methods are used to perform product evaluation by using existing products, like for example the one described by Romero et al. in [7]. These tests are based on the comparison of some new design concepts built by means of some physical prototypes and a successful existing product.

When possible physical prototypes are substituted by virtual prototypes which are less expensive and faster to build. Physical prototypes have not been completely substituted in the product development process because while some aspects, as aesthetic aspects and perceived visual quality, can be today effectively assessed using virtual prototypes, all those aspects related to the use of the products (ergonomics, usability, physical interaction and response) still require physical prototypes.

The initial use of Virtual Prototypes was constrained by the available technologies, which were mainly limited to the visualization. Today a large number of virtual prototyping applications are still based on the visual representation of products [8]. This happens despite the availability on the market and in the research context of technologies that would allow user interaction through multiple sensory channels, like haptic technologies [9], and auditory displays [10].

Virtual Prototyping based on pure visualization has not demonstrated of having the same effectiveness for any kind of evaluation, for example for ergonomic and usability evaluations. In order to effectively substitute physical prototypes with virtual ones, it is necessary to have a virtual model that has the same characteristics of, and behaves like the corresponding physical one for what concerns the aspects we are interested to simulate and validate. It seems that a multimodal approach, i.e. the use of the combination of various sensory channels, in the interaction with Virtual Prototypes is essential in order to get an evaluation procedure that really allows us to effectively and accurately test and validate the various characteristics and usability aspects of future products. For example, Bordegoni et al. [11] have demonstrated the effectiveness of using a multimodal approach based on the combination of vision, sound and touch in performing a design review activity of an interactive product. The aim is to allow the designer to perform a design review together with customers, and refine the haptic feedback of interactive components of a washing machine.

The development of a high fidelity Virtual Prototype is actually costly in terms of time and effort required for the development, and of the technology necessary for the implementation. Among others, haptic technology for industrial applications is particularly expensive. For this reason, there are examples of applications that have been implemented using low cost technology. Belluco et al. [12] describe a system for the assembly of mechanical components performed using two hands, and based on the use of a 6-DOF haptic device and a Wii remote control device. In [13] de Haan et al. describes the use of the Wii Balance Board as a low cost input device for VR environments exploration.

In the research activity described in this paper we try to find a good compromise between the level of fidelity of the prototype and the effectiveness of the users' tests that can be performed, and the overall cost of the virtual prototype implementation.

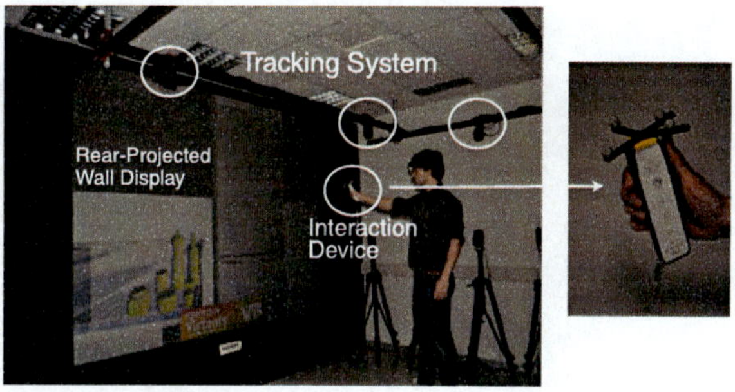

Fig. 15.2 The hardware setup consists of a rear-projected wall display, an optical trackingsystem and a Nintendo Wiimote

15.3 Virtual Prototype of a Selected Consumer Product

The case study that has been selected for this research consists of a refrigerator that has been provided by the Indesit Company (www.indesitcompany.com). The basic idea was to implement an interactive Virtual Prototype (iVP) of the refrigerator and of some objects that are usually stored in the refrigerator (cans, bottles, etc.). The user's task is inserting and placing the virtual objects into the various compartments of the refrigerator. The aim of the company is to check the user comfort during the task execution. A requirement of the application is that the virtual objects are easily and intuitively manipulated by means of an interaction device that also provides a kind of feedback when the objects collide with parts of the refrigerator.

15.3.1 Hardware and Software Components

The hardware setup used in the implementation of the iVP is illustrated in Fig. 15.2 and consists of:

- a rear-projected wall display Cyviz (www.cyviz.com) for scale stereoscopic visualization of the product;
- a Wiimote controller by Nintendo (www.nintendo.com/wii);
- an optical tracking system by ARTracking (www.ar-tracking.de) used to compute the user's point of view position and orientation in real-time. The AR-Tracking system is even used to acquire the position and orientation of the Wiimote.

The Wiimote control device is used as input and output device. It is used to grasp and manipulate virtual objects, and to convey feedback to the user about his actions. We decided not to use the Wiimote own built-in sensors, i.e. the accelerometers and the camera with the IR tracking system, in order to have the same reference axis for both the virtual prototype and the objects that are connected to the Wiimote. This allows us to maintain the same proportions and distances occurring during the interaction with the real product.

Fig. 15.3 The visual environment showing the refrigerator and the objects on the table

Regarding the use of the Wiimote as input device, it is tracked and some buttons are used to input commands. We decided to use only two buttons in order to keep the application simple and limit the cognitive overload. The ‘A’ button (in front of the device) is used to open the refrigerator doors, and the back button is used to grasp the objects.

The controller is also used as an output device that returns to the user the perception of a collision occurring between the object that he is grasping and the refrigerator components through a vibration.

Regarding the software setup, the virtual prototype has been developed by using the 3DVIA Virtools (www.virtools.com). The geometrical model of the refrigerator has been provided by the Indesit Company, and was developed by using the Siemens SolidEdge CAD tool (www.plm.automation.siemens.com). The model has been translated into 3DXML file format through the use of 3DVIA SolidWorks (www.solidworks.com). This format can be imported into the Virtools development environment.

15.3.2 Implementation and Use of the Virtual Prototype

The first step of the implementation of the virtual prototype regards the creation of the visual environment that consists of the fridge, a table and some objects initially put on the table. A red transparent block is visualized in the virtual scene and is used as the avatar of the user’s hand that moves accordingly into the virtual environment. Figure 15.3 shows the visual environment. The refrigerator is initially closed, and all the objects are on the table.

In order to lighten the CAD model of the refrigerator, so as to guarantee a smooth interaction with the virtual objects at run-time, some simplifications on the CAD model have been made. A simplification consists of eliminating those parts that are never displayed during the interaction, and even some details in the part that do not concur to the interaction. This procedure was necessary since the geometrical model provided by the manufacturer was complete and detailed.

Fig. 15.4 Feedback about the collision with parts of the refrigerator is returned in two different ways: the object changes color into red, and the Wiimote starts vibrating

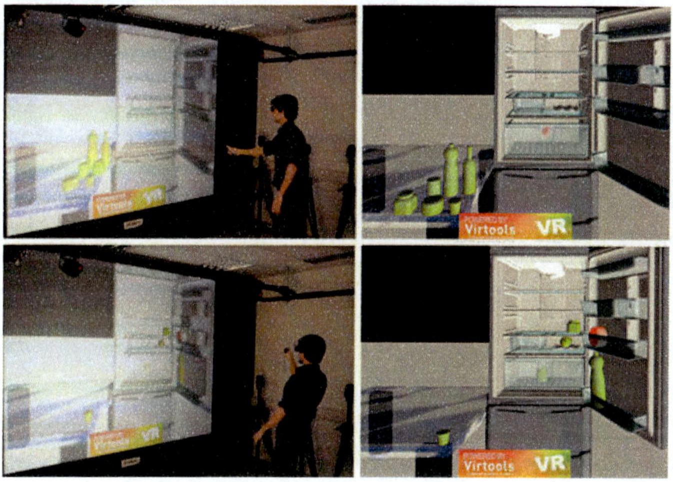

Fig. 15.5 Use of the application: to open the door the user must push the A button when the avatar of the hand is colliding with it. When objects collide they change color and a vibration is communicated to the user

When one of the objects that are on the table is grasped, the red block (avatar of the user hand) disappears, and the movement of the hand is rendered through the movement of the virtual object. A feedback about the collision of the grabbed object with the internal parts of the refrigerator is returned to the user both visually, by turning the color of the object into red (as illustrated in Fig. 15.4), and haptically, through a vibration of the Wiimote. When an object is released by the user, it automatically falls down and stops when a collision is detected and then the equilibrium is reached.

The use of the application is illustrated in Fig. 15.5. Some objects are initially put on the table located on the left hand side of the fridge. The fridge at the beginning is closed. The user is asked to open one of the doors and to open the

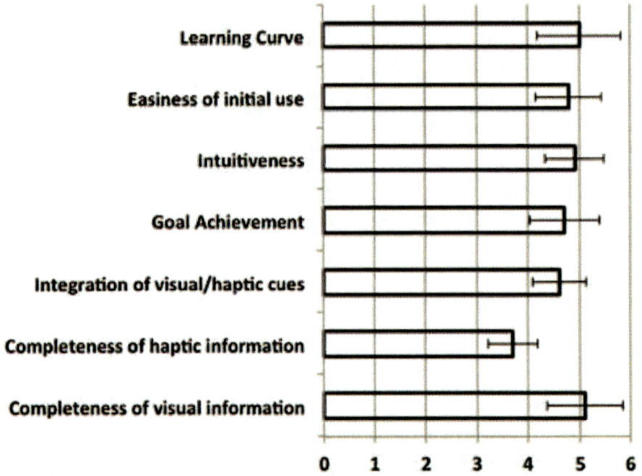

Fig. 15.6 Results of user study

drawer. Then he is asked to grab the objects, one after the other, and put them in a particular position inside the refrigerator compartments, or inside the drawer.

In order to open/close the doors the avatar of the hand must collide with them, and the *A* button of the Wiimote must be pushed. In order to grasp and hold one of the objects the *B* button must be kept pushed until the user decides to release it.

15.4 User Study

We performed a preliminary testing activity to evaluate the interaction modality with the virtual product.

Ten users, 8 male and 2 female, aged between 22 and 36, have been asked to complete a task with the interactive Virtual Prototype of the refrigerator. Users were not familiar with VR technologies even if 6 of them had previous experience with the use of the Wiimote controller. The task consisted in putting some selected objects into specific positions inside the refrigerator, within a given time: in particular two bottles and one glass jar in the door shelf rack.

Before starting the test, the participants were given a demonstration of 5 min about the interaction with the virtual fridge.

After the test they were asked to fill in a questionnaire including a subjective evaluation about the interaction with the virtual prototype of the fridge, by answering to several questions. A 6 point scale evaluation was used to collect participants' rating for each question, ranging from 0 (bad) to 6 (very good), where 3 means acceptable.

Figure 15.6 summarizes the results of the questionnaires. In particular the following aspects are of major interest:

- users judged the application easy to use. It was easy to understand how to interact with the virtual objects since the beginning, and it was also easy to

complete the task. Furthermore the data about the learning curve demonstrates that the application becomes even simpler after some trials;
- visual information is considered sufficiently complete despite the use of non-real objects like the red block that simulates the user's hand, and the little attention paid to the quality of the rendering;
- some details are still missing in the haptic rendering, since vibration is not enough to represent collisions;
- integration between visual and haptic cues during object-refrigerator collisions is considered good;
- finally the application is considered intuitive.

The user's tasks are recorded by the application. Specifically, we record the trajectory of the user's hand when manipulating the virtual objects, for subsequent studies about ergonomics features of the product (accessibility, comfort, etc.).

15.5 Discussion and Conclusion

The paper has described the implementation of an interactive Virtual Prototype (iVP) of a commercial refrigerator to use for a preliminary analysis of the user interaction with industrial consumer products. The interactive Virtual Prototype can be used to test aspects concerning product usage, such as usability and ergonomics, since the very beginning of the design process, while only a preliminary idea of the product is available. The same approach can be used even to test design variants when some decisions have already be taken in the product development process or to run some experiments on existing products by comparing them with others.

The iVP is based on a balance between fidelity and cost. Fidelity is proportional to the quality of hardware/software technologies available and even the time that is spent for implementing the iVP [14]. Since one of the needs of industries is reducing the time and the cost necessary to put the product on the market, named time-to-market, sometimes a good compromise between fidelity and cost is necessary.

A preliminary testing activity has been performed in order to evaluate the effectiveness of the approach. The good results have demonstrated that sometimes it is not necessary big effort in order to perform testing activities originally made on physical mockups on the virtual ones. This is true despite some weak points that have emerged regarding the completeness of the haptic cues. In fact, pure vibration is not sufficient to return to the user a realistic sense of touch, even if integrated and completed with visual cues. Commercial force feedback haptic devices have a limited workspace to be used for such an application and the cost is not comparable with a device used for gaming purposes. Then even if in principle their use would increase the fidelity of the iVP, the related cost, including purchasing and development, would be higher.

The data recorded about the users' task will be used to evaluate some aspects of the refrigerator usage, as reachability, comfort and dimensions.

References

1. Cross N (2008) Engineering design methods: strategies for product design, 4th edn. Wiley, New York
2. Laughery R (2005) Evaluation of human work. Chapter: Simulation and modelling for analysing human performance, CRC Press, pp 219–238
3. Lawson B (2005) How designers think, fourth edition: the design process demystified, 4th edn. Architectural Press, Amsterdam
4. Rodgers P, Pattersoon A, Wilson D (1996) A formal method for assessing product performance at the conceptual stage of the design process. In: Advances in formal design methods for CAD: proceedings of the IFIP WG5. 2 Workshop on formal design methods for computer-aided design, June 1995, Springer, pp 137–164
5. Badler N, Phillips C, Webber B (1993) Simulating humans: computer graphics animation and control. Oxford University Press, Oxford
6. John W, Helen H, Wendy M (2005) Participatory ergonomics. CRC Press, 2011/04/21
7. Romero M, Costa F, Andreoni G, Mazzola M, Vargas J, Conenna L (2009) Proactive ergonomics in refrigerator concept development. In: Kurosu M (ed) Human centered design. Lecture notes in computer science, vol 5619. Springer, Berlin, pp 627–634
8. Zorriassatine F, Wykes C, Parkin R, Gindy N (2003) A survey of virtual prototyping techniques for mechanical product development. Proceedings of the institution of mechanical engineers, Part B: Journal of engineering manufacture 217(4):513–530
9. Hayward V, Ashley O, Hernandez MC, Grant D, Robles-De-La-Torre G (2004) Haptic interfaces and devices. Sens Rev 24(1):16–29
10. Kramer G (2000) Auditory display: sonification, audification and auditory interfaces. Addison-Wesley Longman Publishing Co., Inc., Boston
11. Ferrise F, Bordegoni M, Lizaranzu J (2010) Product design review application based on a vision-sound-haptic interface. In: Nordahl R, Serafin S, Fontana F, Brewster S (eds) Haptic and Audio Interaction Design. Lecture Notes in Computer Science. vol. 6306 Springer Berlin/Heidelberg, pp. 169–178
12. Belluco P, Bordegoni M, Polistina S (2010) Multimodal navigation for a haptic-based virtual assembly application. ASME Conference Proceedings 49088:295–301
13. De Haan G, Griffith EJ, Post FH (2008) Using the wii balance board as a low-cost vr interaction device. In: Proceedings of the 2008 ACM symposium on virtual reality software and technology, New York, VRST'08, ACM, pp 289–290
14. Bordegoni M (2011) Product virtualization: an effective method for the evaluation of concept design of new products. In: Bordegoni M, Rizzi C (eds) Innovation in product design: from CAD to virtual prototyping. Springer

Chapter 16
CPR Module with Variable Chest Stiffness in High Fidelity Mannequins

K. Kanakapriya and M. Manivannan

Abstract Cardio Pulmonary Resuscitation (CPR), especially chest compression, should be performed promptly on victims of cardiac arrest for improved morbidity. CPR is performed in 2 steps, chest compression and chest ventilation. But, commercially available CPR mannequins do not accurately mimic the viscoelastic properties of the human chest, resulting in poor CPR performance. We propose a CPR mannequin which exhibits a chest with increasing stiffness as the depth of compression is increased; the increased bio-fidelity of this mannequin should improve CPR training efficacy. This mannequin also measures depth and rate of compression, the air volume during chest ventilation, and compares the measured values against acceptance criteria that can be tuned to different standards.

16.1 Introduction

Cardio Pulmonary Resuscitation (CPR) performed on the victim of Cardiac Arrest ensures intact brain function until spontaneous blood circulation can be restored. CPR consists of 3 steps [1], first clear the airway to allow exchange of gases between the lungs and the atmosphere, next begin chest compressions at the rate of 100 compressions per minute to a depth of 4 cm; Stop after 30 compressions and ventilate the chest by administering 2 artificial breaths, ensuring the chest rises with each breath. The last 2 steps are repeated for 5 cycles before checking for a spontaneous pulse. CPR is performed both inside the hospital and outside. With

K. Kanakapriya · M. Manivannan (✉)
Biomedical Enginnering Group, Department of Applied Mechanics,
IIT Madras, Chennai, India
e-mail: mani@iitm.ac.in

A. Chakrabarti (ed.), *CIRP Design 2012*, DOI: 10.1007/978-1-4471-4507-3_16,

increasing number of cardiac diseases and concomitant increase in cardiac arrest [2], it is prudent for lay persons also to learn to perform CPR.

CPR is often unsuccessful, especially when performed by lay persons [3]. Apart from the general tendency to confusion in an emergency, performing CPR requires training and practice. The training is complicated by two separate factors, one is the shockingly large force required to attain a 4 cm depth of compression and the other is the speed of compression, which amounts to nearly 2 compressions every second!

Medical Mannequin simulators have been shown to improve acquisition and retention of knowledge [4, 5]. Accordingly many mannequins including the popular Laerdal's ResusciAnne [6] have been developed specifically for training in CPR. But these mannequins suffer from two disadvantages, one is they have a chest with constant stiffness; i.e., in each case the chest is modeled by a single spring giving a single stiffness rate. The human chest, on the other hand is viscoelastic, with stiffness increasing with depth of compression [7]. The other disadvantage is the artificial breath parameters are not tunable, either the volume of air blown into the lungs is not measured or it has a preset pass/fail value which cannot be changed. But, the time and volume of ventilation varies between different agencies, AHA emphasizes CPR, deems ventilation unnecessary especially when performed by a layperson [1], 0.5–2 liter otherwise; The Military specifies a ventilation volume of 0.5–4 liter [6] during each artificial breath; European Resuscitation Council guidelines is 0.4–0.6 liter.

Our design goal is to improve the bio-fidelity of the mannequin chest to improve CPR learning, incorporate pass/fail criteria for verifying CPR skills learned and make the ventilation pass/fail criteria tunable to accommodate various standards.

16.2 Methodology

16.2.1 Chest with Variable Stiffness

Grueben et al. [7] measured the force, displacement, velocity and acceleration during CPR on 16 patients and 3 mannequins with a system specifically designed for the investigation of CPR [8], in a clinical setting. The force applied during CPR was limited to 400 N by visual feedback from the force/displacement recording system to minimize the chances of chest trauma.

The force applied during CPR is a sum of the elastic force F_e and the damping force F_d. Both the elastic force F_e and damping force F_d varied with the depth of compression; The elastic force is a 4th degree polynomial in compression depth while F_d is a linear function and its contribution to the total force applied during CPR is low. The elastic force F_e has also been approximated with a four segment linear spline.

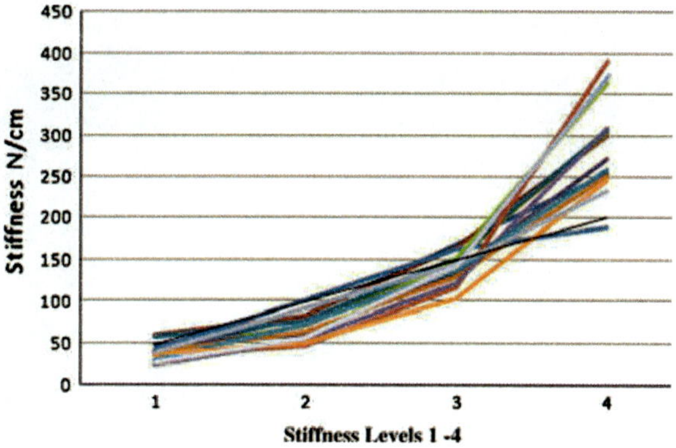

Fig. 16.1 The model stiffness estimate to fit the elastic force data in Gruebel et al. study [7]

$$F_e = \begin{cases} k_1 \quad for\ 0 \le F_e < 20N \\ k_2(x - x_1) + 20 \quad for\ 20 <= F_e < 100N \\ k_3(x - x_2) + 100 \quad for\ 100 <= F_e < 300N \\ k_4(x - x_3) + 300 \quad for\ 300 <= F_e \end{cases}$$

Where $x_1 = 20/k_1$, $x_2 = x_1 + 80/k_2$, $x_3 = x_2 + 100/k_3$, the elastic spline parameters, k_1, k_2, k_3, k_4 provide the piecewise linear approximation of the chest stiffness in four ranges.

We are interested in simplifying the data from Gruben et al. into a set of realisable design. Accordingly we view the data in terms of a set of 4 elastic spline parameters calculated to fit their data set (Fig. 16.1). As is evident the highest stiffness value k_4 ranges from 190 to 389 N/cm. while the initial lowest stiffness, k_1 ranges from 21 to 60 N/cm. The data indicates the human chest shows a wide variation both in stiffness and rate of stiffness change with depth of compression.

Wide ranging stiffness is easily realised in terms of springs: springs in series whose stiffness add up as depth of compression increases, stacked springs of increasing stiffness or a conical spring with varying diameter and/or pitch. The springs had to fit inside the chest of the First Aid mannequin already being built. This put a constraint of a maximum of 16 cm on the height of the spring and a minimum of 5 cm on the inner diameter of the spring. Since the batch size for the spring manufacturer was limited to 2–3 sets, only cold rolled steel springs were available at reasonable cost; Each set of 4 constant pitch springs costs about 120 Rs, the conical spring cost 150 Rs. The total cost of the mannequin worked out to 30,000 Rs while the imported ResusciAnne mannequin without tunable acceptance criteria and changeable chest stiffness costs 1,50,000 Rs. The mannequin cost would be lower once the designs are finalized and transferred to cheaper processors and mannequin bodies.

Fig. 16.2 Mannequin body with 4 concentric springs mounted on the chest

Table 16.1 Four springs of varying heights mounted concentrically

Spring #	1	2	3	4
Diameter cm	5	6	7	8
Height cm	15.5	15.0	13.5	11.5
Stiffness N/cm	37	30	136	146
Cumulative Stiffness N/cm	37	67	203	349
Compression depth at which activated cm	0	0.5	2.0	4.0

16.2.1.1 Springs in Parallel

Four springs of varying heights are mounted concentrically; At the start of compression, the first spring with stiffness k1 is active; When the depth of compression is 0.5 cm, both the first and second spring become active with a total stiffness of $k_1 + k_2$; At 2 cm depth of compression, the first, second and third springs become active with a total stiffness of $k_1 + k_2 + k_3$; At 4 cm depth of compression all the four springs become active with a total stiffness of $k_1 + k_2 + k_3 + k_4$ Figs. 6.2, 6.3, Table 16.1;

16.2.1.2 Conical Spring

A conical spring starts with a low stiffness, a small initial force provides a noticeable depth of compression, but, the stiffness increases with applied load necessitating a large force to produce the required depth of compression. This is because of the variable stiffness of the spring;

As seen from the spring formula, spring coils with the largest diameter have the lowest stiffness. These are the first to deflect; after a certain load they bottom out, now coils with a smaller diameter deflect, but they have a higher stiffness, hence requiring a larger force to produce the same deflection.

$$k = \frac{Gd^4}{8nd^3}$$

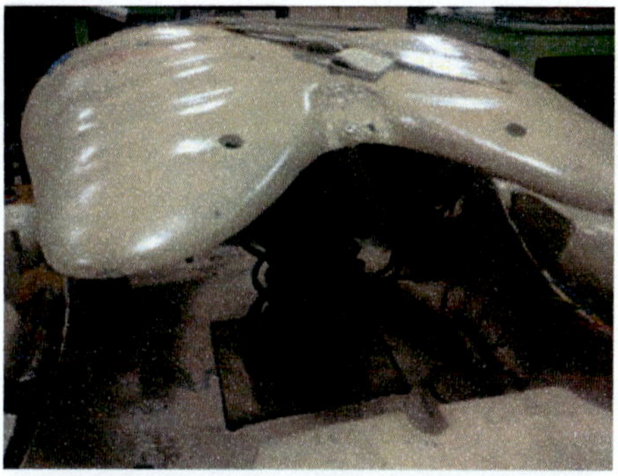

Fig. 16.3 Chest mounted on the body of CPR mannequin

Fig. 16.4 Conical spring developed for CPR mannequin

Where d is the coil diameter, D is the spring mean diameter, n is the number of active coils and G is the modulus of rigidity Fig. 16.4.

A MATLAB simulation was used to generate the force displacement curve, the simulation assumed the spring is made up of a number of single turn springs of equal pitch, with each successive single turn spring having decreased diameter and hence increased stiffness. The pitch has to be changed from turn to turn to obtain desired force displacement curve. Friction between the coils as a result of bottoming out has been neglected. This would add to the increasing stiffness.

16.2.1.3 Stacked Springs

A manufacturer who could make a conical spring in a batch of 2 springs at a reasonable cost could not be found. The design has hence been converted to a set of stacked springs of varying stiffness. As seen from the spring formula, the varying stiffness effect can also be realised by changing the pitch of the spring. The coil diameter, spring diameter and the pitch were modified to generate a set of 3 stacked springs of varying stiffness similar to that from the conical spring Table 16.2.

Along with the springs specified in Table 16.2, Belleville washers of various outer diameter, 55 mm inner diameter and 10 mm height can used to get different stiffness ranges for the same depth of compression.

Table 16.2 Stacked springs of varying stiffness

Spring #	1	2	3
Coil diameter mm	4	4	6
Diameter mm	54	54	58
Height mm	45	45	70
Pitch mm	12	30	17
Stiffness N/cm	40	102	154

In addition to the compression depth, the compression rate and the placement of the hand for performing CPR are also monitored.

16.2.2 Chest Displacement Measurement

Linear Potentiometers caused reliability problems when operated at the nearly 2 Hz frequency required for CPR. The mannequin chest houses the embedded processors and sensors and actuators for other physiological phenomena like respiration and heartbeat. Capacitive or inductive displacement sensors could interfere with these circuits. Contact less displacement systems like optical systems and ultrasonic systems were investigated and we finally settled on a light based displacement measurement setup.

A Light Emitting Diode (LED) is mounted on the sternum, i.e., the top plate sitting on the spring, an Light Dependent Resistor (LDR) is mounted on the base plate, i.e. the base on which the spring is mounted. The light incident on the LDR increases as the compression depth increases and the resistance of the LDR decreases. The change in resistance is mapped to the compression depth.

16.2.3 Flowmeter to Measure Ventilation Volume

Air for ventilating the mannequin's lungs is directed from the mouth of mannequin into its chest by a rubber tube. The flowmeter is mounted at the chest end of the tube to measure ventilation volume. Two different systems have been realised to measure the air flow velocity: one is a rotation based flowmeter and the other is transistor based. The flow rate measured by either system is integrated by the processor to obtain the ventilation volume.

16.2.3.1 Rotation Type Flowmeter

An axial fan is mounted at the end of the windpipe inside the chest of the mannequin. LED and LDR are mounted on either side of the mannequin, whenever the Fan Blade passes the LED, light incident on the LDR is cut off completely, increasing its resistance. The value and frequency of change in resistance is mapped to the air velocity.

16.2.3.2 Transistor Type Flowmeter

This is just a sturdier version of the Hot Wire Anemometer, instead of cooling a heated thin, delicate wire, a transistor [12] is cooled. The temperature differential between two transistors Q1 and Q2 is maintained at about 50° so that they have the same base voltage. The ventilation airflow over the heated transistor Q1 cools it, causing a voltage drop. The sensor circuit drives the voltage back up again by generating a current to heat the transistor Q1. The current required to retain the temperature differential is mapped to the air flow velocity.

Two mannequins have been built so far, one with the rotation flowmeter and the other with the transistor-based flowmeter. The transistor based flowmeter is more sturdy and easier to install than the rotation type flowmeter. A linear actuator mounted in the chest, causes the chest to rise if the minimum required ventilation volume is provided. This provides the visual cue for correct ventilation.

16.2.4 System Interaction

The sensors and actuators are connected to an embedded processor which communicates with a computer via TCP/IP. The computer has a JAVA based GUI which allows the various criteria for successful CPR performance to be edited in admin mode. The CPR module itself is launched via the GUI.

One version of the mannequin uses the Rabbit RCM5700 processor. The sensors in this case communicate with the processor via a 32 to 1 Multiplexer. The processor continuously polls the Multiplexer; once a sensor becomes active, the corresponding program is activated on the computer, which in turn provides the trainee with visual/audio cues regarding the training goals Fig. 16.5.

16.3 Discussion

The 16 volunteers in the seminal study to measure CPR Force [7] showed huge variations in stiffness, so Chest Compression training can also be viewed as learning to gauge a compression depth of 4 cm irrespective of the increasing force required to achieve it; We have come up with a mannequin whose stiffness is a stepwise linear function of compression depth. We can achieve different stiffness rates by changing out one set of springs for another, and this may be the best method to learn to gauge a depth of 4 cm irrespective of the resistance offered.

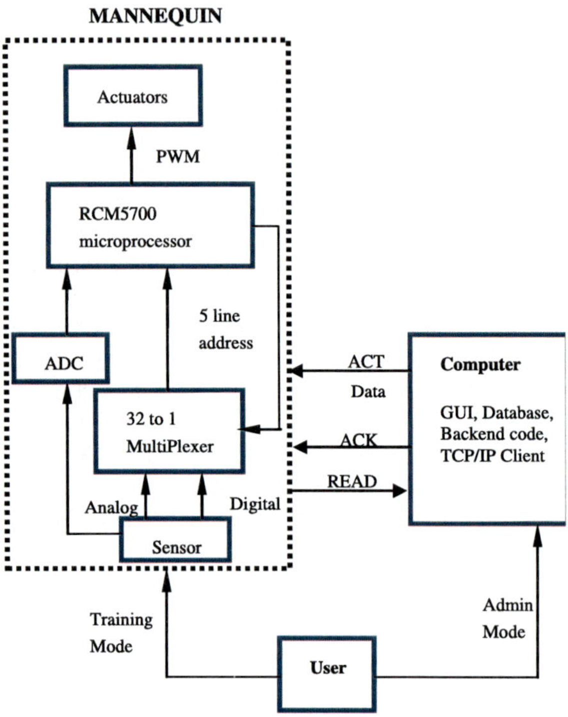

Fig. 16.5 System interaction between the mannequin which houses the CPR module, the main computer and the user

It is possible that the reason for poor CPR performance is the fear [10] one feels when applying so much force on a human being as opposed to a mannequin, still reinforcing the necessity of a 4 cm depth perception might help improve performance. Also, with this mannequin design, springs can be swapped out to achieve different stiffness rates as recommended by Baubin et al. [11].

In most mannequins including ResusciAnne the ventilation air flows into a bag; the bag expands and presses one or more push button depending on the inflow volume. The push buttons are preset for one set of pass/fail criteria. Changing the pass/fail criteria would require hardware rework and calibration. In our case the ventilation volume is measured as velocity and then integrated to get the actual volume; the central processor compares this volume against preset pass/fail values to determine whether the ventilation volume was adequate. Changing the pass/fail volume criteria is a simple setting change in the admin mode. The system can be run in admin, testing and training modes.

16.4 Summary

A CPR module for a high fidelity mannequin with tunable pass/fail criteria has been developed. The efficacy of learning CPR on this module is yet to be established on the field. Future versions of the CPR module will also incorporate a variable damper in the form of a Magneto Rheological Elastomer or Electric Rheological Damper. The next step would be to develop a smart material stiffness element with tunable stiffness range, allowing a single set up to mimic various Stiffness versus Depth curves.

References

1. AHA (2005) American heart association guidelines for cardiopulmonary resuscitation and emergency cardiovascular care part 7.2: management of cardiac arrest circulation, part 3: overview of CPR 2005, 112:IV-58-IV-66
2. Gupta R (2005) Burden of coronary heart disease in India. Indian Heart J 57:632–638
3. Geddes LA, Boland MK, Pervin KT, Vitter J (2007) Chest compression force of trained and untrained rescuers, Cardiova Eng 8:219–224
4. Barry Issenberg S, McGaghie WC, Hart IR et al. (1999) Simulation technology for health care professional skills training and assessment. JAMA 282(9):861–866
5. Cooper B, Taqueti VR (2004) A brief history of the development of mannequin simulators for clinical education and training. Qual Saf Health Care 13(Suppl. 1):i11–i18
6. FM21-11 (2011) First aid for soldiers, retrieved June 1st 2011 from http://www.medtrng.com/Fm21_11/fm2111.htm
7. Gruben KG, Romlein J, Halperin HR, Tsitlik JE (1993) Sternal force-displacement relationship during cardiopulmonary resuscitation. J Biomech Eng 115:195–201
8. Gruben KG, Romlein J, Halperin HR, Tsitlik JE (1990) System for mechanical measurements during cardiopulmonary resuscitation in humans. IEEE Trans Biomed Eng 37(2):204–210
9. Woodward S (2002) Transistor linearly digitizes airflow, EDN, Sep 19, 104
10. Ødegaard S et al (2007) Chest compressions by ambulance personnel on chests with variable stiffness: Abilities and attitudes. Resuscitation 74(1):127–134
11. Baubin MA, Gilly H, Posch A, Schinnerl A, Kroesen GA (1995) Compression characteristics of CPR manikins. Resuscitation 30(2):117–126
12. Resusci® (2011) Anne basic and skillguide, retrieved June 1st 2011 from http://www.laerdal.com/docid/1022079/Resusci-Anne-Basic-and-SkillGuide

Chapter 17
Evaluation of the Accuracy of an Accelerometer Response Generated by Axial Impact Loading

Gauri Ranadive, A. Deb and Bisheshwar Haorongbam

Abstract The basic objective in the present study is to show that for the most common configuration of an impactor system, an accelerometer cannot exactly reproduce the dynamic response of a specimen subject to impact loading. Assessment of the accelerometer mounted in a drop-weight impactor setup for an axially loaded specimen is done with the aid of an equivalent lumped parameter model (LPM) of the setup. A steel hat-type specimen under the impact loading is represented as a non-linear spring of varying stiffness, while the accelerometer is assumed to behave in a linear manner due to its high stiffness. A suitable numerical approach has been used to solve the non-linear governing equations for a 3 degrees-of-freedom system in a piece-wise linear manner. The numerical solution following an explicit time integration scheme is used to yield an excellent reproduction of the mechanical behavior of the specimen thereby confirming the accuracy of the numerical approach. The spring representing the accelerometer, however, predicts a response that qualitatively matches the assumed load–displacement response of the test specimen with a perceptibly lower magnitude of load.

G. Ranadive · A. Deb (✉) · B. Haorongbam
Centre for Product Design and Manufacturing, Indian Institute of Science, Bangalore, India
e-mail: adeb@cpdm.iisc.ernet.in

G. Ranadive
e-mail: gauri_2984@yahoo.co.in

B. Haorongbam
e mail: bisheshwar_h@yahoo.com

A. Chakrabarti (ed.), *CIRP Design 2012*, DOI: 10.1007/978-1-4471-4507-3_17,

17.1 Introduction

A key consideration in a design process is to evaluate the performance of an object and its components under impact. An accelerometer [1] measures the vibration, or acceleration of a structure. The force caused by vibration or a change in a motion causes the mass to "squeeze" the piezoresistive material that changes resistance that is proportional to the force exerted upon it [1]. Since the charge is proportional to the force, and the mass is a constant, then the charge is also proportional to the acceleration. Authors in [2, 3] have explained design and construction of such accelerometers.

However, there is no study available in present literature to quantify the accuracy of accelerometer in a given impact set-up. Here explicit central difference approach is explored to solve governing differential equations of a lumped parameter model representing given impact test set-up. Impact system comprises of impactor, piezoresistive accelerometer and load cell which is connected to impactor head which hits the steel specimen under the impact test. The specimen represents a steel hat-type rail section encountered in car body structures. A given load-displacement response obtained in an actual impact test is assumed to be the true behavior of the specimen. The non-linear behavior of specimen is analyzed in a piece wise linear approach and implemented using code in MATLAB, producing an excellent correlation between input i.e., dynamic response and computed response of specimen, under impact. For the first time, to the authors' best knowledge, comparison between input and computed response of an accelerometer is discussed and quantified in terms of root mean square value (RMSE). An accelerometer is mounted on the vibrating surface to sense the motion and acceleration values are highly dependent on the way in which it is mounted on the surface [4]. Mounting considerations of accelerometers and its effect on the values of measured acceleration is also discussed in this paper.

17.2 Piezoresistive Accelerometers

An accelerometer is a device that measures the physical acceleration experienced by an object i.e., the type of acceleration associated with the phenomenon of weight experienced by a test mass that resides in the frame of reference of the accelerometer device. The piezoresistive effect which is widely used in accelerometer principle states that an electrical resistor changes its resistance when it experiences a strain and deformation. A piezoresistive accelerometer incorporates a crystal semiconductor beam that works as a spring element. A spring element i.e., piezo-resistors can also be made of metal or ceramic material [2, 3]. The crystal carries a seismic mass. Several piezo-resistors are placed in the crystal's body and they are physically an integral part of the beam. By completing a

Wheatstone bridge around the piezo resistors of the accelerometer, a linear relationship between acceleration and voltage can be derived [1].

Accelerometer used in the present study is Endevco 7264B [5]. It is a very low mass piezoresistive accelerometer weighing only 1 g. This accelerometer is designed for crash testing and similar applications that require minimal mass loading and a broad frequency (40 kHz) response.

17.3 Impact System Test Set-up

According to the approach discussed in this paper, the experimentally obtained force–time response is assumed to be a true response for an axially impacted tubular steel component.

In Fig. 17.1 is shown the lower portion of a drop-weight impact testing system with the impactor resting on a crushed steel hat section specimen after being dropped from a given height. For recording load-time history, a column-type load cell is mounted between the main body of the impactor and a rigid impactor head. Additionally, a lightweight piezo-resistive accelerometer [5] is mounted centrally on the main body of the impactor. The impactor slides vertically under gravity on guide rods with minimal friction and acquires velocity proportional to square root of drop height as it strikes a specimen placed on a rigid base. The guide rods add to safety during the impact event. Locating the sensors within the impactor parts ensures versatility in testing; for example, impact perforation of plates with an indenter fitted under the impactor head can be performed in addition to axial impact testing. The load cell and accelerometer responses are digitized through a high speed NI (National Instruments)-supplied data acquisition card at a peak sampling rate of 250 kHz and processed using a Butterworth filter for suppressing high frequency noise in a Lab VIEW environment.

17.4 Lumped Parameter Modeling

Figure 17.2 is a conceptual model of impact testing system with three vertical degrees-of-freedom with springs 1, 2 and 3 representing the specimen, the load cell and the accelerometer respectively. Figure 17.3 represents an LPM corresponding to the Concept of Fig. 17.2. This model is used to simulate the impact testing of a given specimen and for predicting the accelerometer response. The force-displacement curve obtained in an experiment for a given specimen geometry (i.e. cross-section and length), impactor mass and drop height is shown in Fig. 17.4 and is nonlinear because of its varying slope i.e., instantaneous stiffness k_1 which is a function of displacement x_1 although k_2 and k_3 can be assumed as constant slopes of linear force–displacement responses of load cell and accelerometer respectively.

The governing equations can be simplified without sacrificing accuracy by assuming a piecewise linear representation of the force–displacement response of

Fig. 17.1 Image of impactor along with load cell and accelerometer mounted on it resting on a specimen after being dropped from a height

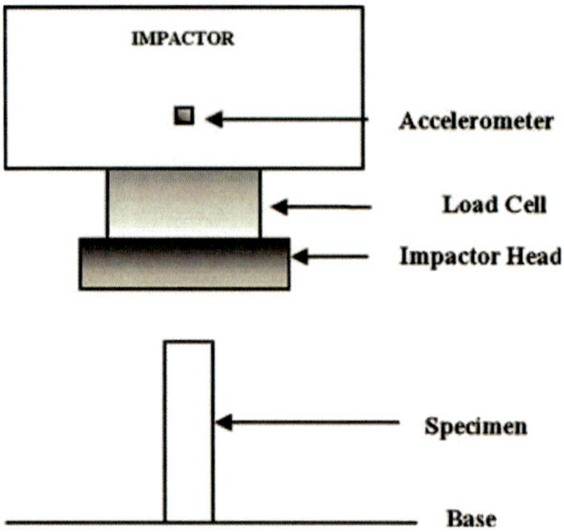

Fig. 17.2 Conceptual model of impact testing system

spring 1 as shown in Fig. 17.5 and increasing the number of linear segments. It is noted from Fig. 17.5 that the force f in the spring (i.e., the specimen) for any displacement x_1 is:

$$f = F_j + k_1^j (x_1 - X_j) \tag{17.1}$$

Where F_j is the force in Spring 1 and X_j is its shortening at the beginning of the jth linear segment in Fig. 17.5, and k_1^j is the constant slope of the jth segment.

With the above assumption, the governing differential equations of the 3-DOF spring-mass system (i.e. the LPM) of Fig. 17.3, in the matrix form, are obtained as:

$$\tilde{M}\,\ddot{\bar{x}} + \tilde{k}_j\,\bar{x} = \bar{R}_j \tag{17.2}$$

The solution of the above Eq. (17.2) can be initiated by imparting an initial common downward velocity (i.e. $\dot{x}_1$, $\dot{x}_2$ and $\dot{x}_3$ being equal to the impact velocity at time $t = 0$) to m_1 m_2 and m_3. Where,

Fig. 17.3 A lumped parameter representation of conceptual impact testing system

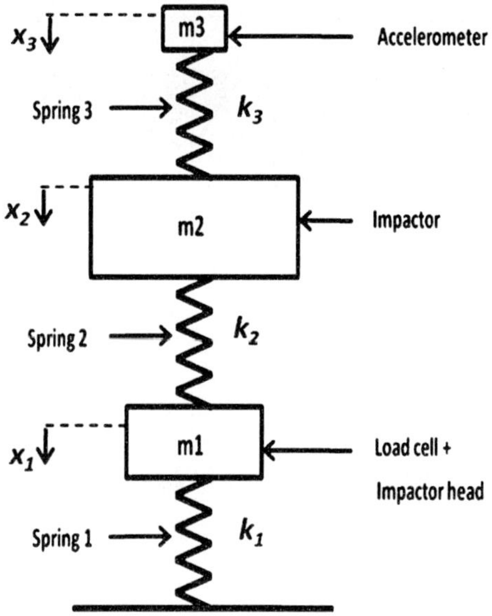

Fig. 17.4 Force-displacement behavior of specimen (Spring 1) obtained in impact test

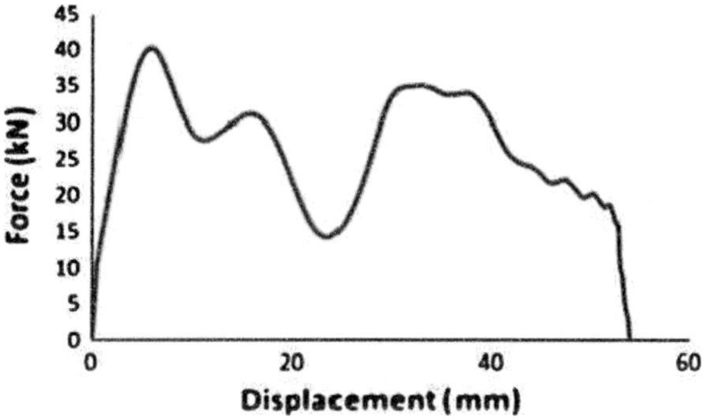

$$\tilde{M} = mass\,matrix = \begin{bmatrix} m_1 & 0 & 0 \\ 0 & m_2 & 0 \\ 0 & 0 & m_3 \end{bmatrix} \tag{17.2a}$$

$$\ddot{\tilde{x}} = acceleration\,vector = \begin{Bmatrix} \ddot{x}_1 \\ \ddot{x}_2 \\ \ddot{x}_3 \end{Bmatrix} \tag{17.2b}$$

$$\tilde{k}_j = stiffness\,matrix = \begin{bmatrix} k_1^j + k_2 & -k_2 & 0 \\ -k_2 & k_2 + k_3 & -k_3 \\ 0 & -k_3 & k_3 \end{bmatrix} \tag{17.2c}$$

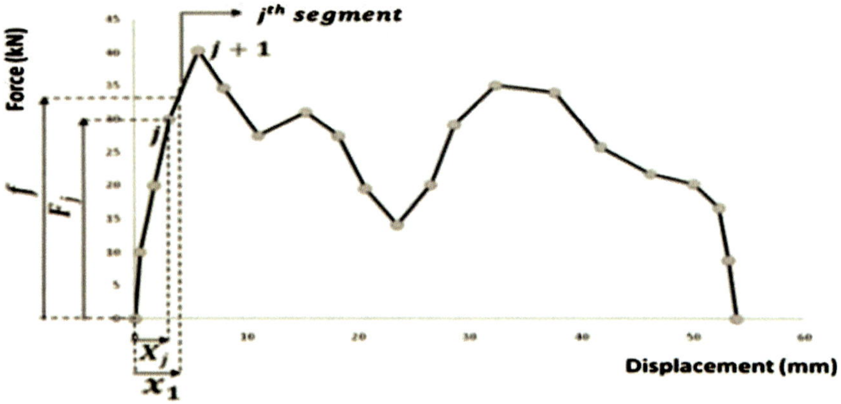

Fig. 17.5 Piecewise linear representation of the force-displacement curve of Fig. 17.4

$$\bar{x} = Displacment\ vector = \begin{Bmatrix} x_1 \\ x_2 \\ x_3 \end{Bmatrix} \tag{17.2d}$$

$$\bar{R}_j = Force\ vector = \begin{Bmatrix} k_1^j X_j - F_j \\ 0 \\ 0 \end{Bmatrix} \tag{17.2e}$$

m_1 = mass of impactor head and local cell
m_2 = mass of main body of the impactor
m_3 = mass of accelerometer
x_1 = displacement of m_1
x_2 = displacement of m_2
x_3 = displacement of m_3
$\ddot{x}_1$ = acceleration of spring 1
$\ddot{x}_2$ = acceleration of spring 2
$\ddot{x}_3$ = acceleration of spring 3

17.5 Numerical Solution and Analysis

In the present study, Eq. (17.2) is solved by numerical integration by marching forward in time with a small uniform time increment Δt. Explicit central difference method is adopted here [6]. This method is known to be conditionally stable and here error in computation can be reduced with decreasing time step Δt.

For Central Difference method, following approximations have been considered here:

$$\dot{\bar{x}}_t = \frac{1}{2\Delta t}(\bar{x}_{t+\Delta t} - \bar{x}_{t-\Delta t}) \tag{17.3}$$

$$\ddot{\bar{x}}_t = \frac{2}{\Delta t^2}\left(\frac{1}{2}\bar{x}_{t+\Delta t} - \bar{x}_t + \frac{1}{2}\bar{x}_{t-\Delta t}\right) \tag{17.4}$$

Using Eqs. (17.3) and (17.4) to calculate displacement at $(t - \Delta t)$ initially,

$$\bar{x}_{t-\Delta t} = \bar{x}_t - \dot{\bar{x}}_t \Delta t + \frac{\Delta t^2}{2}\ddot{\bar{x}}_t \tag{17.5}$$

Substituting Eq. (17.4) into Eq. (17.2),

$$\widetilde{M}\bar{x}_{t+\Delta t} = [2\widetilde{M} - \Delta t^2 \widetilde{K}_j]\bar{x}_t - \widetilde{M}\bar{x}_{t-\Delta t} + \Delta t^2 \widetilde{R}_j \tag{17.6}$$

Using Eqs. (17.5) and (17.6), the displacements of the springs in Fig. 17.3 can be obtained at any incremental (i.e., current) time $(t + \Delta t)$ knowing the responses at the previous instant of time t. Matrix $\widetilde{M}$ is diagonal and the system of three linear simultaneous equations represented by Eq. (17.6) need to be solved at only previous instant of time, the current time integration method is therefore called as 'explicit'. This numerical algorithm was implemented in the form of a MATLAB script, after every increment Δt and necessary computation (starting with $t = 0$ and $j = 1$), it is checked in which segment of the force–displacement curve in Fig. 17.4 will x_2 lie; accordingly the value of k_1 (i.e. k_1^j in Eq. (17.2)) for that j^{th} segment is chosen for next incremented time.

17.6 Simulation Results

The input data for the simulation is:

1. The behavior of Spring 1 (i.e. nonlinear k_1) is defined by the force–displacement curve shown in Fig. 17.4 with a discrete set of points;
2. $k_2 = 1000\,\frac{\text{kN}}{\text{mm}}$ (Obtained experimentally);
3. $k_3 = 63\,\frac{\text{kN}}{\text{mm}}$ (Based on the information on natural frequency and mass of a piezoresistive accelerometer provided by the manufacturer);
4. $m_1 = 30$ kg (Mass of impactor head and load cell);
5. $m_2 = 130$ kg (Mass of main body of impactor);
6. $m_3 = 0.02$ kg (Mass of accelerometer);
7. An impact velocity of 4.5 m/s provided as an initial velocity to each of the masses m_1, m_2 and m_3.

Calculation of critical time step for simulation is of prime importance for central difference method. Initial stiffness k_1 for spring 1 is given by Eq. (17.7a):

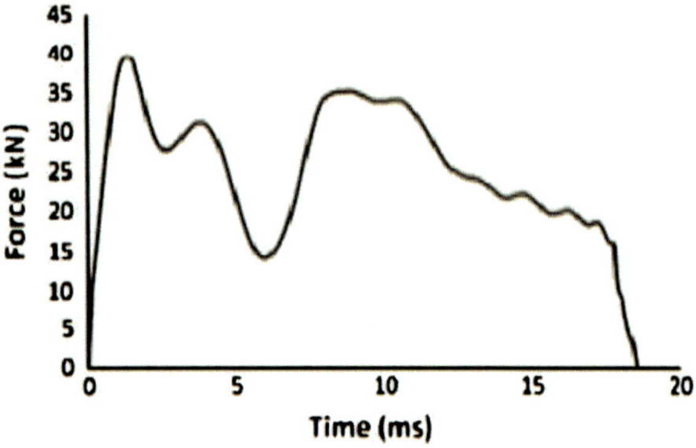

Fig. 17.6 Computed force time history of specimen (Spring 1)

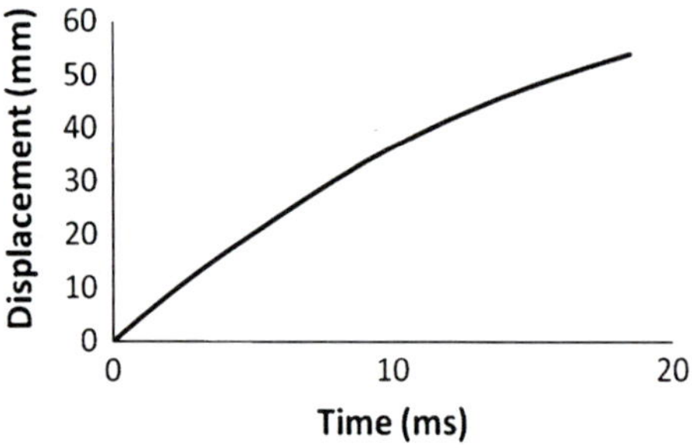

Fig. 17.7 Computed displacement time history of specimen (Spring 1)

$$k_1 = \frac{f_2 - f_1}{d_2 - d_1} \tag{17.7a}$$

$$m = m_1 + m_2 \tag{17.7b}$$

$$\omega = \sqrt{K_1/m} \tag{17.7c}$$

$$\text{Time period} = \frac{2\pi}{\omega} \tag{17.7d}$$

$$\text{time step } \Delta t = \text{Time period}/\mathrm{n_{time}} \tag{17.7e}$$

Where,

f_1, f_2 = *initial force values of input behavior of test specimen in* Fig. 17.4
d_1, d_2 = *initial displacement values of input behavior of test specimen in* Fig. 17.4
ω = *angular frequency*
n_{time} = *arbitary number of time intervals assumed*

It is noted that the mass of the specimen is a fraction of the impactor mass and can be lumped with mass m_1 in the current model with little effect on the results presented here. The time step Δt for analysis is calculated as 0.0043 ms (millisecond) using Eqs. (17.7a–17.7e). The maximum computation time, T_f, is given as

25 ms, however, the analysis is automatically terminated when m_1 reaches a velocity close to zero i.e. when it comes to rest.

The computed force–time response of Spring 1 (i.e., the test specimen) is shown in Fig. 17.6. The displacement–time response of mass m_1 is given in Fig. 17.7. By eliminating time from these two figures, the force-displacement response of Spring 1 can be plotted. This computed response is compared with the input behavior of spring 1 in Fig. 17.4, which is the experimental data representative of an actual crash test. As seen from Fig. 17.8, there is excellent correspondence between input and computed responses. Present numerical simulations are able to reproduce the input behavior closely, confirming the accuracy of computations. Since the force–displacement relation of the actual impact test is given as the input to the model, stiffness of specimen, whether low or high would not have any effect on the simulation results.

The basic objective of the study is to evaluate the accuracy of accelerometer response which is mounted on the impactor. Figure 17.9 shows comparison between force–displacement behaviour of spring 1 and spring 3, representing stiffness of accelerometer in LPM in Fig. 17.3. It can be said from Fig. 17.9 that:

1. Peak load of computed response (39.03 kN) is slightly lower than that of input (40.3 kN).
2. The shapes (patterns) of both the responses are matching.
3. Accelerometer response is oscillatory (actual acceleration measured during the impact test also follows the same behaviour).
4. Accelerometer is usually designed as a single DOF spring mass system. Accelerometer usually has a high natural frequency and its useful range up to 6 % of the natural frequency i.e., 2.4 kHz. A simple lumped system is a good representation of the accelerometer.

For conditionally stable central difference method, the accuracy of computed response is greatly affected by time step (Δt). Here error in simulation is calculated by difference between the input and computed response. In order to quantify the error, the new metric is defined as root mean squared error (RMSE) and can be given as:

$$RMSE = \sqrt{\frac{\sum_{i=1}^{N} \left(f_{computed,i} - f_{input,i}\right)^2}{N}} \tag{17.8}$$

where, the quantity within the parentheses in (17.8) is the error at any ith displacement point between any computed spring force, $f_{computed,i}$, and the reference input force, $f_{input,i}$, for Spring 1. The number N in (17.8) is the total number of discrete points at which error is calculated and should be large enough such that RMSE has a stable value.

For the current simulation, time step value (Δt) is 0.0043 ms. For the comparison between input behavior of spring 1 and computed response in Fig. 17.8, RMSE value is 0.1728 and for comparison between input behavior of spring 1 and

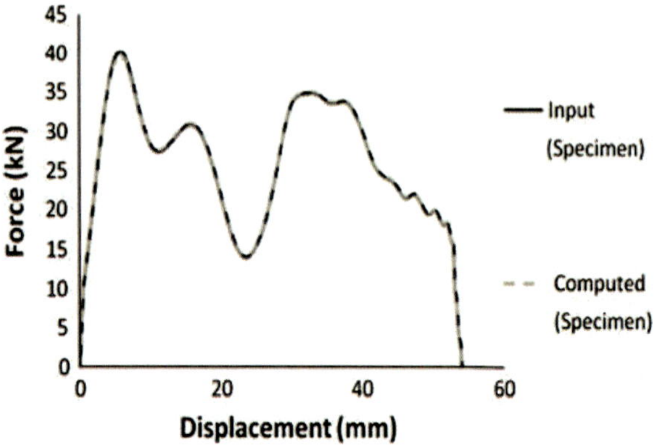

Fig. 17.8 Comparison between input force-displacement behavior of Spring 1 and computed response

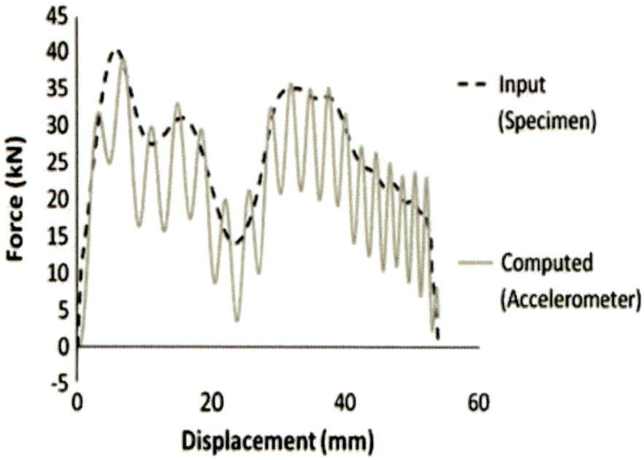

Fig. 17.9 Comparison between input force–displacement behavior of spring 1 and computed response of accelerometer

computed response of accelerometer (i.e. spring 3) in Fig. 17.9, RMSE value is 7.3027.

The RMSE value is very low for small values of (Δt). Time step (Δt) chosen in this simulation here is 0.0043 ms and the RMSE value seems to be minimum (0.1728) and satisfactory in terms of computed responses. Therefore choosing correct time step is very essential for explicit time integration method.

17.7 Accelerometer Mounting Considerations

An accelerometer senses the motion of a surface to which it is attached, producing an electrical output signal precisely analogous to that motion. The ability to couple motion, (in the form of vibration or shock), to the accelerometer with high fidelity, is highly dependent upon the method of mounting the instrument to the test surface [4, 7]. It is important that the mounting surface of the accelerometer be tightly

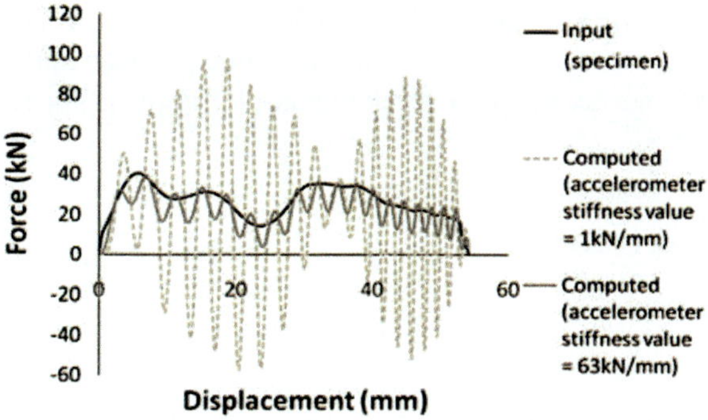

Fig. 17.10 Comparison between responses of accelerometer for different mounting stiffness values and input behavior of specimen

coupled to the test surface to ensure the duplication of motion, especially at higher frequencies. There are different types of method for mounting:

1. Stud mounting (used in current impactor system)
2. Adhesive mounting (mounting the accelerometer to thin sheet metal or to other surfaces where drilling a mounting hole is not allowable)
3. Electrical Isolation bases (to electrically insulate the housing of an accelerometer from the test surface)
4. Magnetic coupler (to attach accelerometers to ferromagnetic surfaces)

It is very important that the mounting surface should be flat for the best frequency response [7, 8]. Also mounting material should be very rigid because stiffness value of mounting material can affect the accelerometer response. Figure 17.10 is the comparison between input behavior of test specimen and computed accelerometer response at different stiffness values of 1 and 63 kN/mm. This stiffness is a combined stiffness of accelerometer and mounting assembly of accelerometer. As seen from Fig. 17.10, if the stiffness value is too low, the accelerometer response has lots of oscillations and has no correspondence with input response.

17.8 Conclusion

The present study is an attempt at quantifying the accuracy of response of a piezoresistive accelerometer under impact testing conditions when accelerometer is mounted on the impactor. Impact test set-up is represented by novel lumped parameter model approach. Non-linear behavior of the test specimen is assumed to be piece-wise linear to solve 3 DOF governing equations. Using explicit time integration scheme, central difference method, it is shown that the dynamic response of specimen can be accurately reproduced by simulation, if time step (Δt) selected is sufficiently small. The comparison between input and computed response is quantified in terms of the RMSE (root mean squared error) values. Effect of the change in the mounting stiffness due to the incorrect method of mounting of an accelerometer to the vibrating surface is also considered and discussed here. The present simulation procedure can thus be useful in ascertaining the adequacy of an accelerometer for a given impact testing configuration.

References

1. Khazan AD (1994) Transducers and their elements: design and applications, 1st edn. Prentice Hall, USA
2. Lynch JP, Partridge A, Law KH (2003) Design of piezoresistive MEMS-based accelerometer for integration with wireless sensing unit for structural monitoring. J Aerosp Eng. doi:10.1061/(ASCE)0893-1321 (16)
3. Partridge A, Reynolds JK, Chui BW (2000) A high-performance planar piezoresistive accelerometer. J Microelectromech Sys. doi:10.1109/84.825778
4. Dytran Instruments Inc, Dytran tech education, accelerometer mounting considerations
5. Endevco data sheet for model 7264B piezoresistive accelerometer, Meggitt sensing systems, Endevco
6. Choi KK, Kim Nam-ho (2004) Structural sensitivity analysis and optimization 1: linear systems. Springer, USA
7. Cristalli C, Paone N, Rodríguez RM (2006) Mechanical fault detection of electric motors by laser vibrometer and accelerometer measurements. J Mech Syst Sig Proc. doi:10.1016/j.ymssp.2005.11.013
8. Tran Duc Tan, Nguyen Tien Anh (2011) Three-axis piezoresistive accelerometer with uniform axial sensitivities. In: International conference on intelligent systems, modeling and simulation ISMS2011, Cambodia, pp 395–399

Chapter 18
Behaviour Simulation in Computer Aided Product Concept Sketching

Prasad S. Onkar and Dibakar Sen

Abstract In the product conceptualization phase of design, sketches are often used for exploration of diverse behaviour patterns of the components to achieve the required functionality. This paper presents a method to animate the sketch produced using a tablet interface to aid verification of the desired behaviour. A sketch is a spatial organization of strokes whose perceptual organization helps one to visually interpret its components and their interconnections. A Gestalt based segmentation followed by interactive grouping and articulation, presented in this paper, enables one to use a mechanism simulation framework to animate the sketch in a "pick and drag" mode to visualize different configurations of the product and gain insight into the product's behaviour.

18.1 Introduction

Concept sketch characteristics: A product sketch is a set of strokes; a stroke in turn is typically a simple curve. The form of an object represented in a sketch is not explicit as in a physical prototype or a computer model; it is subject to the interpretation of the observer who subjectively and spontaneously groups certain strokes and ignores some. The interpretation is influenced strongly by the Gestalt principles of perception as well as the skill with which the sketch is created. The product concept sketches, on the other hand, visually represent a working principle through

P. S. Onkar · D. Sen (✉)
Centre for Product Design and Manufacturing, Indian Institute of Science,
Bangalore 560012, India
e-mail: dibakar@cpdm.iisc.ernet.in

P. S. Onkar
e-mail: prasad@cpdm.iisc.ernet.in

A. Chakrabarti (ed.), *CIRP Design 2012*, DOI: 10.1007/978-1-4471-4507-3_18,

spatial arrangement of different components, created often as an assembly of strokes. Interestingly, the actual form of the components is just indicative here; representation of a causal relationship is the primary focus. Thus, understanding of a concept involves complex cognitive activity of *perceiving* the form of the components, *discovering* the physical interconnections and *enumerating* local and global motion implications of the connections through mental simulation of the product in action.

Sketching plays an important role in the conceptual design process [1–5]. Traditional pen and paper based sketching process has several limitations [1]. Sketchpad of Sutherland [6] is an early line drawings interface which revolutionized the interaction with the computers and motivated further development; viz. electronic cocktail napkin [7] which was developed for creating concepts for architectural designs. Other gesture based domain specific applications are [8–10].

Sketch understanding: In 2D, it is a challenge to interpret the shapes created using freehand strokes. Use of spatio-temporal patterns of strokes, called gestures, for quick generation of standard objects or operations have been reported in [11–14]; the strokes per say are not retained in the sketch. Specific spatial patterns constitute sketching of domain specific symbols [15, 16] and form parts of schematic sketches. Since symbols are standard, interpretation of these sketches is well structured. In product sketches, strokes may pertain to a shape or annotation which are perceptually distinct but computational segregation and interpretation is still a challenge. Saund [17] proposed techniques to identify closed paths in conceptual architectural sketches. Gestalt [18] based techniques have been used to identify features in sketches [17–19] and simplify line drawings [20]. Validity and efficacy of functional segmentation using spatial and temporal parameters have been demonstrated in [19]. The present work includes an extension to that.

Sketch based analysis: In Product design, components generally have relative motion between them. Concept sketches need validation for their functional appropriateness [21]. This was recognized by Babbage [22] long back while studying motion of machinery. He developed a novel notational scheme to represent relative motion of components over a full-cycle operation. In ASSIST [23] simple parts are sketched and their dynamic behaviour is simulated using physical laws. In "Feasy" [24], the strength viability of the concept is estimated using a commercial FEA package, for which geometry, forces, and boundary conditions are derived from the sketches.

Thus, it is observed that there are tools available for form exploration in conceptual design and sophisticated analysis tools for the detailed design. But tools available for behaviour exploration in the early stages of design are scarce and no systematic methodology for this purpose is available. Thus *concept evaluation* depends solely on the designers' ability to do accurate *mental simulation*!

Motivation: Although the relative movements of the components, which constitute the behaviour of the conceived/proposed system, are deterministic because it is governed by the principles of kinematics. It is difficult to visualize them through mental simulation for moderately complex assemblies. For product designers it poses a greater challenge because they are not typically kinematicians! Analogy is a common method of idea exploration. But, it is not intuitive to foresee the implications of the extensions

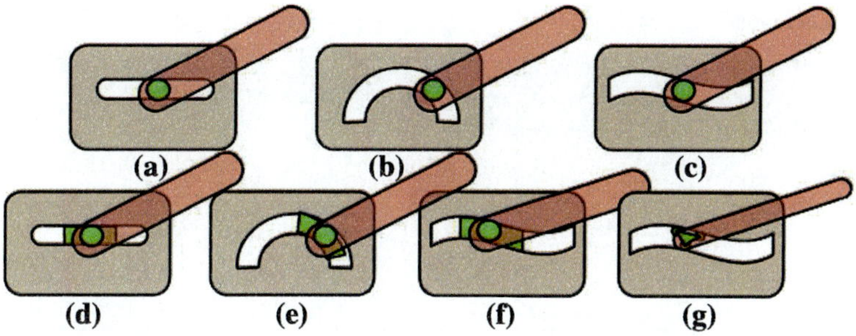

Fig. 18.1 Analoguous exploration of kinematic relations (**a**) pin in a linear slot (**b**) pin in a circular slot (**c**) pin in a curvilinear slot (**d**) rectangular slider in linear slot (**e**) arc slider in a circular slot (**f**) curvilinear slider in a curvilinear slot (**g**) roller supported slider in a curvilinear slot

of an idea in an *analogous situation* in kinematics. Here, Figs. 18.1a–c shows a circular pin in a linear, circular and curvilinear slot respectively. The pin enjoys two degrees of freedom (d.o.f.), namely one rotation and one sliding in the slot. Figures 18.1d and e are analogous to Figs. 18.1a and b wherein a rectangular and an arc slider supports the pin in corresponding slots; in both these cases, the slider has only one d.o.f.; only sliding in the slot. This mobility is not available for any shape other than linear and circular. Hence, situation in Fig. 18.1f is not a valid derivative of the analogical reasoning on 18.1a–e. To make a body move in a curvilinear path, much more elaborate arrangement (e.g. Fig. 18.1g) would probably be needed! *This physical fact is not easily derived from mental simulation.* If a sketch is created in a computer, kinematic simulation could identify that a concept using Fig. 18.1f is not good. Thus it is useful to facilitate the designers through kinematic simulation.

Objective: Conventional kinematic simulation systems need domain knowledge and expertise in modelling the articulation of the components. This paper presents a systematic procedure, through the "pick and drag" modality to intuitively and interactively animate the sketch. A perceptual segmentation scheme identifies the components automatically in a computer mediated sketch data; they are then interactively articulated through an intuitive and approximate prescription of joints which in effect defines a mechanism. However, unlike conventional mechanism modelling and analysis, where *precise motion attributes* are under investigation, the scheme proposed here is aimed at *observational* (not numeric) validation of the desired behaviour. This, we believe, would relieve the designers of mental workload and enable them to create novel functional concepts with more confidence and agility.

18.2 Requirements for Behaviour Exploration

In light of the complex cognitive process in sketch understanding discussed in Sect. 18.1, the sketching environment should support freestyle stroke generation and concept validation through the phases of perception, cognition and simulation.

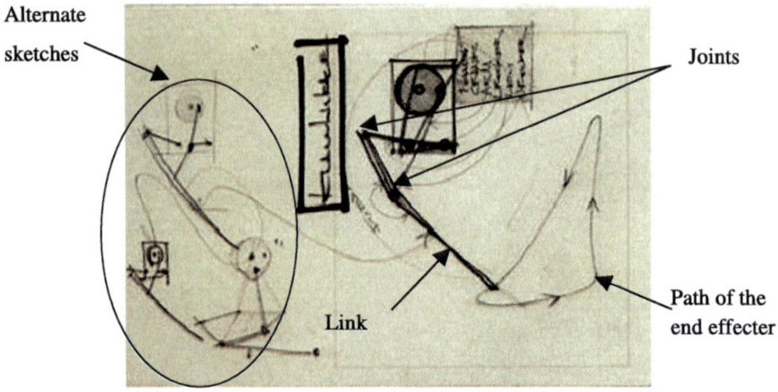

Fig. 18.2 Sketch representing a device tracing a path

It is observed that designers draw auxiliary sketches to explore and then emphasize the accepted parts by drawing darker strokes [19]. The sketch in Fig. 18.2[1] also follows a similar pattern. This is an example of a concept sketch which pertains to an exercise to come up with a mechanism which traces a given path. However, it is *difficult to evaluate* to what extent the sketched device actually traces the given sketched path. Presently the designers use commercially available mechanism simulation software to verify motion. This requires the *crisp* geometry and precise dimension of the components which is inappropriate in the early stage. It also requires the procedural steps to create the elements and simulate the motion which are not intuitive for a product designer.

In conventional product sketching scenarios, it is observed that, strokes for different components of the product are created in arbitrary order and the level of details is commensurate with the local information density. The product is sketched as an articulated system wherein articulations are either embodied or annotated. The embodied articulations contain the information about the relationship between the components. The concept is sketched in the assembled configuration and the relation between the components and input–output motions are indicated by sketch strokes. As shown in Fig. 18.2, connections are represented by drawing small circles and input and output motions are represented by arrow marks.

Therefore, a method is required to quickly simulate the motion based on the components sketched to provide better insight into the motion behaviour of the sketched product *without digressing much from the product sketching activity and continuity of the associated thought process*. Towards that, the following requirements are deemed important for conceptual sketch based behaviour exploration.

- To support normal sketching procedure. The system would identify the components matching with the designers intensions automatically.
- To support intuitive method to indicate connections between the components.

[1] http://mv122011.files.wordpress.com/2011/05/mv_wp_2vm_mechanismstudy3.jpg.

- To support a feel of direct interaction by providing a "pick and drag" mode of animation, using any component in the articulated sketch.
- To support indication of violation of environmental constraints.

18.3 Stroke Grouping

The form of an object represented in a sketch is not explicit as in a physical prototype or a computer model; it is subject to the interpretation of the observer who subjectively and spontaneously groups the certain strokes and ignores some. The interpretation is influenced strongly by the Gestalt principles of perception as well as the skill with which the sketch is created. Use of Gestalt principles for functional segmentation is reported in [17–19]. The methods in [19] show that the sketch strokes grouped based on the spatial, temporal and pressure patterns reveal characteristic features in the sketches and it matches the designers' perception. In this work, we extend the clustering algorithms presented in [19] to find meaningful features as discussed below.

Algorithmic grouping: Gestalt laws of perceptual organization [18] helps in perceiving the composition. Van Sommers [25] identified the starting point as an important measure providing information about the drawing process for non professionals. Here, first the clusters of start points and end points are found. The distance of a point from a cluster is defined as the minimum of its distance from all the points in the cluster. Distance between two clusters is defined as the minimum distance between the points in the two clusters. If the distance between two entities, point and clusters, is less than a given value, they are grouped into one. The strokes associated with the points in a given cluster are identified as a cluster of strokes. Bunch of strokes are identified having common start and end point clusters representing a component in the sketch.

To verify the grouping methodology, a designer was asked to sketch different elements used in mechanisms in conventional styles in a single sketch using our interface. He sketched a fixed pivot, binary, ternary, quaternary links and slider in black. The grouping algorithms identified the sets of strokes defining different components satisfactorily as shown in Fig. 18.3 in different colours.

Interactive Grouping: In conceptual design, sketches contain details of the product apart from those relevant for the behaviour. Also, all the elements pertaining to a single component need not be perceptively recognizable. The sketched components are not as well defined as in case of geometric and multi-body modelling. This makes product concept sketches more complex; thus, stroke groups identified by clustering algorithms may contain multiple groups pertaining to the same component. To take care of this knowledge driven process, an interactive grouping facility is provided. The groups pertaining to the same component are selected interactively by the designer, which are then merged as a single component. In Fig. 18.4a, both the hinge location are single entity representing the

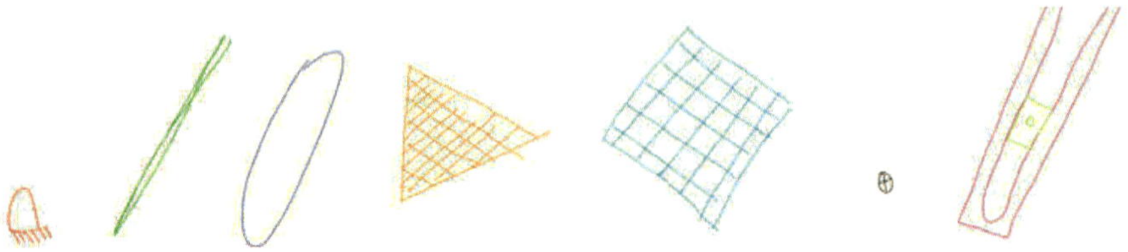

Fig. 18.3 Different types of components identified by the algorithm

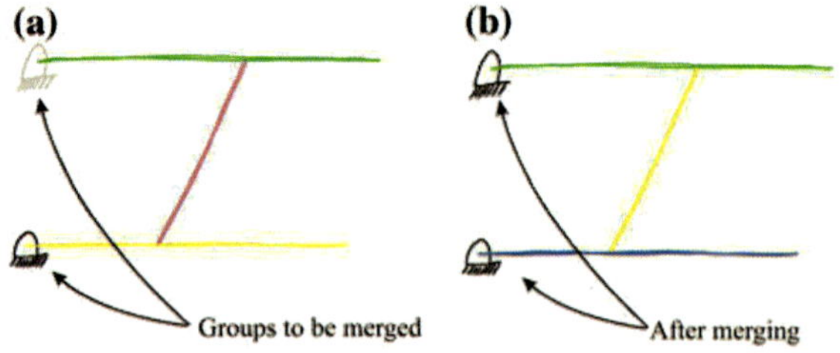

Fig. 18.4 Sketch showing interactive merging **a** before merging two components shown in different colors **b** after merging both are colored same

fixed link from the kinematics point of view; but, the algorithmic grouping identified them as two separate elements (shown in different colours). After identification of these elements, Fig. 18.4b shows them as a single entity sharing a common colour. The group merging will only associate the groups with one another and the identity of individual group is maintained to enable demerging at a later stage, if needed. The process is repeated if more components related to different links are to be grouped.

18.4 Motion Simulation

The actual form of the components in a product concept is just indicative; representation of a causal relationship is the primary focus. This causality is determined by the nature of physical interconnections among the components. Motion simulation involves reliable visualization of the motion of the components in space. Although animation software has been used for concept visualization, it needs knowledge and expertise in both software and kinematics to create such animations. Using the components as links and interconnections as joints, behaviour simulation can be performed through kinematics. Component identification has been discussed in the previous section. The primary challenge in motion simulation is intuitive description of the interconnections (joints) and ease of operation of a physical prototype.

Identification of Joints: Joints define the relation between motions of two links. In mechanism design joints are explicitly identified on the links at a predefined location called as markers. The notion of marker is abstract and

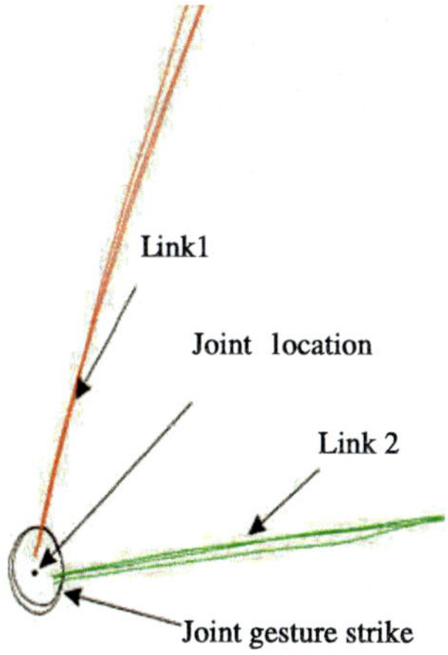

Fig. 18.5 Joint identification from stroke groups in gasture based method

mathematical; it has no physical meaning. It helps in systematic and rigorous definition of the type of relative motion between two components. It requires domain knowledge. At conceptualization stage, the designers indicate interconnections using shapes in typical embodiments. Two methods of identifying revolute joints through the interactive sketching interface have been implemented. Figure 18.5 shows an example in which the joint needs to be defined between two stroke groups shown in different colours. First, the designers interactively select the stroke groups which correspond to two components to be interconnected and then explicitly identify the joint location by specifying an arbitrary coincident point which is not restricted to be within the bounds of the sketched links. This helps the designers to explore motions even before embodiment. The second approach is based on the observation that in concept sketches designers identify revolute joints by drawing small circles at the junction of the two links. Correspondingly in this interface, designers define joints by drawing the circular gesture strokes for revolute joint. The location of the joint is estimated at the mean of the points on the gesture stroke as shown in Fig. 18.5. The Euclidian distance of the strokes in different groups from this mean point is computed and the groups which are within a predefined threshold get connected at this point by a revolute joint. If there are more links within the threshold, it creates a multiple joint. The extension of the interface for other joint types is being explored by the authors. It is found that the first method is more tedious and time consuming but accurate. The second method is more intuitive and works well when the sketch is not too cluttered.

Interactive Simulation: In physical prototypes motion exploration is unstructured, wherein the designer holds and moves any component of choice; in computer models, simulation is strictly structured wherein fixed link, input pair and input functions are predefined. Hence in the present context, although the representation of the articulation is kinematic, simulation needs to support unstructured exploration. In this work, we used an in-house mechanism modelling library that enables closed form kinematic analysis of planar mechanisms as well as inverse kinematics of

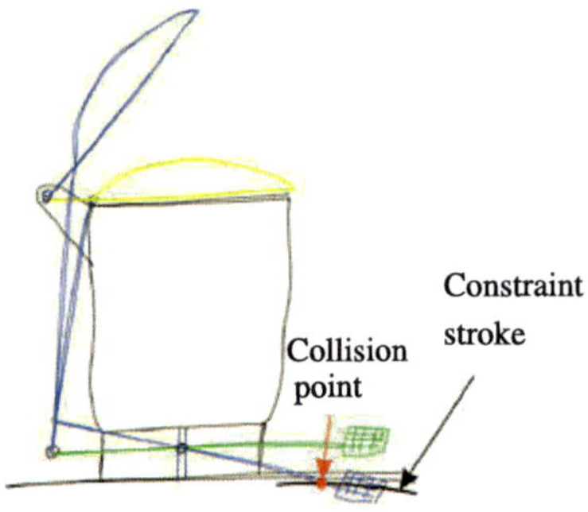

Fig. 18.6 Sketch simulation showing collision of components with constraint curve

arbitrary planar mechanisms using the concept of modular kinematics. An explicit method of selecting the fixed link is provided in the interface. This helps the designers to select any of the stroke groups as fixed link and study the behaviour of the connected components. Another important parameter is input pair. This is the physical location where the input force or torque is given. This is required by the kinematic kernel for simulating the behaviour of the connected components. The kind of input given to the kernel to simulate the motion may not be directly available from the sketch itself. In conceptual design, the designers are not concerned about location of input actuator. To enable unstructured motion exploration, we use inverse kinematic based technique which computes the corresponding configuration of the mechanism based on the stylus position. When the designers click on a component and drags the stylus, the inverse kinematic module calculates the corresponding configuration for the present location of the stylus. This provides a realistic feel of actually moving a link of a physical prototype.

Constraints Verification: Designers generate product concepts to satisfy the design requirements. These requirements impose certain constraints which need to be respected by the designers. For example, in case of design of foot operated dust bin (Fig. 18.6), the motion of the pedal should not go below ground line and at the same time the lid has to open sufficiently. These kinds of constraints are difficult to verify in the static sketches. Designers often draw elements, for example, the environment details, which are not related to the kinematic behaviour of the system but are potential obstructions to the motion. To support this kind of evaluation, a constraint verification check is provided which finds the proximity of the sketch components with the constraints and highlights the violation of the constraints. In this implementation, a coloured dot is displayed at the point of collision.

18.5 Design Experiments

Design experiments are conducted to validate the methods implemented in the interface and to evaluate the performance of the tool to support the conceptual design process. The application is developed in C++ for Wacom Cintiq® 21UX

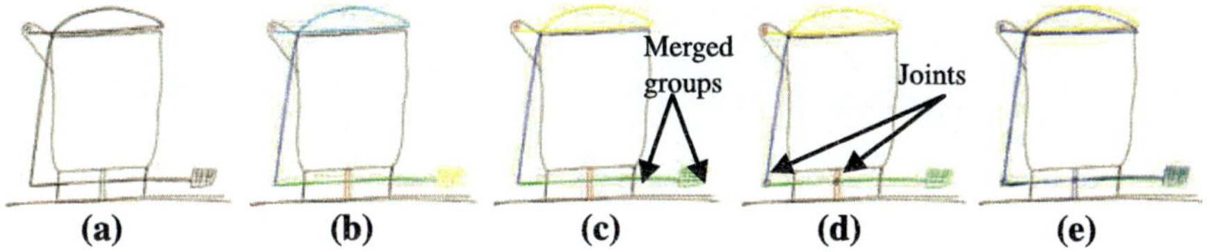

Fig. 18.7 Stages in motion exploration **a** sketch created by the designer **b** algorithmic grouping result shown in different colors **c** interactive merging by designers **d** connections and fixed link identified **e** sketch ready for motion exploration

touch screen tablet using OpenGL, wintab and bbTablet APIs. Two graduate design students participated in the experiments. A brief introduction of the application is given designers. Each designer is given a distinct problem. The designers are asked to explore the concepts using our sketching interface. Once the sketch is sufficiently done for describing the concept (Fig. 18.7a), the designer opts for algorithmic grouping in which the individual components are identified (Fig. 18.7b). If any inadequacy in grouping is observed, they compensate it by interactively selecting the components to be grouped (Fig. 18.7c). A circular gestural stroke at the junction of the link groups is drawn for indicating a joint location (Fig. 18.7d). Then he selects the ground link and starts the simulation by *picking* a sketched component and *dragging* the stylus. This causes the underlying mechanism to move. The articulated model is displayed overlaying on the original sketch (Fig. 18.7e), enabling the designer to observe the behaviour of the concept.

Figure 18.8 shows the snapshots taken at discrete time intervals showing the animation of the two product concepts explored. The moving components are coloured and static ones are in black. Figure 18.8(I) belongs to the sketch of an excavator bucket. It can be observed that even in extreme position of operation the components do not intersect each other. The designer can explore how much would the crank rotate for moving the bucket to a close position. It also gives a subjective correlation between the crank and bucket rotations in different phases. These types of inferences are difficult to derive from static sketches. The simulation can be extended for the arms of the excavator to evaluate the full excavator arm in sketch phase itself. It can also be used to find the optimum configuration for transportation and packaging.

Figure 18.8(II) shows the different stages of a foot operated dustbin. One can explore how much the pedal has to be pressed to open the lid completely. The Fig. 18.8II(e) shows that when the lid is completely open, the pedal is penetrating the ground-line. The designer can alter the position of the footrest to avoid the penetration. The designers can also explore by varying the joint location to refine the motion to satisfy the requirements.

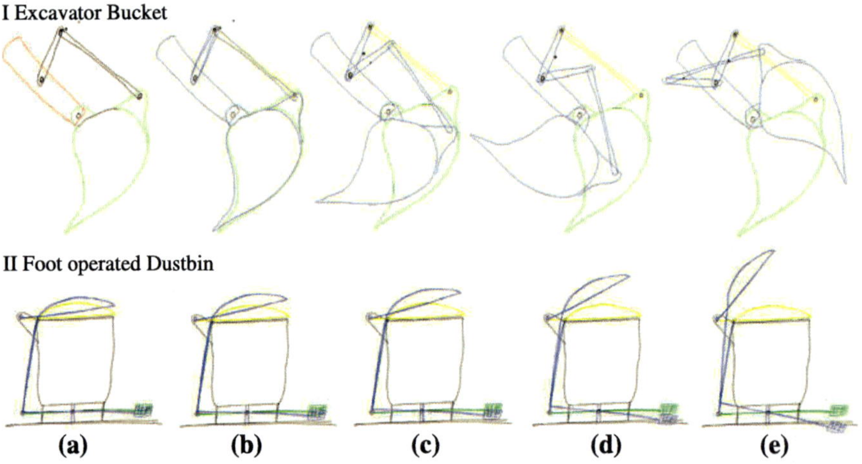

Fig. 18.8 Different stages of simulation. (I) Excavator bucket. (II) Foot operated dustbin (**a–e**) different configurations obtained by moving one component superimposed with original sketch

18.6 Conclusions

The work presented a sketch animation methodology to aid mental simulation towards direct concept verification. This reduces both the mental workload of the designer and uncertainties associated with guessing how the concept works, without the necessity of creating detailed geometric and multi-body modelling. This is enabled by stroke grouping algorithms to identify functional components in the sketch automatically, intuitive methods for component articulations and a direct manipulation scheme for motion simulation. This brings the experience closer to physical prototyping without losing the versatility of over-stroked concept sketches and continuity of the sketching process.

References

1. Fish J, Scrivener S (1990) Amplifying the mind's eye: sketching and visual cognition. Leonardo 23(1):117–126
2. Goel V (1991) Sketches of thought. The MIT press, Cambridge, MA, USA
3. Goldschmidt G (1991) The dialectics of sketching. Creativity Res J 4(2)
4. Cross N (1999) Natural intelligence in design. Des Stud 20:25–39
5. Schon DA, Wiggins G (1992) Kinds of seeing and their function in designing. Des Stud 13(2):135156
6. Sutherland IE (1964) Sketchpad: a man-machine graphical communication system. In DAC'64: proceedings of the SHARE design automation workshop, 6.329–6.346
7. Do E, Gross MD (1996) Demonstrating the electronic cocktail napkin: a paper-like interface for early design. Conferrence on human factors in computing systems CHI'96, pp 5–6

8. Sezgin TM, Davis R (2007) Sketch interpretation using multiscale models of temporal patterns. IEEE Comput Graphics Appl 27(1):28–37
9. Hammond T, Davis R (2006) Tahuti: a geometrical sketch recognition system for UML class diagrams. In ACM SIGGRAPH, NY
10. Kara LB, Stahovich TF (2004) Sim-U-Sketch: a sketch-based interface for simulink. Proceedings of AVI-2004, pp 354–357
11. Jorge JA, Fonseca MJ (2000) A simple approach to recognise geometric shapes interactively. In GREC '99, London, pp 266–276
12. Diehl H, Mller F, Lindemann U (2004) From raw 3D-sketches to exact CAD product models—concept for an assistant system, sketch based interfaces and modelling. UK
13. Zeleznik R, Herndon K, Hughes J (1996) SKETCH: an interface for sketching 3D scenes. Proceedings SIGGRAPH, pp 163–170
14. Igarashi T, Matsuoka S, Tanaka H (1999) Teddy: a sketching interface for 3D freeform design. In SIGGRAPH'99, pp 409–416
15. Kara LB, Stahovich TF (2005) An image-based, trainable symbol recognizer for hand-drawn sketches. Comput Graph 29(4):501–517
16. Olsen L, Samavati F, Sousa MC, Jorge JA (2009) Sketch-based modeling. Surv Comput Graph 33:85–103
17. Saund E (2003) Finding perceptually closed paths in sketches and drawings. IEEE Trans Pattern Anal Mach Intell 25(4):475–491
18. Wertheimer W (1923) A source book of gestalt psychology, chapter laws of organization in perceptual forms. Routledge, London, pp 71–88
19. Onkar P, Sen D (2010) Functional segmentation of strokes for product sketch understanding. Int J Shape Model 1:9–38
20. Barla P, Thollot J, Sillion FX (2005) Geometric clustering for line drawing simplification In ACM SIGGRAPH
21. Pahl G, Beitz W, Wallace K (1996), Engineering design: a systematic approach. Springer, Berlin
22. Babbage C (1826) On a method of expressing by signs the action of machinery. Phil Trans R Soc London 116(1/3):250–265
23. Alvarado C, Davis R (2001) Resolving ambiguities to create a natural computer-based sketching environment IJCAI, pp 1365–1374
24. Murugappan S, Ramani K (2009) Feasy: a sketch-based interface integrating structural analysis in early design. Proceedings of ASME IDTC/CIE conference
25. Van Sommers P (1984) Drawing and cognition. Cambridge university press, UK

Chapter 19
Non-Linear Signal Processing Techniques Applied on EMG Signal for Muscle Fatigue Analysis During Dynamic Contraction

Ram Kinker Mishra and Rina Maiti

Abstract In the field of ergonomics, biomechanics, sports and rehabilitation muscle fatigue is regarded as an important aspect for study. Work postures are basically dynamic in nature. Classical signal processing techniques used to understand muscle behavior are mainly based on spectral based parameters estimation, and mostly applied during static contraction. But fatigue analysis in dynamic conditions is of utmost requirement because of its daily life applicability. It is really difficult to consistently find the muscle fatigue during dynamic contraction due to the inherent non stationarity time-variant nature and associated noise in the signal along with complex physiological changes in muscles. Nowadays, different non-linear signal processing techniques are adopted to find out the consistent and robust indicator for muscle fatigue under dynamic condition considering the high degree of non stationarity in the signal. In this paper, various nonlinear signal processing methods, applied on surface EMG signal for muscular fatigue analysis, under dynamic contraction are discussed.

19.1 Introduction

Electromyographic (EMG) signal is an evaluation tool in the field of sports science, physical therapy, biomechanics, ergonomics, occupational health and rehabilitation. There are three applications which dominate the use of the surface EMG signal in these fields: its use as an indicator for the initiation of muscle activation, its

R. K. Mishra (✉) · R. Maiti (✉)
CPDM, IISc, Bangalore, India
e-mail: ramkinker@cpdm.iisc.ernet.in

R. Maiti
e-mail: rmaiti@cpdm.iisc.ernet.in

A. Chakrabarti (ed.), *CIRP Design 2012*, DOI: 10.1007/978-1-4471-4507-3_19,

relationship to the force produced by a muscle, and its use as an index of the fatigue processes occurring within a muscle. Muscle fatigue is defined as inability of produce the required power output for a given stimulus. This condition is characterized by a notable change in physiological and biochemical processes which includes recruitment of larger motor unit, reduction in muscle conduction velocity, alteration of blood flow, decreased mean power frequency of EMG signal, increased hydrogen ion and other metabolites etc. The study of muscle fatigue has two important applications. First, it can be used to identify weak muscles. The most famous application is in the analysis of low back pain patients. Second, it can be used to prove the efficiency of strength training exercises [1], [2]. Surface EMG technique for monitoring the fatigue using different computation methodologies has drawn attention of many researchers to under-stand more insights about the events occurring inside the muscle. Different authors use different objective criteria to indirectly quantify or identify fatigue related phenomena. For the study of surface EMG and its quantification there is a variety of signal processing tools available. Among them spectral analysis technique is the traditional technique most commonly used for muscle fatigue assessment [3], [4]. But spectral technique is reliable only during the static muscle contraction not during dynamic contraction. The reason behind it is the non stationarity induced in the signal during dynamic contraction. Stationarity of the signal is the basic requirement for the spectral analysis techniques. But in daily routine dynamic muscle contraction are most common and thus is a bigger matter of concern. In order to quantify surface EMG during dynamic condition various non-linear signal processing techniques have been used. In present study, a review of different non-linear and non-stationary signal processing methodologies related to fatigue estimation along with their merits and demerits are presented.

19.2 Characteristics of EMG

Muscular fatigue is generally analyzed on recorded surface EMG (1) during static contraction or (2) during dynamic contraction, discussed below. During isometric muscle contraction, the joint angle and muscle length are constant. So, recorded EMG signal during this contraction can be assumed to be wide sense stationary signal. Fatigue condition can be identified as a change in power spectrum result, given in Fig. 19.1 [5]. This assumption is not true during dynamic muscle contraction, so dynamic field postures are simulated in a set of static conditions and fatigue is estimated [6].

In the field, the physical working postures are generally dynamic in nature, where both joint angle and muscle length vary. During dynamic contraction, because of physical (from the skin–electrode interface) and physiological constraints, causes the non-stationarity in the EMG signal, which makes it challenging to reliably estimate the fatigue indices. Various authors have attempted to estimate fatigue indices assuming following different conditions.

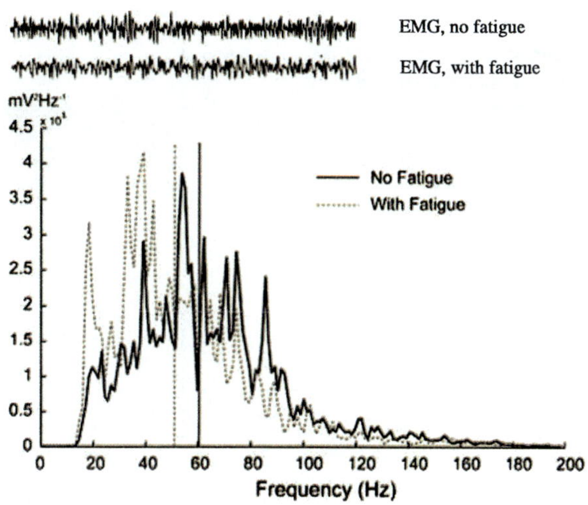

Fig. 19.1 Comparative change in power spectrum during fatigue while doing static contraction along with EMG signal [5]

19.3 Non-Linear and Non-Stationary Signal Processing Techniques

Random discharge of active motor units imposes non-stationarity of the EMG signals, whereas non-linear characteristics are caused by functional interference between different muscles, changes of signal sources and paths to recording electrodes, variable electrode interface etc. All the classical methods use different assumptions before processing the signal. To better understand the system dynamics, different nonlinear time series analysis methods have been employed as alternatives in determining EMG-based fatigue indices.

19.3.1 Recurrence Plot

In a chaotic system, at any time instant, the phase space trajectory will follow roughly equal to the phase space area [7]. Therefore, quantification of recurrences and line segments of the phase space trajectory in recurrence plot (RP) method is used to capture nonlinearity of a non-stationary system.

First step in generating Recurrence plot, involves transformation of a uni-dimensional time series data, X into a multidimensional form as follows:

$$X_i^m = (X_i, X_{i+d}, X_{i+2d}, \ldots\ldots X_{i+(m-1)d}) \tag{19.1}$$

Where, d denotes time delay, m is the embedded dimension. These parameters should be selected with care otherwise non-optimal embedding parameters can cause result in discrete diagonal lines with smaller blocks [8]. Then a symmetric matrix of distances (e.g., Euclidean distances) can be constructed by computing

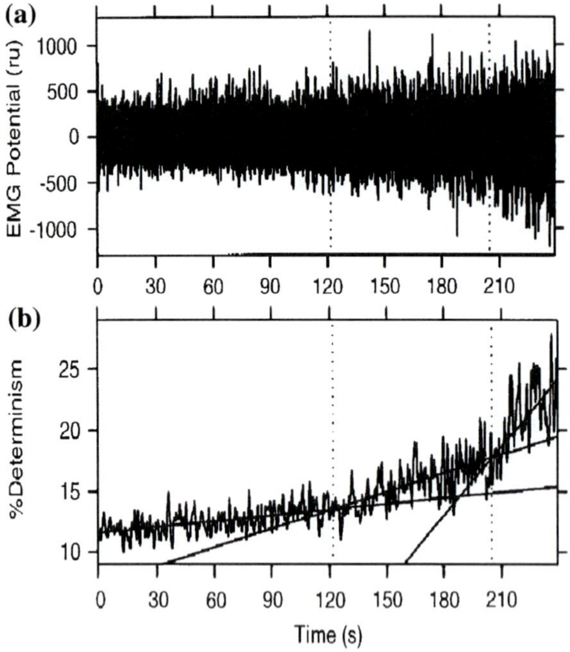

Fig. 19.2 **a** EMG signal obtained for 238 s. during isometric contraction for endurance test. **b** % determinism of the signal increased with progression of fatigue linearly fitted in 3steps[10]

distances between all pairs of embedded vectors. In the recurrence plot, the color of the pixels corresponds to the magnitude of data values (either actual or based on threshold) in a two-dimensional array and the coordinates correspond to the locations of the data values in the array. From the graphs, percent of determinism (%DET) is calculated as the ratio of number of selected (active) points on diagonal lines (>2) and total number active points. Percentage of recurrence (%REC) is calculated as the ratio of total number of active points and total number of points. Farina et al. have shown that %DET and %REC are good indicators in determining the muscle fatigue [8], whereas Yiwei et al. [9] have emphasized on %DET in determining muscle fatigue, similarly shown in Figs. 19.2 and 19.3.

19.3.2 Entropy

Entropy is generally used to quantify the randomness and complexity in signal. In particular, it is used to characterize non-periodic, random phenomena and indicates the rate of information production in relation to the dynamic systems [11]. Especially, when dealing with surface EMG of muscle fatigue, previous non-linear techniques shows a wide statistical variation in result, which causes difficulties in proper identification of muscle fatigue. Another problem is the unavoidable noise associated with this signal. To solve these problems, Pincus [12] developed

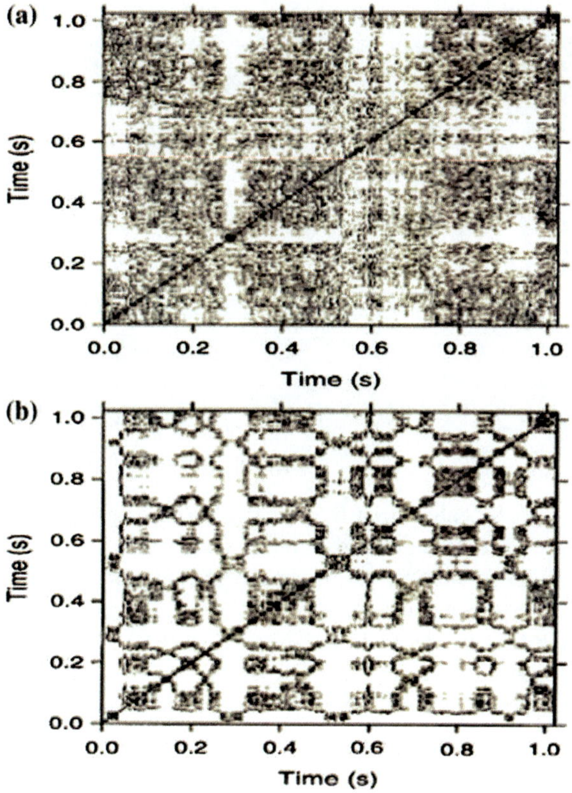

Fig. 19.3 Recurrence Plot for **a** first 1.024 s and **b** last 1.024 s of above EMG signal. RP of last epoch shows increased periodicity in signal with increased determinism [10]

approximate entropy (ApEn) to measure the system complexity, which is applicable to noisy and short datasets.

Algorithm for calculating approximate entropy (ApEn) is shown below:

From a time series {u(i): $1 \leq i \leq$ N} a vector sequence, x_i^m is formed defined as:

$$x_i^m = \{u(i), u(i+1)\ldots u(i+m-1)\}, \quad \text{where } 1 \leq i \leq N-m+1 \tag{19.2}$$

Where Distance, d_{ij}^m between x_i^m and x_j^m is defined as

$$d_{ij}^m = d\left[x_i^m, x_j^m\right] = \max_{kc\ (0,\mathrm{m}-1)} |u(i+k) - u(j+k)| \tag{19.3}$$

Probability for similarity between the two different vectors x_i^m and x_j^m can be calculated as:

$$C_r^m(i) = \frac{1}{N-m+1}\sum_{j=1, j\neq i}^{N-m+1} \backslash\mathrm{Uptheta}(d_{ij}^m - r) \tag{19.4}$$

Where, Θ is the Heaviside function:

$$\Theta(z) = \begin{cases} 1 \, if \, z \leq 0 \\ 0 \, if \, z > 0 \end{cases} \tag{19.5}$$

Tolerance, r is defined as: $r = k.std\ (t)$ Where, k is a constant $k > 0$ and std (.) represents the standard deviation of the time series. Logarithmic average over all the vectors of probability, $C_r^m(i)$ calculated as:

$$\Phi^m(r) = \frac{1}{N-m+1} \sum_{i=1}^{N-m+1} \ln[C_r^m(i)] \tag{19.6}$$

ApEn for different epoch signal can be defined as:

$$ApEn(m, r, N) = \Phi^m(r) - \Phi^{m+1}(r) \tag{19.7}$$

ApEn calculation is basically based on Heaviside function, which may not be sensitive enough to the minor changes in the signal complexity. Therefore, the fatigue condition can be separated from non-fatigued state, but Hongbo et al. [13] have shown that ApEn is insensitive to the change in muscle fatigue, and they have used *Fuzzy Approximate Entropy (fApEn)* to improve the result. A comparative result is given in Fig. 19.4. *fApEn* is based on the similar algorithm as of ApEn except it uses ambiguous boundaries. The soft and continuous boundary of a fuzzy membership function makes the fApEn statistics decrease smoothly and monotonically when there is a slight increase in the tolerance r.

Distance, D_{ij}^m between x_i^m and x_j^m is determined by a fuzzy membership function on normalized data by removing the effect of baseline, where r is having similarity degree,

$$D_{ij}^m = u(d_{ij}^m, r) \tag{19.8}$$

Comparing of *fApEn* with *ApEn*; *fApEn* showed better monotonicity, relatively more consistency and more robustness to noise while characterizing signals with different complexities.

19.3.3 Huang-Hilbert Transform

Huang et al. [14] have proposed the Huang-Hilbert transform method (HHT) as a new tool for the analysis of nonlinear and non-stationary data. Unlike the Fourier transform, which is predicated on a priori selection of basis functions that are either of infinite length or have fixed finite widths, Empirical Mode Decomposition (EMD) decomposes a signal into finite basis functions called the intrinsic mode functions (IMFs) (fission process). EMD assumes that data have many different

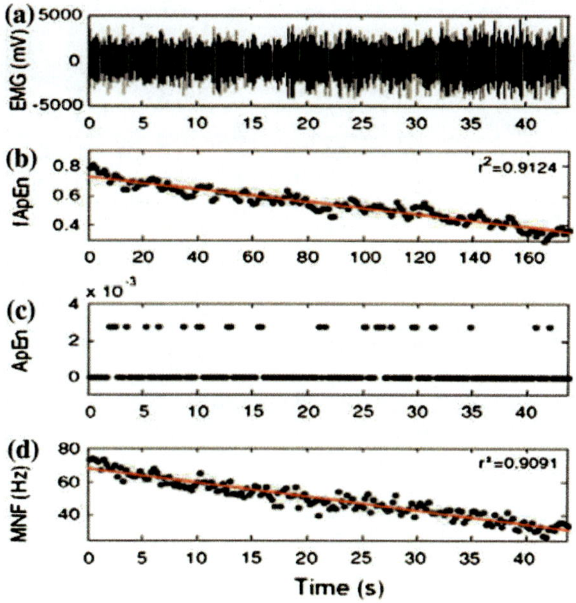

Fig. 19.4 **a** Actual EMG signals, **b** the estimation of fApEn, **c** ApEn **d** MNF. The EMG signals were recorded from the biceps during the static isometric contraction from non-fatigue to exhaustion state [12]

coexisting modes of oscillation, one superimposing on the others. EMD decomposition and separates a time series into a finite number of its individual characteristic oscillations (as shown in Fig. 19.5) in order to define a meaningful instantaneous frequency (as shown in Fig. 19.6). After EMD of time series, $x\ (t)$ can thus be expressed as follows:

$$x(t) = \sum_{j=1}^{n} C_j(t) + r_n(t) \qquad (19.9)$$

Where, n is the number of IMFs, $C_j(t)$ are the IMFs and $r_n(t)$ is the final residue.

The first component has the smallest time scale, which corresponds to the fastest time variation of data. Since the decomposition is based on the local characteristic time scale of the data to yield adaptive basis, it is applicable to non-linear and non-stationary data in general and in particular, fatigue EMG data considered in the following section. The second step of the HHT is Hilbert transform (HT). HT of the time series, x(t) is the convolution with 1/πt, Shown as:

$$\mathrm{H}[x(t)] = x(t) * \frac{1}{\pi t} = \frac{1}{\pi} P \int_{-\infty}^{\infty} \frac{x(t')}{(t - t')} dt' \qquad (19.10)$$

Where, P indicates the Cauchy principle value. Then we define the mean instantaneous frequency, MIF (j) of $c_j(t)$ with m data points as the weighed mean

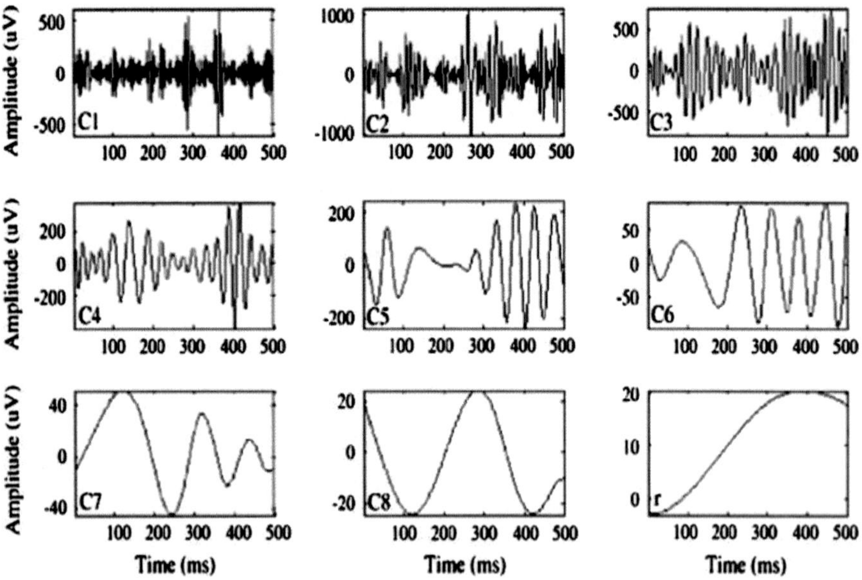

Fig. 19.5 9 different sets of IMFs of EMG data with 500 ms analysis window. The 9th mode is residue, which was not considered during Hilbert transform [15]

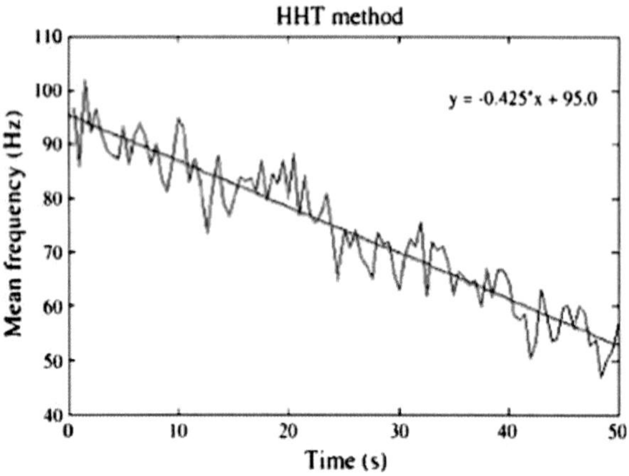

Fig. 19.6 Time courses of the mean frequency derived from Hilbert-Huang transform. The analysis window was 500 ms [15]

of instantaneous frequency, $\omega_j(t)$ and amplitude, $a_j(t)$ of hilbert spectrum computed as:

$$MIF(j) = \frac{\sum_{i=1}^{n} \omega_j(i) a_j^2(i)}{\sum_{i=1}^{n} a_j^2(i)}. \tag{19.11}$$

The mean frequency of the original signal is defined by:

$$MIF = \frac{\sum_{j=1}^{n} ||a_j||MIF(J)}{\sum_{j=1}^{n} ||a_j||} \tag{19.12}$$

By means of the combination of the amplitude and the derivative of the phase (i.e. the instantaneous frequency) of each component, it is possible to obtain the resulting amplitude, time, and frequency representation of the original series:

$$x(t) = Real \sum_{j=1}^{n} a_j(t) \exp\left(\mathrm{i} \int \omega_j(t) dt)\right) \tag{19.13}$$

HHT is applicable on non-linear and non-stationary signal and it can capture the non-linear dynamics in a better way than the power spectrum analysis especially for fatigue estimation during dynamic contraction [15].

19.3.4 Interpretation of Results

Muscular fatigue is characterized by a complex combination of physiological and biochemical process induced by physical exercise. During dynamic contraction, alteration of mean power frequency is not universally accepted conclusion. Different authors adopt different methodologies to estimate muscular fatigue during dynamic contraction in separate ways. Using Recurrence Quantification Analysis (RQA), researchers assume EMG signal as nonlinear deterministic chaotic system and try to capture the increase in motor unit synchronization (from the nature of the recurrence plot) and reduction in conduction velocity (from the slope of the determinism) related to fatigue in an effective way. Approximate entropy, Fuzzy Approximate Entropy and Huang-Hilbert Transformation work on non-linear and non-stationary signal based on local time scale input data at any point, therefore the output does not suffer the problem of spectral leakage, unlike power spectral analysis. The result of Approximate Entropy often corrupt with inherent non-linear dynamics of EMG data during dynamic contraction along with due to noise, short data length etc. as it access the complexity and/or regularity based on Heaviside function However, Fuzzy Approximate Entropy is more robust than Approximate Entropy to estimate inherent dynamics and stochastic behaviors (Fig. 19.4). Different authors highlight that during dynamic muscle contraction, the non-stationarity dynamics is better captured by Huang-Hilbert Transformation from the slope and higher resolution of the time–frequency analysis. In comparison, Fig. 19.6 shows a clear and gradual decline in mean frequency than observed by power spectrum analysis result (Fig. 19.1). According to Srhoj-Egekher et al. [16], during dynamic contraction the local maxima of median frequency are directly correlated with the number of active muscle fiber, and it declines during fatigue [16]. In conclusion, future work will include a broader comparison of these methods to other new and established fatigue indices.

19.4 Conclusion

Efficient signal processing techniques have made it possible to detect fatigue from EMG signal with a limited degree of reliability during static contraction. In order to improve the reliability of fatigue indicator different non-linear signal processing techniques like Recurrence plot and Hilbert-Huang transform have shown promising results. Therefore, use of above mentioned techniques for non-linear and non-stationary surface EMG signal will provide another dimension for the muscle fatigue analysis. These techniques can be helpful to find out the most consistent and robust indicator for muscle fatigue during dynamic contraction.

References

1. De Luca CJ (1997) The Use of surface electromyography in biomechanics. J App Biomech 13:135–163
2. McQuade KJ, Dawson J, Smidt GL (1998) Scapulothoracic muscle fatigue associated with alterations in scapulohumeral rhythm kinematics during maximum resistive shoulder elevation. JOSPT 28(2):74–80
3. De Luca CJ (1984) MyoElectrical manifestations of localized muscular fatigue in humans. Crit Rev Biomed Eng 11(4):251–279
4. Paiss O, Inbar GF (1987) Autoregressive modeling of surface EMG and its spectrum with application to fatigue, IEEE, BIOMED ENGG, BME-34, 10:761–770
5. Dennis T, He H, Kuiken TA (2010) Study of stability of time-domain features for electromyographic pattern recognition. JNER 7:21
6. David ED, McClure PW, Karduna AR (2006) Effects of shoulder muscle fatigue caused by repetitive overhead activities on scapulothoracic and glenohumeral kinematics. J Electromyogr kinesiol 16:224–235
7. Chandre C, Wiggins S, Uzer T (2003) Time–frequency analysis of chaotic systems. Physica D 181:171–196
8. Farino D, Fattorini L, Felici F, Filligo G (2002) Nonlinear surface EMG analysis to detect changes of motor unit conduction velocity and synchronization. J Appl Physiol 93:1753–1763
9. Yiwei L, Markku K, Zbilut JP, Webber CL Jr (2004) EMG recurrence quantifications in dynamic exercise. Biol Cybern 90:337–348
10. Ikegawa S, Shinohar A, Fukunaga T, Zbilut JP, Webber CL Jr (2000) Nonlinear time course of lumbar muscle fatigue using recurrence quantifications. Biol Cybern 82:373–382
11. Richman JS, Moorman JR (2000) Physiological time-series analysis using approximate entropy and sample entropy. Am J Physiol Heart Circ Physiol 278:H2039–H2049
12. Xie H, Guo J, Zheng Y (2010) Fuzzy approximate entropy analysis of chaotic and natural complex systems: detecting muscle fatigue using electromyography signals. Annals of Biomed Engg 38(4):1483–1496
13. Pincus SM (1991) Approximate entropy as a measure of system complexity. Proc Nati Acad Sci USA 88:2297–2301
14. Huang NE et al (1998) The empirical mode decomposition and the hilbert spectrum for nonlinear and non-stationary time series analysis. Proc Royal Society A: Math, Phys Eng Sci 454:903–995
15. Xie H, Wang Z (2006) Mean frequency derived via Hilbert-Huang transform with application to fatigue EMG signal analysis. Comput Methods Programs Biomed 82:114–120

16. Srhoj-Egekher V, Cifrek M, Medved V (2011) The application of Hilbert–Huang transform in the analysis of muscle fatigue during cyclic dynamic contractions. Med Biol Eng Comput 49:659–669

Part IV
Global Product Development and PLM

Chapter 20
Improvement of Product Design Process by Knowledge Value Analysis

Yang Xu, Alain Bernard, Nicolas Perry and Florent Laroche

Abstract Nowadays, design activities remain the core issue for global product development. As knowledge is more and more integrated, effective analysis of knowledge value becomes very useful for the improvement of product design processes. This paper aims at proposing a framework of knowledge value analysis in the context of product design process. By theoretical analysis and case study, the paper illustrates how knowledge value can be calculated and how the results can help the improvement of product design processes, such as deciding which knowledge to choose and what to do next.

20.1 Introduction

In this world of globalization, more and more enterprises consider knowledge management (KM) process as an important part, if not the only part, of their production activities [1, 2]. Meanwhile, how to pay deeper attention to the crucial competence "knowledge" is becoming a strategic approach in production management, and product design processes are linked more and more tightly with knowledge [3].

When considering knowledge management in product design activities, how to evaluate knowledge has always been a challenging problem. Which knowledge is more "useful" and thus can add more value to the product? What knowledge to be acquired in the next step of design? The answers of such questions may greatly improve design activities, so the following sections will focuses on this related issue: how to calculate and analyze knowledge value in product design processes and help this improve.

Y. Xu · A. Bernard (✉) · N. Perry · F. Laroche
Ecole Centrale de Nantes, LUNAM Université, IRCCYN UMR CNRS 65971 rue de la Noë
BP 92101, 44321 Nantes, France
e-mail: alain.bernard@irccyn.ec-nantes.fr

A. Chakrabarti (ed.), *CIRP Design 2012*, DOI: 10.1007/978-1-4471-4507-3_20,

20.2 How to Evaluate Knowledge Value

20.2.1 Product Development Process

We may describe the product development process as the following procedure: a product starts from its initial state and arrive at a required state (final state), and a task T is supposed to be accomplished to realize this product evolution from that initial state P_0 to the final state P_n. The product development process can be described by a series of state changes. Given an initial state P_0, the product development process can be characterized by a sequence of product states «$P_0 \rightarrow P_1 \rightarrow P_2 \rightarrow \cdots \rightarrow P_n$», where P_1 is the product state when t_1 is accomplished, P_2 is the product state when t_2 is accomplished, etc., and when task T is accomplished, the product comes to its final state P_n.

Task T can be defined as follows.

Definition 1. Task T is represented by a weighted directed graph $G(T) = (H, A, \Omega)$, where:

- H is a set of tasks, whose elements are the task T, the non-atom tasks t_m and the atom-tasks at_n, i.e., $H = \{h_i\} = \{T, t_1, t_2, \ldots, t_m, at_1, at_2, \ldots, at_n\}$;
- A is a set of directed arcs α_{pq}, i.e. h_p and h_q are linked by α_{pq}, from h_p to h_q;
- Ω is a set of weights ω_{pq} which are assigned to each arcα_{pq}.

Particularly, the sub-tasks which do not have successors are named atom-tasks, noted as at_i.

The reason that T is characterized by a graph, not a tree, is that there may be several sub-tasks which are not independent and they may have one or several sub-tasks in common.

20.2.2 Knowledge Value

Supposing that knowledge K is necessary to accomplish the task T and knowledge fragment k_i is needed to accomplish sub-task t_i, thus, k_i is the solution for the sub-task t_i, and knowledge K can be considered as a set of solutions which together can accomplish the task T. Obviously, a knowledge fragment k_i can be a person, a book, a plan or any type of solutions provided, and knowledge characterization in detail can be referred from the works of (Xu and Bernard [4]).

Based on this proposal, some questions may be raised. Can knowledge K accomplish the task T completely? If knowledge K can only solve a part of the task T, which part is solved? What knowledge fragments k_i have to be added in order to solve the remaining parts? How to choose the knowledge fragments k_i to accomplish the unsolved sub-tasks?

In order to answer these questions, some hypotheses are first presented:

Hypothesis 1. The atom-tasks are noted as at_i, and all atom-tasks correspond to an explicit answer "yes" or "no" which shows whether it can be solved or not. In other words, the atom-tasks cannot be solved partially.

Hypothesis 2. The principles of task decomposition are as follows.If the task$T$ is decomposed into T_1, T_2, …, T_n, we have:

a. $T \subset (T_1 \cup T_2 \cup \cdots \cup T_n)$

(The combination of the sub-tasks should cover the original task T)

b. $T \not\subset T_i$

(Any sub-task T_i cannot cover the original task T)

c. The task T is decomposed with weights, noted as:

$$T : \omega_1 T_1 + \omega_2 T_2 + \ldots + \omega_n T_n, \text{ and } \sum_{i=1}^{n} \omega_i = 1$$

(The weights indicate the importance of the sub-tasks to the original task, for example, if the design of a car focuses more on speed improvement, then the sub-task of speed improvement will have a higher weight than the sub-task of cost diminution)

The value of knowledge K_i to the task T_i is noted as $V(T_i, K_i)$. This notation indicates that knowledge is always in context, in other words, knowledge evaluation is linked with specific tasks. Knowledge value thus varies according to different tasks. For example, given the same knowledge fragment "to adjust the height of a chair", it could have a high value to the task "to consider the ergonomics" and have a low value to the task "to control the cost". The value of knowledge K to the atom-task at_i is defined as follows.

Definition 2. $V(at_i, K) = \begin{cases} 1, & at_i \text{ can be solved by } K \\ 0, & at_i \text{ can be solved by } K \end{cases}$

Based on the two hypotheses and Definition 2, knowledge value can be measured by the procedure as follow.

Procedure for knowledge value measurement:

Step 1: All the value of knowledge K to the atom-tasks is obtained according to Definition 2.

Step 2: For any $h_i \in H$, find all the (h_i, h_j) and their associate ω_{ij}, then:

$$V(h_i, K) = \sum_j \omega_{ij} \cdot V(h_j, K)$$

The procedure shows that from Step 1 we can obtain all the $V(at_i, K)$ and from Step 2 we can obtain $V(T, K)$. When $V(T, K) \neq 1$, it means there are one or several sub-tasks which are not accomplished, so additional knowledge is necessary to make $V(T, K) = 1$. During this process of knowledge addition, both explicit knowledge and tacit knowledge might be needed. Usually, explicit knowledge comes from databases, publications, rules, etc. and tacit knowledge comes from experience, expertise, wisdom, judgment, etc.

If K_i can solve at_i and at_i is linked to T by a sequence of arcs with weights of $\omega_1, \omega_2, \ldots, \omega_m$, then

$$V(T, K_i) = \prod_{u=1}^{m} \omega_u \cdot V(at_i, K_i) = \omega_{at_i} \cdot V(at_i, K_i)$$

This calculation process is realized by a calculator called CAL-KNOW, which is used in a case study introduced later.

If two knowledge fragments K_1 and K_2 are both available, $V(T, K_1)$ and $V(T, K_2)$ can be calculated and compared. Generally, knowledge that has a higher value is usually chosen. As collaborative networks are regarded as a critical success factor to achieve product innovation [5], it is always useful to choose the most valuable knowledge to be exchanged and shared.

20.3 Knowledge Evaluation in Product Design Process

During product lifecycle design, which can be defined as a sequence of tasks [6], both tacit and explicit knowledge may be required to accomplish the tasks at_i, so these two kinds of knowledge can add value to the knowledge of design K and thus make knowledge evolution [7, 8].

Here are the main steps to take during the procedure of knowledge evaluation in supporting product design.

1. To decompose of the product development process into simpler processes, in other words, to realize the decomposition of the task T into atom-tasks at_i.
2. To evaluate the value of the existing knowledge using the evaluation model introduced in the previous section.
3. If not all the atom-tasks are solved, find out which at_i should be solved next.
4. Add appropriate knowledge, explicit and/or tacit, to accomplish at_i.
5. Do Step 3 and Step 4 repeatedly until all atom-tasks are solved.

In the product design process, knowledge may add value to products and product may also make knowledge more valuable [9], and such mutual value adding process can be explicitly described and controlled using our knowledge evaluation model.

20.3.1 Case Study

This paper has chosen a case of chair design, which is a part extracted from the product lifecycle of a chair. The concentration is implemented on the phase of design as it is a key phase where major decisions are made concerning knowledge. In this example, the task « design a chair » should be accomplished in order to make the product (chair) evolves in the development process. Figures 20.1 and 20.2 illustrate how the task is decomposed. Although the decomposition lacks in completeness, for example, several tasks such as market study, packaging and logistics matters, particular optimization, etc., are neglected, it can serve as an adequate demonstration.

Based on the criteria obtained from experience in product design, the principle task « design a chair » is decomposed into four sub-tasks.

The weights ω_i are given by the experts of different roles who have different points of view in design activities. Table 20.1 shows the weights given to each sub-task by experts of different roles. In order to determine a weight, we have taken into account the results given by a group of experts for each given role. How to improve the results of collecting and analyzing the weight values given by different people is another complicated topic, which needs further research on statistical techniques, human behaviors, etc. In this paper, we simply regard the weight value as is the average of the weights proposed by all the experts assigned in each group.

Here are some illustrations about Fig. 20.1:

- «Perception test» and «To consider the psychological comfort issues» can be solved by questionnaire surveys.
- «Ergonomic studies» mainly focus on examining the degree of fatigue of different parts of the body (muscle, bone, joint, etc.) of a person who sits in the chair for a period of time or by simulations.
- «Tests of the material attributes»may include the thermal conductivity (in winter, people do not like to sit in a chair with a surface of iron, because it's too cold), the sensation of the material (for example, smooth or rough, soft or hard), etc.
- «To consider the aesthetics of the chair» considers the intrinsic beauty of the chair, which depends on the cultural and social context. In other words, for a same chair, it may vary from beautiful to disgusting due to different tastes of people from different countries or groups.
- «To consider the adaptiveness in the context of use» considers whether the chair matches the environment of use. For example, in a fast-food restaurant, sofas are not suitable to the environment although they are very beautiful.
- «Architectural design» is considered before the design in details.
- For the assignments of the values of the weights ω_{B1} and ω_{B2}, they depend on whether the designer takes optimization into account. Table 20.2 shows two examples in determining ω_{B1} and ω_{B2}. In an extreme situation, when a designer assigns $\omega_{B1} = 100\ \%$, it means the designer will simply look for a solution in a database of archived designs.
- The tasks «To consider the mechanical holding issues» and «To consider the stability» have the same sub-task«To consider the positions of gravity centers». Such situation that several tasks may have a same sub-task in common is acceptable according to Definition 1 which defined the Task T as a graph.

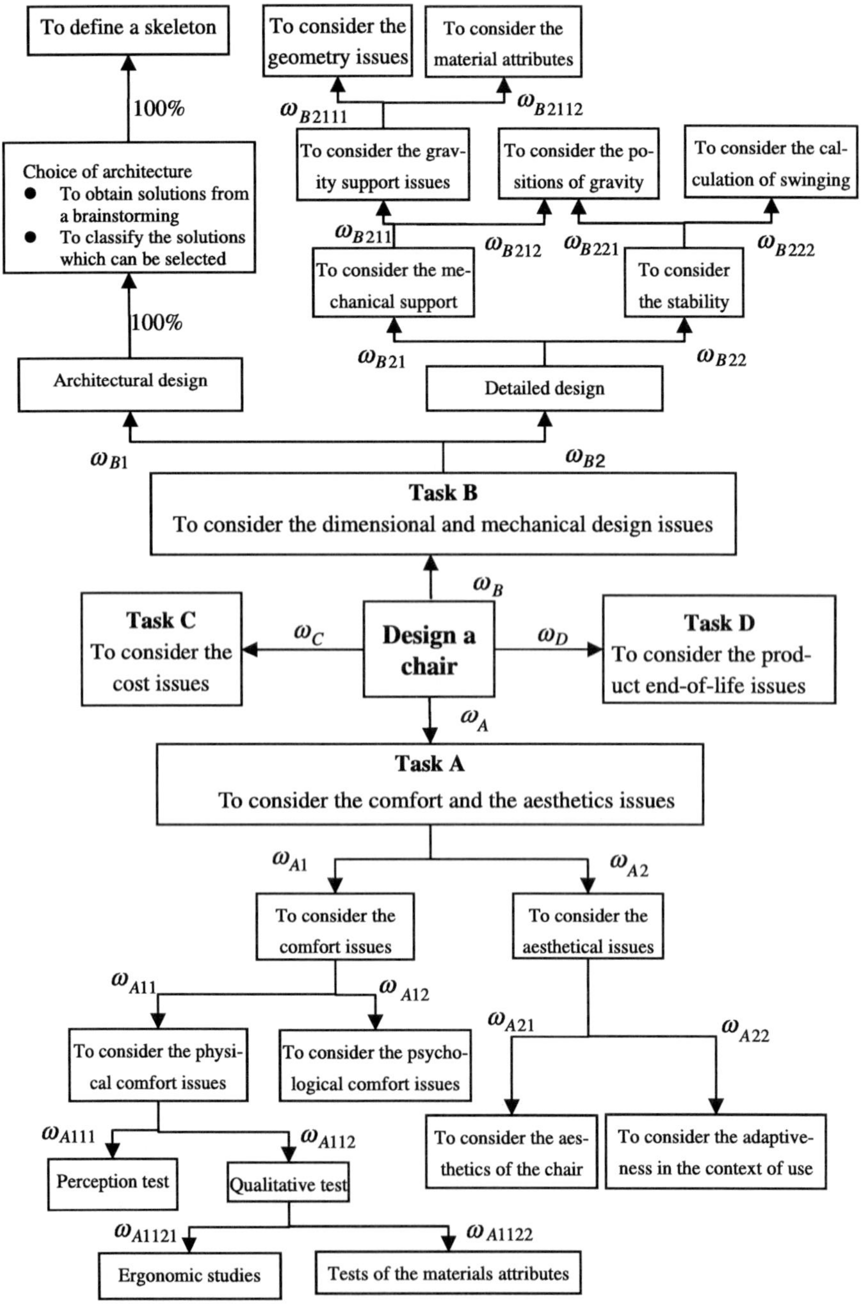

Fig. 20.1 The decomposition of the task « design a chair » (*Part I*)

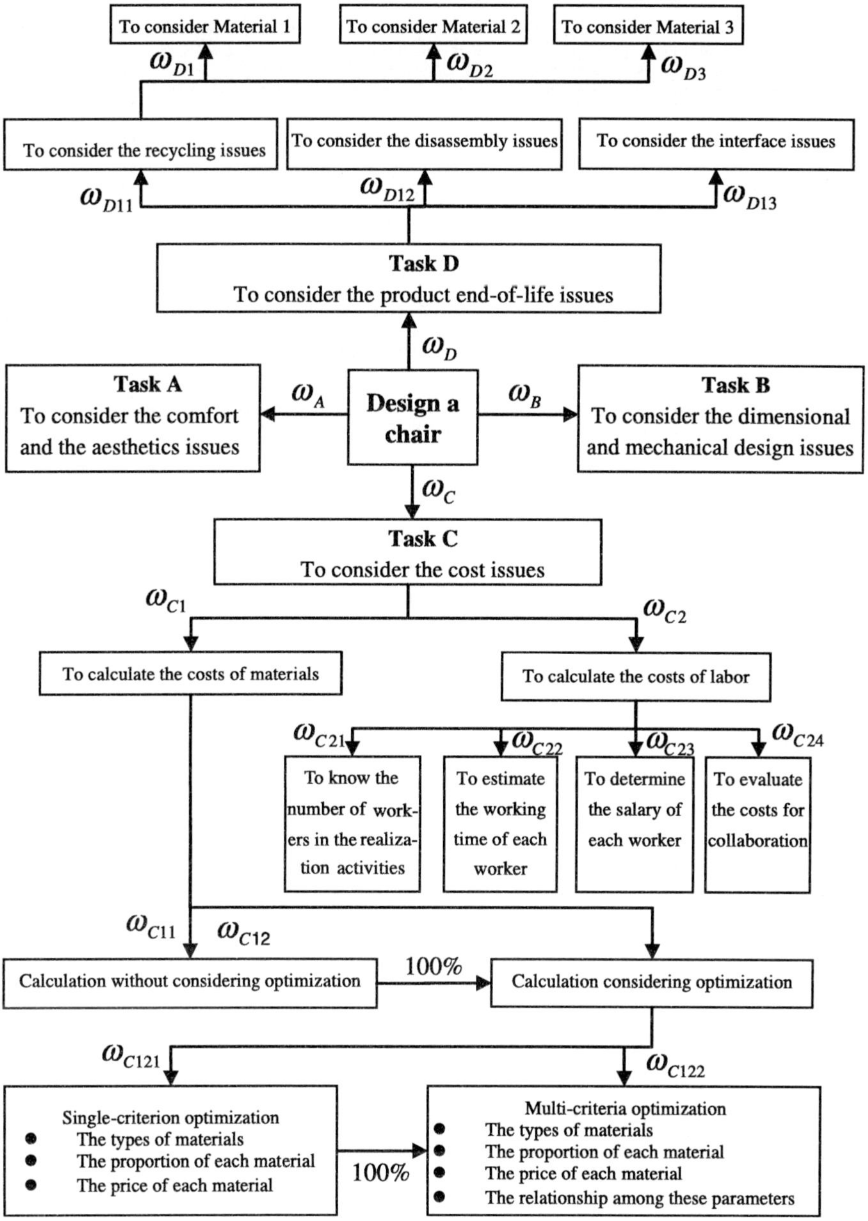

Fig. 20.2 The decomposition of the task « design a chair » (*Part II*)

Table 20.1 The values of weights

Experts of different roles	ω_A Comfort/aesthetics (%)	ω_B Dimension/mechanics (%)	ω_C Costs (%)	ω_D End of life (%)
Client	50	10	30	10
Designer	10	50	30	10
Manufacturer	0	30	50	20
Seller	30	10	40	20
Transporter	0	60	30	10
Recycler	0	0	30	70

Table 20.2 The values of the weights ω_{B1} and ω_{B2}

	ω_{B1} (%)	ω_{B2} (%)
If the designer pay much attention in optimization issues during the design process	30	70
If the designer does not wish to spend too much time in searching for optimization solutions for task B	50	50

Table 20.3 The values of the weights ω_{C1} and ω_{C2}

	ω_{C1} (%)	ω_{C2} (%)
If the chair is designed to be produced in large quantities	80	20
If the chair need a custom design with a small amount of production expectation	20	80

- Here are two weights which have the value « 100 % ». They mean that the tasks linked by an arrow of a weight of «100 %» are «equal». In this case, when people have accomplished «to define a skeleton», they have accomplished the «architectural design» at the same time.
- In order to determine the values of the weights ω_{C1} and ω_{C2}, the context of design should be considered, in other words, they depend on the amount of production of the chairs provided by customers. Table 20.3 gives two examples. In the condition that the chair is designed to be produced in large quantities, the cost of materials has a weight of greater importance. When it is a case of custom design, the weight of materials is lower. The client is willing to pay the extra cost for differentiation even if the materials used are more expensive.
- If several tasks have the relations of inclusion, an arrow with a weight of "100 %" is used. Design optimizations are often made retrospectively by taking into account new knowledge [10].
- Why the arrow from the task "Single-criterion optimization" to the task "Multi-criteria optimization" has a weight of "100 %"? Obviously, when people can perform the task of "Multi-criteria optimization", they are able to accomplish the task of "Single-criterion optimization". In other words, these two tasks have

a containment relationship. In case when two tasks have a containment relationship, an arrow of a weight of "100 %" is used. Optimizations of the design are often made retrospectively, taking new knowledge into account, [11].

- Management of product end-of-life and recycling are critical issues in environment treatment for manufacturing enterprises so they should be considered in product lifecycle design [12, 13]. The task «To consider the recycling issues» needs knowledge about the possibilities of recycling the materials used.
- The number of materials to be considered is not limited to three, and it may differ from case to case. In other words, this number depends on how many principal types of materials are used to build the chair.
- The three weights ω_{D11}, ω_{D12} and ω_{D13} are determined by several factors of the chair, for example, the proportion of each material used, the cost of each material used, etc.
- The task «To consider the disassembly issues» evaluate whether the designed chair can be disassembled. The easy disassembly of a product will facilitate the recycling of material used and the reuse of different parts of the chair.
- The task «To consider the interface issues» mainly considers the reuse issues of different parts of the chair. For example, if a chair has a leg broken, instead of throwing away the chair and replacing it by a new one, people can simply substitute the broken leg. But in order to realize the substitution of the broken leg, the interface between the leg and the body of the chair should be well designed. In such cases, the design of the interface should be given special attentions.

In real cases tested, each time a solution (knowledge fragment) with a higher value is chosen, and from the list of unaccomplished atom-tasks, we could find out easily which tasks should be accomplished next. Every time that K reaches a state that can solve one more task, its value increases.

When knowledge reaches its final state, its value may not always be 100 %, but it is not critical if people are already satisfied with its current value. In the given example, if we do not have to accomplish the task of "To calculate the cost of labor", knowledge can remain in a state that its value is not 100 %. In such cases, people have to take some risks when they are going to the next stage of the product lifecycle.

20.4 Conclusions

Knowledge evaluation is a key issue in knowledge management, and this paper has presented a knowledge evaluation model in product lifecycle design. The model integrates the process of knowledge evolution and product development, and the mutual effects between knowledge and product are analyzed. Based on the theoretical definitions and models, this paper illustrates how knowledge value can be assessed by studying a specific case. In the applications of product lifecycle design, knowledge values calculated by the model can serve as important factors in a decision making system that decides which knowledge to choose and what to do next. The model could serve as a framework to describe the knowledge related activities and could be a useful tool for managing knowledge in product lifecycle design and support.

Interesting perspectives may include deeper analysis about the optimization issues of weights and dynamic product development processes.

References

1. Nonaka I, Takeuchi H (1995) The knowledge creating company. Oxford University Press, New York
2. Bernard A, Tichkiewitch S (2008) Methods and tools for effective knowledge life-cycle-management. Springer, Berlin
3. Perry N, Bernard A, Bosch-Mauchand M, LeDuigou J, XU Y (2011) Eco global evaluation Cross benefits of economic and ecological evaluation. In: Proceedings of the 18th CIRP International conference on life cycle engineering, Braunschweig, Germany, 681–686
4. Xu Y, Bernard A (2010) Knowledge value chain: an effective tool to measure knowledge value. Int J Comput Integr Manuf 23(11):957–967
5. Perry N, Candlot A, Schutte C (2010) Collaborative knowledge networks emergence for innovation. Factors of success analysis and comparison. J Decis Syst 19(1):75–91
6. Nacsa J, Bueno R, Alzaga A, Kovács GL (2005) Knowledge management support for machine tool designers using expert enablers. Int J Comput Integr Manuf 18(7):561–571
7. Bernard A, Xu Y (2009) An integrated knowledge reference system for product development. CIRP Ann Manuf Technol 58(1):119–122
8. Xu Y, Bernard A (2010) Measurement of enterprise knowledge by state characterization. Expert Syst 27(5):374–387
9. Xu Y, Bernard A (2011) Quantifying the value of knowledge within the context of product development. Knowl Based Syst 24(1):166–175
10. Sébastian P, Chenouard R, Nadeau J-P, Fischer X (2007) The embodiment design constraint satisfaction problem of the BOOTSTRAP facing interval analysis and genetic algorithm based decision support tools. Int J Interact Des Manuf 1(2):99–106
11. Del Prete A, Mazzotta D, Anglani A (2010) Design optimization application in accordance with product and process requirements. Adv Eng Softw 41(3):427–432
12. Bufardi A, Sakara D, Gheorghe R, Kiritsis D, Xirouchakis P (2003) Multiple criteria decision aid for selecting the best product end of life scenario. Int J Comput Integr Manuf 16(7–8):526–534
13. Ueda K, Nishino N, Nakayama H, Oda SH (2005) Decision making and institutional design for product lifecycle management. CIRP Ann Manuf Technol 54(1):407–412

Chapter 21
Risk Minimized Procurement in Low Wage Countries

Thomas Zentis and Robert Schmitt

Abstract International product procurement is a risky business, one that can, in extreme cases, threaten the survival of a business. This is mainly due to the fact that the total costs of purchasing activities are insufficiently calculated and the company is inadequately oriented towards international procurement. For this reasons, two methods have been developed, with which companies can efficiently safeguard their procurement in international markets. The first method serves as a systematic forecast of purchasing costs. The second is a new risk identification and assessment method tailored to procurement needs. Project findings show that risks can be identified far better with the aid of these new user-friendly methods.

Abbreviations

TCO	Total cost of ownership
SME	Small and medium-sized enterprises
FMEA	Failure mode and effect analysis
RPN	Risk priority number
TCS	Total cost of supply
QRC	QuickRiskCheck

T. Zentis (✉)
Department of Production Quality and Metrology, Fraunhofer Institute for Production Technology IPT, Steinbachstr. 17, 52074 Aachen, Germany
e-mail: thomas.zentis@ipt.fraunhofer.de

R. Schmitt
Laboratory for Machine Tools and Production Engineering (WZL), Department for Metrology and Quality Management, RWTH Aachen University, Steinbachstr 52074 Aachen, Germany
e-mail: r.schmitt@wzl.rwth-aachen.de

A. Chakrabarti (ed.), *CIRP Design 2012*, DOI: 10.1007/978-1-4471-4507-3_21,

21.1 Introduction

Success in the procurement of bought-in parts often means treading new paths in order to get ahead of the competition. This might entail cultivating new and unknown suppliers or tapping completely new markets. Every company necessarily takes some risks here, but these will only have a negative impact if they are not calculated with sufficient care. Only 28 % of businesses are satisfied with the quality of the products and services they purchase. Defective parts and delivery delays are often cited as reasons for dissatisfaction [1]. Smaller companies take a particularly high risk when purchasing abroad. With their limited human resources and lack of experience in worldwide procurement they are often unable to make an a priori estimate of the total costs of this type of procurement activity [2]. To be successful, businesses must deploy risk management methods covering all procurement activities. But comprehensive risk management often entails a great deal of effort and expense to take into account all risks and handle them properly. Too much effort is spent on low-risk parts 10 procurement processes is only rarely undertaken. The central challenge here is to minimize all critical risks while at the same time ensuring efficient procurement.

To this end, two methods have been developed within a research project for enabling companies to plan and organize their international procurement effectively while keeping risks as low as possible and ensuring product quality.

21.2 Current Approaches to Determine Procurement Costs and Identify and Assess Technical Risks

21.2.1 Methods to Determine Procurement Costs

A widespread method for determining the costs of acquisition is the Total Cost Of Ownership (TCO). Originally, it was developed for the IT industry in 1987 by the Garter Group. Beside the simple costs for acquisition any additional costs and expenses of the purchase process are determined, for example, the travel expenses for auditing the supplier, costs for transportation, installation and maintenance or the expenses for the development and qualification of suppliers [3–7].

Therefore, TCO is a possibility to determine the suitability of a supplier and can be a basis for supplier selection [8, 9]. The disadvantages of TCO are its high complexity and therefore the difficult implementation, as well as the high effort for the determination and maintenance of costs data [6, 8, 9]. The user is frequently confronted with the problem that the costs data are not precise enough or are not available at all. This delays using the method and makes it more difficult to apply, as well as negatively influencing the validity of the result. The consequence is, that the success of the method strongly relies on the level of commitment of both management and the employees [10]. Moreover, with TCO possible procurement risks related to the supplier cannot be identified. Furthermore, the TCO method is currently limited mainly to lifecycle costs of bought-in parts and does not extend to the followup costs of strategic decisions made during procurement activities. In addition, the TCO method does not demonstrate the efficiency in handling that small and medium-sized enterprises (SME) with little experience in the field of international procurement.

The literature refers to a variety of methods that are very similar to the method of TCO. These approaches target specific procurement costs beside the pure purchasing price [5]. One of these methods is Life-Cycle Costing [11, 12], in which costs of supply before signing the acquisition contract are unintended. Another method is Zero-Based Pricing [13, 14], which especially considers the suppliers structure of costs. The method of Cost-Based Supplier Performance [13, 14] only focuses on costs apart from the supplier costs and the Cost-Ratio method [15] determines the cost-pushing factors only (e.g. poor quality, delivery delay, etc.).

21.2.2 Methods to Assess Technical Risks

Due to the high complexity of a company, that is inherent in the vast number of different system states and alternative possibilities of actions, a complete capture and analysis off all possible risks is not feasible. Therefore, a limitation of risks is necessary which can be supported by using different methods [16].

One of these methods is the Failure Mode and Effect Analysis (FMEA), a widespread and effective approach for risk limitation. The objective of an FMEA is the examination of products and processes during the early stages of development regarding potential errors and to initiate early measures to prevent failures. This is carried out by a risk analysis during the integrated development and planning processes [17–19]. The implementation of a FMEA can be realized in five steps: (1) creation of system elements and their structures, (2) identification of functions and their structures, (3) failure and risk analysis, (4) risk assessment and (5) the final optimization according to the identified and by Risk Priority Number (RPN) structured failures [20]. The RPN assesses risks in three categories: (1) The Occurrence (O) related to the probability of the failure mode and cause, (2) the Severity (S) to measure the seriousness of the effects of a failure mode and (3) the Detection (D), which is generated on the basis of likelihood for detecting the failure mode before the product is used by a customer. The rating is scaled from 1 to 10 for each category [17]. The result of an FMEA requires a differentiated interpretation, because it only reflects the estimation of the experts involved [19]. Another problem is the ordinal scale for each of the three RPN indices, because the distance between the values is not clarified by a distance function. The RPN is just a product of three independent variables and not an absolute measure of risk [17]. It should be noted that an FMEA allows no objective quantification of risks and it is not a possibility to find solutions to eliminate failure causes. Especially SME are not able to afford FMEA's expenditures.

Recent research addressed some of these problems and focused the improvement of the traditional FMEA. For example Petri nets were used to analyze multiple failure effects [21], logarithmic scales were used for RPN [22] or Monte Carlo Simulation were applied for simulation on RPN [23].

21.2.3 Central Deficits and Need for Action

Current approaches to global sourcing, in particular to total cost calculation and risk management, do not cover all the needs of small to medium-sized enterprises that are interested in aligning their procurement activities with global markets, or do not address these needs sufficiently. Although there are rough concepts and methods available, there is still a need for pragmatic recommendations for how to proceed and for methods that can readily be applied to the day-to-day business of procurement. TCO calculations, for example, should be used to support the search for suppliers with respect to assessing their long-term suitability. But since to date, there is no total cost of ownership model available for purchasing bought-in parts on global markets, companies have to accept a high level of uncertainty in their cost estimates. The biggest challenge here is to guarantee the most complete examination of costs possible, while however minimizing the effort needed to deploy the method in question. Especially smaller businesses need methods with which they can minimize the risks in their procurement activities. This would reduce the time and effort entailed by procurement activities without suffering the quality of procurement. Risk management approaches and methods in use up to now are framed for the most part on a high level of abstraction and are therefore not adequately applicable to actual procurement activities. Even using well-known risk management methods does not give companies any secure basis for deciding on supplier's suitability.

21.3 Methodical Approach

The first major focus of the research project was to develop a pragmatic method for calculating the Total Cost of Supply (TCS) (see Sect. 21.3.1). The aim here is to enable SMEs in particular to globally compare all suppliers from the perspective of cost and to efficiently predict the consequential costs of a procurement decision. The method used the strengths of the Total Cost of Ownership approach but contains additional cost categories for procurement activity. The results of the costs analysis are used to determine the maximum purchase price that should be paid for a bought-in part that, in comparison with other suppliers, entails the lowest lifecycle costs for procurement activity.

The second focus of research was to develop a risk management method tailored to procurement needs (see Sect. 21.3.2). Therefore proven (e.g. FMEA)

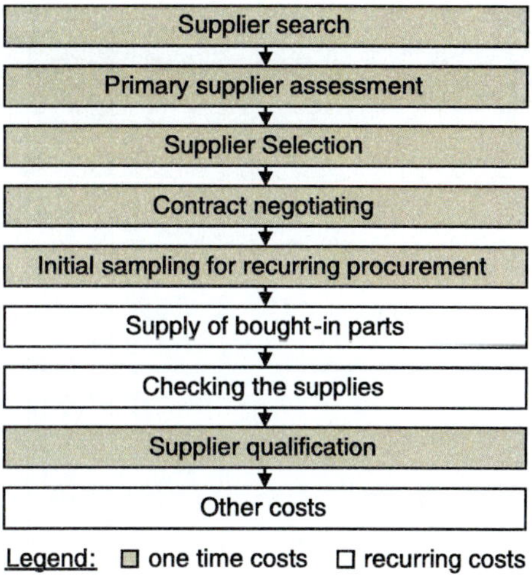

Fig. 21.1 Cost categories through procurement lifecycle

and innovative risk management methods are fine-tuned along the special conditions of procurement and reduced to the essential elements needed to provide pragmatic assistance to SMEs.

21.3.1 Total Cost of Supply

The basis for risk reduction in international procurement is provided by a high degree of transparency in the forecast of procurement costs, also taking into account the consequential costs of failures. If procurement costs are not forecast completely, this can threaten a company's survival, because the consequential costs of failures are often several times as high as the actual costs for purchasing bought-in parts.

The Total Cost of Supply (TCS) method enables production companies to carry out a targeted calculation of the total costs of procurement activities. The predicted procurement costs then comprise not only customs fees and the actual costs of bought-in parts and their transport, but also take into account the entire procurement lifecycle (see Fig. 21.1).

This cycle begins with the effort and expense involved in searching for a supplier and ends with the termination of the supply relationship. Other costs include taking into account opportunity costs, for example missing out on discounts offered by alternative suppliers, as well as the non-conformity costs that arise when a bought-in part does not meet the required specifications. In this case it is necessary to quantify e.g. the effort needed to search for back-up suppliers on short notice or the damage claims possibly asserted by a customer for delivery delays. Taking into consideration the probability that such cases may arise, a company can then make a decision on an appropriate procurement market based

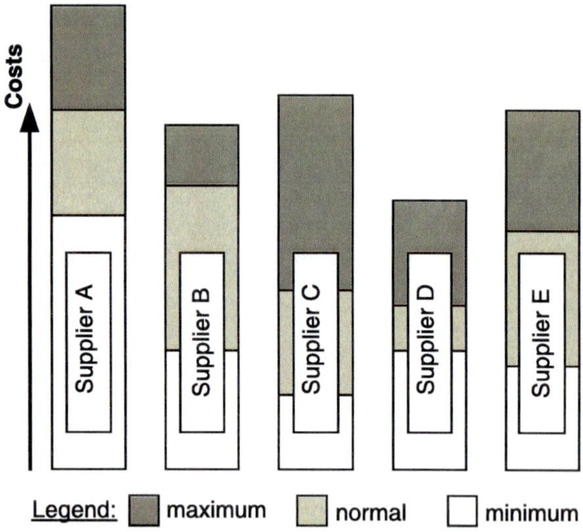

Fig. 21.2 Cost overview of each supplier

on quantifiable data that enables a long-term estimation of costs and thus reduces the risk of any unexpected costs arising. Since the probability of something occurring can as a rule only be roughly estimated, the normally incurred costs are used in the TCS method, along with the minimum and maximum possible costs. The range between minimum and maximum costs then reflects the financial risk of a procurement activity. In addition, the method enables a direct comparison to be made between several suppliers with the corresponding cost ranges. For a less time consuming application of the TCS method, users have the option of roughly estimating various parts of the procurement process costs, for example the cost of searching for a supplier. This abbreviated estimation is only of benefit, however, when the user already has an exact picture of the costs or can provide a precise estimate based on past experience. To improve a company's ability to plan for variable costs of procurement, the TCS method distinguishes between onetime and recurring cost categories. This breakdown of the total costs for a supplier helps the user to rate suppliers because the amount of recurring costs and hence the impact on the company's overall liquidity might otherwise be underestimated (see Fig. 21.2). In this Context the TCS allows the rework of recurring costs by a feedback loop.

Initial results from practice show that the use of the Total Cost of Supply method markedly increases cost transparency. In order to test the effectiveness of the method, employees from several companies first estimated procurement costs using the old methods, and then the same bought-in parts were rated with the TCS method. The real costs determined via the TCS method were maximal five times higher than those originally identified. If non-conformity was also included in the mix, such as for later express deliveries due to defective parts or additional visits to the supplier, this difference was even at maximum ten times higher.

After all costs from each supplier are detailed demonstrated the user gets a good basis for supplier selection. It is now up to the user how risky his procurement activities should be.

On the one hand he can make a decision for a potential cheap supplier, which also represents a high risk of maximum costs (see supplier C in Fig. 21.2). On the other hand he can select a supplier with less minimum procurement costs, but also a less risk of maximum costs (see supplier D in Fig. 21.2).

The results from practice also show that the »Supply of bought in parts« demonstrates the biggest cost driver in procurement activities. These recurring costs are not only driven by the costs for the physical bought-in part. Big cost drivers are also the effort for the long logistics distance as well as possible costs for loss of production if the transport of bought-in parts delay or even fail. Through the practical application of the TCS possible costs for loss of production, for example reasoned by unpunctual transport, were identified as the biggest risk for unprofitable procurements and potential cost drivers.

21.3.2 QuickRiskCheck

The crucial prerequisite for efficiently identifying and rating risks in procurement processes is to focus at an early stage on risk-critical sub-processes. In existing approaches, e.g. FMEA, this important pre-selection is not made, or not made carefully enough, so that frequently the analysis entails a great deal of effort and expense. There is also a danger that critical risks will go undetected or will be rated falsely in relation to other risks. If risks are identified and rated inefficiently, this directly leads to the problem of how to define and select effective measures to reduce them.

To deal with the deficits described here, the QuickRiskCheck (QRC) method was developed, which creates a cascade for the identification and rating of critical risks. This method enables the time-effective definition of measures appropriate for reducing risks. The method is primarily applied to procurement processes, but can also be adapted and used for products or projects. Its procedure is structured into 7 phases (see Fig. 21.3).

The procedure starts off with a rough structuring of the procurement process under examination, dividing it into sub-processes and recording risks that are already known. In the subsequent evaluation, the sub-processes are compared two-by-two and the sub-process with the relatively higher risk of failure is assigned a higher numerical value. The sum is calculated for every subprocess and from the results a ranking is created for the most critical sub-processes. From this ranking, the user himself chooses the sub-processes for which he would like to examine the risks in more detail. The user's experiences therefore come into play in the determination of critical sub-processes, whereby the detailed examination focuses on the most essential subprocesses. The detailed examination begins with a further division of a sub-process into individual steps. The risks within these steps are

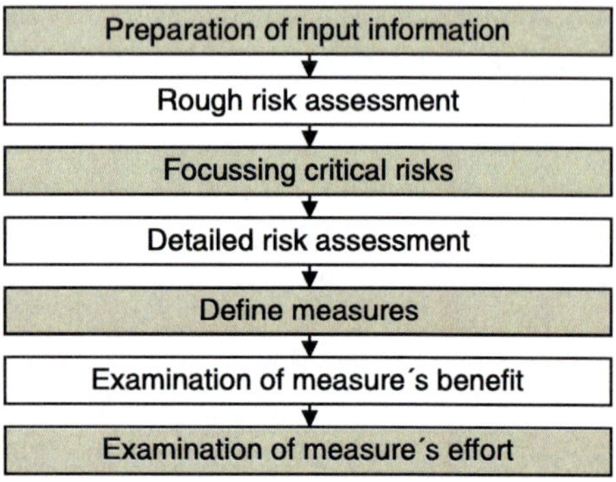

Fig. 21.3 Seven phases of QuickRiskCheck

identified and rated according to the 5 M: Man, Machine, Method, Materials, Maintenance [24].

These risks are rated by the user on a three-step scale from low to medium to high. For risks rated medium or high, the user formulates additional measures. By then neglecting the lower-rated risks, the time taken for risk assessment can be reduced by around one third. The next step is to determine for the measures defined whether they reduce, neutralize or heighten each individual risk.

Here, the QRC method offers the advantage that the impact of a measure is not only evaluated for one particular risk, but across all risks in the individual steps of the sub-process. The fact is that every risk-minimizing measure can in fact increase a different risk. In order to determine the benefit of a measure, the numerical values recorded for the significance of the risk in question are multiplied by the values for the respective impact of the measure. By summing up the values, the risk-minimizing benefits of every measure are then quantified and a ranking of measures is derived. For a complete evaluation of the measures, the effort and expense involved in their implementation must additionally be calculated into the equation. Only in this way does the user have the possibility of selecting measures according to the Pareto principle, according to which around 20 % of the total effort should achieve around 80 % of the total benefit. In determining how much effort is needed to implement a measure, the user indicates the costs of the personnel and resources involved. From these figures the relative ratio of total benefits to costs is calculated and a corresponding diagram is generated for each measure. The user is thus able to visualize how to select the most effective measures according to the above-described Pareto principle (see Fig. 21.4).

In practice, the use of the QRC method has been shown to reduce the time needed to identify and assess process risks while increasing the number of risks identified. To test the method's effectiveness, employees of several companies first identified and rated risks using the old methods, and then repeated the same procedure using the QRC. The time needed to apply the QRC was in all cases approximately 50 % shorter than that required for the procedures previously used. The number of identified risks could be raised by a maximum of 6 times.

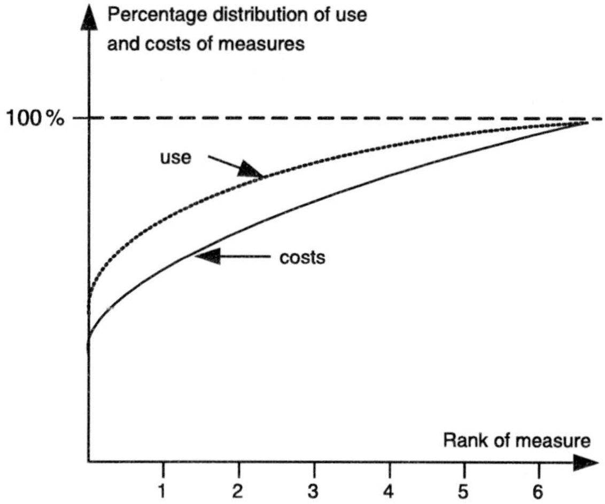

Fig. 21.4 Demonstration of measure's cost-value-ratio

21.4 Summary

In summary, the presented methods safeguard against procurement risks while at the same time increasing the efficiency of procurement.

The methods support businesses in their efforts to safeguard against procurement risks through a comprehensive approach consisting of TCS calculations, and risk classification by QRC. In addition to offering businesses greater security in their procurement efforts, the methods also reduce the amount of effort and costs incurred for supplier management. With more efficient procurement, it is easier for Western European companies to take part in the global market while extending their competitive advantage over worldwide rivals.

Acknowledgments The support of the Federal Ministry of Economics and Technology (BMWi) as part of a program promoting "Industrial Joint Research (IGF)" within the framework of the Confederation of Industrial Research Associations (AiF) and supervision by the Federation for Quality Research and Science (FQS).

References

1. Pulic A (2004) Einkaufsgeschäfte in den Niedriglohnländern boomen. Procurement Lett 4(12):1–12
2. Lazowski F (2004) Global sourcing—an anchor format for successful globalization strategies. ReeseMcMahon, Chicago
3. Aberdeen Group (2003) The global sourcing benchmarking report—balancing supply cost, performance, and risks in an uncertain economy. Aberdeen Group, Boston
4. Kirwin B, Mieritz L (2003) Total cost of ownership as a common denominator. Business Issues 1:8–13

5. Bremen P, Alard R (2008) Total cost of ownership—Gesamtkostenanalyse bei der Beschaffung in China. Beschaffungsmanagement, Heft 3:12–15
6. Degraeve Z, Roodhoft F, van Doveren B (2005) The use of total cost of ownership for strategic procurement: a company-wide management information system. J Oper Res 56:51–59
7. Ellram LM (1995) Total cost of ownership—an analysis approach for purchasing. Int J Phys Distrib Logist Manag 4:4–23
8. Bhutta KS, Hup F (2002) Supplier selection problem: a comparison of the total cost of ownership and analytic hierarchy process approaches. Supply Chain Manag Int J 24:126–135
9. Wildemann H (2009) Total cost of ownership, Leitfaden zur Optimierung der Gesamtkostenposition in Beschaffung, Produktion und Logistik, TCW, München
10. Bremen P, Oehmen J, Alard R (2007) Costtransparend sourcing in China applying total cost of ownership. In: Proceedings of the 2007 IEEE IEEM, 13th IEEE international conference on industrial engineering and engineering management, Singapore, pp 262–266
11. Jackson DW, Ostrom LL (1980) Life cycle costing in industrial purchasing. J Purch Mater Manag 18(4):8–12
12. Fernandez JP (1990) Life-cycle costing, NAPM Insights 6:6
13. Burt DN, Norquist WE, Anklesaria J (1990) Zero base pricing. Probus, Chicago
14. Monczka RM, Trecha SJ (1988) Cost-based supplier performance evaluation. J Purch Mater Manag 24(1):2–7
15. Soukup W (1987) Supplier selection strategies. J Purch Mater Manag 23:7–12
16. Allenspach M (2001) Neue Beratungskonzepte zum integrierten Risikomanagement—was entscheidet über den Erfolg?, Integriertes Risiko-management- Perspektiven einer chancenorientierten Unternehmensführung, St. Gallen, pp 178–195
17. Seung J, Kosuke I (2003) Using cost based FMEA to enhance reliability and serviceability. Adv Eng Inform 17:179–188
18. Korayem M, Irvani A (2008) Improvement of 3P and 6R mechanical robots reliability and quality applying FMEA and QFD approaches. Robotics and Computer-Integrated Manufac turing 24:472–487
19. Pfeifer T (2001) Qualitätsmanagement: strategien, methoden, techniken. Hanser, München
20. Haffner A (2005) Ein Modell zur Bestimmung der monetären Einsparungspotenziale bei der Durchführung einer Fehlermöglichkeits- und Einflussanalyse (FMEA), Dissertation Universität Stuttgart
21. He D, Adamyan A (2001) An impact analysis methodology for design of products and processes for reliability and quality. In: Proceedings of ASME Design Engineering Technical Conference, Pittsburgh
22. Ben-Daya M, Raouf A (1996) A revised failure mode and effective analysis model. Int J Qual Reliab Manag 13:43–47
23. Bevilacqua M, Braglia M, Gabbrielli R (2000) Monte Carlo simulation approach for modified FMECA in a power plant. Qual Reliab Eng 16:313–324
24. Masaaki I (1986) Kaizen the key to Japan's competitive success. Mcgraw-Hill Higher Education, New York

Chapter 22
Methodological Approach to Evaluate Product Adaptations Based on Real Options

G. Lanza and S. Ruhrmann

Abstract Global business activities of companies grow continuously because cost savings must be achieved and new markets must be tapped into. Today's companies face the challenge to either satisfy local markets with their existing product portfolio or to develop locally adapted products. These products must not only be adjusted to the individual requirements of the customer but must also realise cost reduction potentials. Such an investment project in the form of product adaptations has to be calculated and evaluated at an early stage. This paper presents an approach to use the real option analysis in order to include uncertainties and flexibility during the development phase. Using this approach industrial companies are capable to identify the cause and effect relationships between product design and production technology adaptations, to calculate the costs of these adaptations and to evaluate the investment project regarding uncertainties and flexibility.

22.1 Introduction

Not just during the last decades of globalisation but still today companies are facing the strategic questions of how to compete with other companies, how to meet customer requirements and how to perform on a new global and international scale [1]. Globalisation leads companies to locate their production site in best-cost countries, because global production networks promise competitiveness and access to new selling markets [2]. Getting access to new selling markets and meeting customer requirements are the core issues for successful project implementation.

G. Lanza (✉) · S. Ruhrmann
wbk Institute of Production Science, Karlsruhe Institute of Technology (KIT),
76131 Karlsruhe, Germany
e-mail: Gisela.Lanza@kit.edu

A. Chakrabarti (ed.), *CIRP Design 2012*, DOI: 10.1007/978-1-4471-4507-3_22,

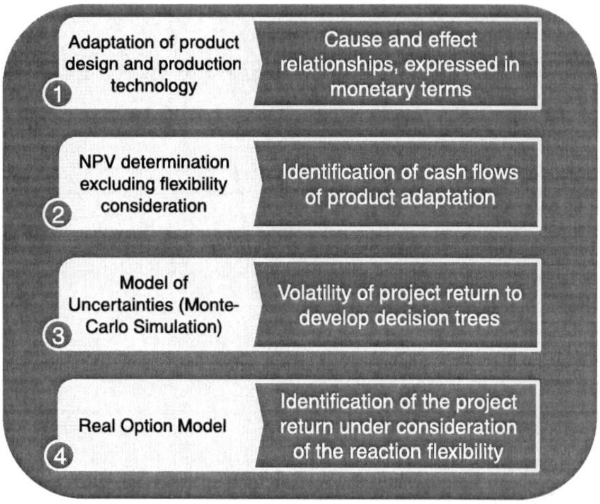

Fig. 22.1 Schematic approach to calculate and evaluate product adaptations based on real options

Therefore, companies make strategic decisions to gain access to new markets by introducing even their existing portfolio or designing adapted products to meet local requirements.

The configuration of the adaptation of product design to domestic markets is one big factor which helps to get additional sales opportunities. The disadvantage of the introduction of the existing product portfolio is the non-fulfilment of the customer needs and corresponding orders. At this point, companies need to assess whether product adaptations are economically beneficial or if the costs would exceed the revenue. It leads to the creation of new investment projects through the definition of new product portfolios for localised markets. The evaluation of the advantage of investment projects is especially useful for the real option analysis because in this case uncertainties and flexibility that are identified during the development phase of product adaptation can be taken into consideration. Before using the real option analysis it is necessary to determine the mutual effects between product development activities for product adaptation and activities for production technology adaptations [3], so called cause and effect relationships.

Reducing uncertainties in the early stages of product generation is one success factor for companies, which is published in studies of the engineering industry [4]. Therefore it is necessary to calculate the costs of the product adaptation in the early stages of product design. Thus, the management is able to estimate the profitability of the investment in a product customisation for a domestic market. The decision to adapt a product design for the purpose of getting access to new selling markets is not a single yes-or-no decision with a certain or uncertain outcome [5]. It is a decision which has to be done repetitively in order to quantify the added value. The consideration of flexibility using the real option analysis yields the desired result.

The wbk Institute of Production Science is developing an approach to estimate the total cost of adapting a product design to a domestic market based on uncertainty and decision flexibility (Fig. 22.1).

22.2 Literature Review

The following paragraph shows the state of the art in the two key elements, i.e. integrated product and process design and the evaluation of investment projects. This is done to value the cause/effect linkages between product development and production technology and also to transform those interdependent effects into the model framework of the real option analysis.

22.2.1 Approaches in the Field of Product Adaption

Penetrating new markets and satisfying the requirements of local customers is achieved by adopting renowned products that are produced in the home country. Abele points out the adaptation of the product technology in addition to adapting the product design because in this categories only the simultaneous performance of adaptation the lead to the greatest possible success. [6]

Hurschler derives the design adaptation from the causes that led to product adaptation. However, the trigger can be located in the procurement market or the sales market. [7] Reinecke presents a method which describes how to adapt products to the conditions of global competition [8]. Therefore product flexibility and production flexibility are seen as key factors of global product adaptation.

Große-Heitmeyer introduces a tool to support the global product development of small and medium-sized enterprises (SME) [9]. The method that is presented concentrates on product structuring that factors in globalisation by taking core competences into account. Structuring takes place on the basis of common design criteria such as manufacturing or assembly. Lanza points out the challenges and opportunities of Design for Low-Cost Country Sourcing. The research goal of supporting production by developing and supplying applicable tools aims to make good use of the opportunities of design that is suitable for low cost sourcing. This concept represents a suitable basis for the adaptation of product construction to site- specific conditions. [10] Koehler categorises technological product against the backdrop of changes on products, processes, resources and cost effects [11]. He distinguishes between primary and secondary effects. Accordingly, the effects on products and processes are seen as inseparable.

22.2.2 Methods for Evaluating Investment Projects

Methods for evaluating investment projects are particularly suitable to assess the expected consequences of investments with regards to quantifiable interest [12]. The Net Present Value method (NPV method) and the decision tree process belong to the traditional procedures. Just as the NPV method, most of the evaluation methods are based on the assumption that forecasts will come true. So these evaluation processes ascertain the benefits of investments only under virtually safe expectations [13].

However, the decision tree process gives the investor the possibility to intervene during the course of the project and enables him to make decisions [1, 14]. By responding to risks and managing the project accordingly, the investor has the opportunity to influence the value of the project after the initial investment. Due to the fact of different levels for investment decisions, the investor gets the opportunities to take actions at each level [15]. In the evaluation different environmental statuses and their probability of occurrence are taken into account [16].

22.2.3 *Deficit in the State of Art*

The deficit of the state of the art is found in the areas of product adaptation and investment analyses under consideration of uncertainties and the flexibility to react. In regard of product adaptation, the above-mentioned authors refer to integrated product and process design, but do not mention any cause and effect relationships between possible adaptation measures and their consequences. Some of the causes can be found in insufficient project management and information asymmetries between different organizational company departments. It is exactly those uncertainties that are not taken into consideration for investment projects looking into product adaptations.

22.3 Integrated Product and Production Design

The following chapter aims to represent an approach to identifying the causal loops for the representation of the many different adaptation measures. The approach also defines how the causal loops can be assigned to the different phases of development.

22.3.1 *Identification of the Causal Loops of Product and Production Technology*

For adapted and cost-optimized products to be successful, integrated product design combined with the design of the accompanying production technology is indispensable. The integrated methodology is used to define the causal loops between product and production design that must be taken into consideration. Working with these causal loops is beneficial as all possible alternatives and their

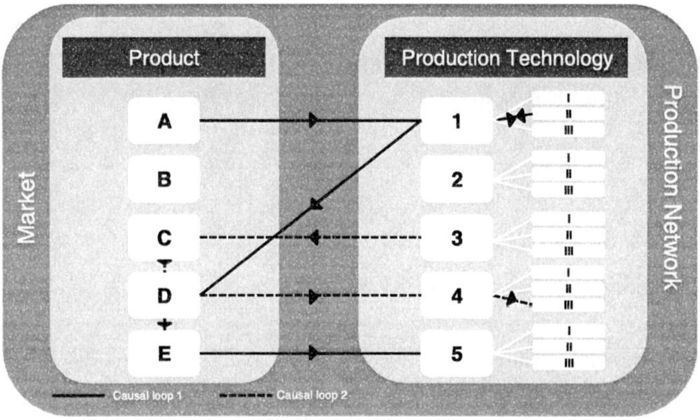

Fig. 22.2 The causal loops of product and production design

potential effects can be considered. For example, a change of the product design can require modifications of the production technology, as can be seen in Fig. 22.2. The modifications of the production technologies can have other consequences, too. Variables A–E, 1–5, i–iii are place markers for potential adaptations. For example, if the material is changed, it can be necessary to change the tool of the machine tool (production technology). This in turn, can affect the handling systems and the assembly processes. Modifications can be made to the shape, the material, the manufacturing process, the product mechanism, the way that quality is inspected, the joining processes, the handling systems and the assembly processes. Additional factors of influence and restrictions resulting from the production network and the market must be taken into consideration, too. The availability of adapted components within the production network affects the total cost, because higher logistics costs are incurred if a product is not produced locally. For each market, different customer requirements must be factored in. For product adaptations to be useful in the first place, it is indispensable to satisfy these customer requirements. In addition, any technology trends that a market has can require product modifications.

These causal loops must be expressed in monetary terms and must be analyzed through the use of binomial models to be able to evaluate the total adaptation expenses. The expression of monetary terms is necessary to quantify the investments which have to be made, if the option is carried out.

22.3.2 The Development Phases of Product Adaptation

New products are generally developed using specific standardized processes for each project. The evaluation methodology of product adaptations allows to take the different project phases into consideration. The product adaptation process can be subdivided into any number of individual process steps. The number of processes depends on the development process of the company and the phases it is composed of. The causal loops of the product adaptation measures can be assigned to the different phases. This step allows the flexibility to take action at different points along the process to be considered for the real options analysis.

22.4 Methodological Approach

This chapter introduces the methodological approach to build up the real option model. First of all, the net present value (NPV) is calculated. Secondly, the project volatility under consideration of uncertainty factors is identified. Finally, the decision trees could be built and the decision makers' options are integrated as a real option.

22.4.1 NPV Determination Excluding Flexibility Considerations

The first step of the method is based on a modified NPV method. It is geared towards identifying the cash flows of the product adaptation. The incoming payments consist in the expected revenue (savings) resulting from the successful adaptation of the product and the successful market launch. The revenues are calculated from the prices that can be asserted for the adapted product and from the volume of demand. It is also useful to know the cost structure of the product because any changes to individual cost items have an impact on revenue. The corporate success and the successful launch of a new product (or of an adapted product in our case) depends on the parts prices that are achievable in the market and on the total cost for the adaptation measures. In this context, adaptation costs include the manufacturing costs, and the additional development expenses. The manufacturing costs are subject to some uncertainties (i.e. overall demand, currency rate fluctuations, material price fluctuations). The factors can have positive or negative effects. For example, fluctuations of material prices, labour costs, exchange rates and prices can affect the manufacturing costs of the suppliers. The required project return is discounted on the cash flows of the revenue, and the cash flows are then added up.

$$\begin{aligned} NPV &= -I_0 - \sum_{t=1}^{T}\left(\prod_{k=0}^{t-1} p_k\right)\frac{I_t}{(1+r_f)^t} + \sum_{t=0}^{T}\left(\prod_{k=0}^{t} p_k\right)\frac{x_t\cdot(E(TO_t)-E(K_t))}{(1+i)^t} \\ &= -I_0 - \sum_{t=1}^{T}\left(\prod_{k=0}^{t-1} p_k\right)\frac{I_t}{(1+r_f)^t} + \sum_{t=0}^{T}\left(\prod_{k=0}^{t} p_k\right)\frac{E(CF_t)}{(1+i)^t} \end{aligned} \tag{22.1}$$

NPV = Net Present Value
I_t = Investment expenditure in period t
x_t = Supply quantity
t = Periods (0–T)
T = Planning duration (number of periods)
CF = Cash flow or savings in period t
TO_t = Turnover of the adapted product
K_t = Variable costs of the adapted product
E() = Expected value
p_k = Success probability of period k
i = Discount rate (capital cost, WACC)
ix_t = Risk-free interest

The products are broken down into their cost structures. This allows to determine the advantages of each of two adaptation options. These two adaptation options are the result of two different causal loops. The cost structure includes material cost, labour cost, overheads and transport cost. Furthermore, the duration of the adaptation project, the calculated demand volume and the project margin are defined. First, the net value of the savings is calculated in the form of contribution margins. The expected contribution margins for each project phase are multiplied by the value of success probability p_k in order to identify the expected value of the contribution margins. Then, the investment quantity of the respective project phase is subtracted from the expected contribution margins. For the calculation of the net present value of the total project, the project return is discounted on the net values of the different project phases to get the figures for the time of the project start, which are then added up. A positive net present value is considered beneficial. The disadvantage of this calculation method is that it neither provides for any alternatives nor takes into consideration the uncertainties in the form of fluctuating input parameters. In spite of that, the calculations must be performed for the identification of cash flow volatility.

22.4.2 Identifying Project Volatility Under Consideration of the Uncertainty Factors

The next step is aimed at identifying the binomial tree while taking the uncertainties of the base value of the present value of the real options into consideration. The base value is the present value of the savings of each period.

The uncertainties can differ widely for the adaptation projects. Therefore, the uncertainties should be identified by the decision-makers prior to an investment project for the specific situation. For example, labour costs will have a bigger impact if a design change requires a higher level of manual labour than if it leads to more automation. The different uncertainties are modeled based on a Monte Carlo Simulation in order to integrate them to the binomial tree. Cox, Ross and Rubenstein have developed the formulae for the positive (u) and the negative (d) value changes per period in the form of a binomial tree. The binomial tree for the present value of the project is created based on the calculated volatility σ of the total project using the formulae of the model developed by Cox, Ross and Rubinstein. [17]

A Monte Carlo simulation is performed in order to bundle the different sources of uncertainties in the form of one variable, i.e. the volatility of the present value of the savings. For each uncertainty variable, a simulation run is performed for the entire project duration, meaning that a new present value of savings will result from each run. Afterwards, the volatility of the project return is determined identifying the continuous return of present value z. Each uncertainty factor equals a random variable. Distribution assumptions are made for each random variable and each simulation run generates estimated present values. Out of it the distribution assumption of the project return is made. The standard deviation of this project return is equivalent to the volatility of the investment project. Based on the assumption that the volatility is constant over the project duration, the volatility could be used at different instant of times. The project return is established based on the quotient of two successive present values of the project [18, 19].

$$z = \ln\left(\frac{PV_1 + FCF_1}{PV_0}\right) \tag{22.2}$$

PV_t = Present value of total savings at end of period
T = Number of observations with t = 0…T
FCF = Free cash flow
z = Return on savings

Consequently, present value PV_0 will either take the value uPV_0 or dPV_0. The probability of an upward development of the present value is equal to p, whereas the probability of a downward development is equal to $1 - p$. This yields a multi-period binomial model.

The binomial tree of product adaptations is an approximated representation of the causal loops. The different paths already include the different costs that would be incurred for the implementation of a specific measure. This makes it possible to integrate the adaptation costs that are incurred into Copeland's method for the calculation of real options.

22.4.3 Identifying the Project Return Under Consideration of the Reaction Flexibility

According to the approach taken by Copeland, the next step would be the creation of a decision tree. The reaction flexibilities would be integrated into the tree. This is a difference to the binomial tree that did not include such flexibilities [18].

Prior to each new project adaptation phase, which can also be considered an investment phase, the decision-maker is presented with several options to choose from. He has the possibility to abort the project if the development of the value of the savings is not beneficial. This would save him from making all future investments into that project. For instance, investment into a machine that is part of a causal loop can be saved. Another option would be to switch to an alternative way of producing the part or to delay certain processes. A project that consists of different pre-defined stages which are implemented one after the other is referred to a compound option. This means that the success of each investment stage depends on the success of the previous stage. If an option is exercised at the beginning of the project, another option is generated for the next period [19]. For the following option to be exercised it is indispensable that the savings exceed the investment made in that phase. The possibility for the decision-maker to exercise the option represents his flexibility to interfere at each stage of the project. If those options are taken into consideration, the decision tree can be created on the basis of the binomial model, and then the project can be evaluated under consideration of the flexibility.

The roll-back method is used to identify the real option value. The evaluation begins with the project end and then gradually covers the different stages one after the other until the start of the project.

The value that the options have at each node of the decision tree is calculated using the following formula [10]:

$$C_t^{u^i d^j} = \max\left[0; P_t \cdot \left(u^i d^j \cdot PV(FCF_t) + \left(\frac{p \cdot C_{t+1}^{u^{i+1} d^j} + (1-p) \cdot C_{t+1}^{u^i d^{j+1}}}{1 + r_f}\right)\right) - I_t\right] \tag{22.3}$$

$C_t^{u^i d^j}$ = Option value in period t after i up and j down movements
P = Risk-neutral probability for up movement
(1 − p) = Risk-neutral probability for down movement
U = Positive change in value
D = Negative change in value
FCF = Savings in t
i = Number of up movements with i + j = t
j = Number of down movements with i + j = t
t = Period with t = 0…T
r_t = Risk-free interest
$PV(FCF_t)$ = PV of savings from period t
p_t = Success probability of period t
I_t = Investment in period t
r_t = Risk-free interest

Finally, the results of the formulae represent the option value, which is the decisive factor to make the next adaptation step or abandon the project. The decision is made out of the ratio of the total of savings of an investment period and the investment required. If the total of savings (weighted with its success probability and the risk-neutral expected value) exceed the investments, the option is exercised, otherwise the project won't be continued.

22.5 Summary

Without doubt, the progression of globalization offers big opportunities to companies that are active at the global level. Competitiveness can be increased or maintained by reducing costs, creating new and innovative products and adapting products to local customer requirements. These local customer requirements differ between different regional markets. Corporate decision-makers develop strategies which do not only satisfy the market demand with the existing product portfolio but which are also targeted at the adaptation of products to local needs. The approach presented in this article looks into the complexity of the way in which products are adapted, the cause/effect relationships of the adaptations and the profitability of these adaptations. The real options analysis is used to evaluate the benefits of an investment project. The adaptation options are first analyzed on the basis of literature research, and are then transferred to causal loops. Finally, they are translated into the methodology of the real options analysis.

We extend our sincere thanks to the Deutsche Forschungsgemeinschaft (German Research Foundation) for supporting this research project LA 2351/11-1 "Wirtschaftliche Gestaltung eines Lieferantenaufbaus" ("Economical Designing of Supplier Qualification").

References

1. Vietor R (2007) How countries compete: strategy, structure, and government in the global economy. Harvard Business School Press, Boston
2. Reinhart G, von Bredow M (2006) Methoden zur Gestaltung und Optimierung von Wertschöpfungsnetzwerken, wt—werkstattstechnik online, Jahrg 7/8:561–565
3. Eversheim W, Schuh G (2005) Integrierte Produkt- und Prozessgestaltung. Springer, Heidelberg
4. Hesmer A, Duin H, Thoben KD (2010) Towards a guideline for the early stage of product development. Proceedings of the 20th CIRP design conference, Nantes, pp 83–91
5. Friedl G (2006) Real options and investment incentives. Springer, Heidelberg
6. Abele E (2008) Global production: a handbook for strategy and implementation. Springer, Berlin
7. Hurschler P, Boutellier R (2008) Design to supply chain: central versus decentral decision taking. International conference on value chain sustainability in Izmir
8. Reinecke M, Nickel R, Overmeyer L (2007) Redesigning products for global competition. In: Proceedings of DET2007: 4th international conference on digital enterprise technology, 19th–21st September, Bath, United Kingdom, pp 160–168
9. Große-Heitmeyer V (2006) Globalisierungsgerechte Produktstrukturierung auf Basis technologischer Kernkompetenzen. IFA J 4

10. Lanza G, Weiler, Vogt S (2009) Design for low-cost country sourcing: motivation, basic principles and design guidelines. Proceedings of the 19th CIRP design conference, Cranfield, pp 229–236
11. Köhler C (2009) Technische Produktänderungen –Analyse und Beurteilung von Lösungsmöglichkeiten auf Basis einer Erweiterung des CPM/PDD-Ansatzes. Universität des Saarlandes, Saarbrücken
12. Weiler S (2010) Strategien zur wirtschaftlichen Gestaltung der globalen Beschaffung. Shaker, Aachen
13. Franke G, Hax H (2004) Finanzwirtschaft des Unternehmens und Kapitalmarkt. Springer, Berlin
14. Hommel U, Lehmann H (2001) Bewertung mit dem Realoptionsansatz—Ein Methodenüberblick. In: Hommel U, Scholich M, Vollrath R (eds) Realoptionen in der Unternehmenspraxis. Springer, Berlin
15. Meise F (1998) Realoptionen als Investitionskalkül: Bewertung von Investitionen unter Unsicherheit. Oldenbourg, München
16. Trigeorgis L (1997) Real options—managerial flexibility and strategy in resource allocation. 2. Aufl. The MIT Press, Cambridge
17. Cox JC, Ross SA, Rubinstein M (1979) Option pricing: a simplified approach. J Financ Econ 7:229–263
18. Copeland T, Antikarov V (2001) Real options a practitioner's guide. Texere, New York
19. Herath H, Park C (2002) Multi-stage capital investment opportunities as compound real Options. Eng Econ 47(1):1–27

Chapter 23
Clustering Regional-Specific Requirements as a Methodology to Define the Modules of a Car Concept

Frank Nehuis, Marcel Ibe, Carsten Stechert, Thomas Vietor and Andreas Rausch

Abstract The increasing globalization in the automotive industry creates worldwide new target groups. Because of that, regional-specific requirements have to be considered in the vehicle development process. This increases the complexity in conception of the vehicle's design. In order to create vehicles economically the depth of production must not go up disproportionately. To develop systematically various factors in the vicinity of a vehicle and further to demonstrate them, in relation to the technical characteristics of a vehicle concept, a model was created in the early stages of the developing process. Finally the characteristics will be clustered to so-called 'prototypes', which also summarize required similar characteristics from the customer. In this case the prototypes represent a compromise between the large numbers of regional-specific alternatives and less economical ones.

23.1 Introduction

In the long term, worldwide over 70 % of people will be living in cities as a result of globalization. The simultaneous need to individual mobility creates regional-specific, ecological and traffic related problems, which has become a factor limiting the city development. Car manufacturers pursue the need to individual mobility of customers, in realizing a large number of different derivatives. This results in

F. Nehuis (✉) · C. Stechert · T. Vietor
Institute for Engineering Design, Langer Kamp 8, 38106 Braunschweig, Germany
e-mail: f.nehuis@tu-braunschweig.de

M. Ibe · A. Rausch
Department of Informatics—Software Systems Engineering, Clausthal University of Technology, Julius-Albert-Street 4, 38678 Clausthal-Zellerfeld, Germany
e-mail: marcel.ibe@tu-clausthal.de

A. Chakrabarti (ed.), *CIRP Design 2012*, DOI: 10.1007/978-1-4471-4507-3_23,

platforms on which several different vehicle alternatives are based on, developed by diverse customized derivatives, [1]. One example for this platform strategy is the PQ35/A5 platform from the undertaking Volkswagen. This platform was used among other things to manufacture the models VW Golf, VW Touran and Audi TT. In addition, unified interfaces have been defined within the platform in order to exchange individual modules. An example for a consistent module is the vehicle's engine, which was used in 10 different alternatives both in the Golf and the Touran. The reduced number of components and the resulting lower cost are the advantages of the platform strategy. Furthermore, the automotive industry tries to sell their portfolio of vehicles in rapidly expanding markets, such as in the BRIC countries. In order to develop a vehicle, consequently, not only the individual customer needs have to be taken into account by defining the platform and the modules, but also the regional-specific ones. As a general rule, regional-specific requirements, e.g. different customer needs, legal regulations, or environmental conditions compared to the European market have not been taken into account when the existing vehicle models were developed, [1]. Thus, for example by climatic conditions in India (dust, high temperature and humidity) squeaky sounds occur when opening the side windows. Such a phenomenon can be attributed to the friction combination between door and door seal, which is influenced differently by the environmental conditions. That is why an unmodified introduction of existing vehicle models in other regions is not possible without a loss of quality. If a car manufacturer wants to maintain his high standard of quality also in other regions, the regional-specific requirements have to be worked out and must be included into the vehicle platform and the modules. At the same time the number of alternatives has to be as low as possible, in order to create an economic viable solution. Therefore, particularly succinct differences in requirements, so-called 'spreads of requirements' have to be identified. Afterwards this spreads of requirements can be clustered and single requirements can be identified.

23.2 State of Research

23.2.1 Requirements in the Product Development Process

A key step to develop a product with a complex structure is to define the requirements of the system or individual objects of the system, in order to fulfill customer's needs and the production environment. The established requirements will be subsequently documented in a list of requirements, [2]. At the beginning of the methodological development process, the task has to be clarified immediately after the task has been formulated. In the course of the developing process, the product to be created has to be concretized. In this way, other information can be inferred and included into the list of requirements, [3]. However, requirements can change during the development process. That is the case, for example, if the fulfilling of a requirement is checked only when production knowledge increases or certain framework conditions and restrictions are not known, [4, 5]. If potential solutions, because of changing requirements, are only excluded late from development process, development resources were used unnecessarily, so that development costs and times rise. Therefore, product requirements have to be determined as early as possible and be secured by information, in order to prevent any changes. For this to happen, methodologies and tools have to be used, which dispose required information of the developing process at an early stage, [6]. Regarding the surrounding of a technical product e.g. checklists can be used to get necessary information about the target region as complete as possible. This enables to fix technical product requirements, which decreases the number of potential changes in later stages of the development process, [7].

23.2.2 Effect of Spreads in Requirements on the Vehicle Development

The term spreads of requirements means different forms of a single requirement, which may occur through various scenarios, [8–10]. Spreads of requirements happen, e.g. if the environmental conditions in which the technical system is used change in different regions. In that case, the requirement has to be changed, for reacting to the varying conditions and restrictions in a region.

If spreads of requirements are identified in a technical system, it has to be considered, from which degree of deviation the differences in the product development should be taken into account. Particularly succinct spreads of requirements, which have to be considered during the development, are collected in a common target system, [11]. In the automotive industry, spreads of requirements within a target system are compensated by varying vehicle derivatives or modules. To consider as much as possible spreads of requirements, because of different customer and regional-specific needs, the highest possible number of different alternatives should be proposed to the customer. On the other side there must be a sufficiently small number of individual alternatives and modules, in order to offer the customer a reasonable vehicle. This means balancing, in which extent single spreads of requirements have to be taken into consideration, during the design of the vehicle concept, [11]. Should no additional alternatives be built, as a result of a minor difference, is it necessary to make a compromise to specify the requirement on the vehicle. For this purpose, the solution space of the requirements is examined and clusters of requirements are made. Therefore, the clustering shows a compromise between the number of variants and economical realization.

23.2.3 Clustering

Clustering is the assignment of data to groups, so-called clusters. The intention is to find such an assignment, that all data within one cluster are similar as possible. To determine which data contains to which cluster, a prototype is defined that represents the cluster. This procedure is called prototype based clustering. One data point is assigned to that prototype that has the lowest distance to the data point. Thereby the definition of the prototype is essential for the quality of the clustering. In the past neural networks have proven highly effective for prototype based clustering. An easy and intuitive approach is the learning vector quantization, presented by Kohonen [12]. A neuron [13] is defined by:

- the input dimension $n \in \mathbb{N}$
- a weight vector $w \in \mathbb{R}^n$
- a threshold/bias $\theta \in \mathbb{R}$
- and an activation function $f : \mathbb{R} \to \mathbb{R}$

When a neuron gets an input $x \in \mathbb{R}^n$, the output is computed with the aid of the weight vector, the bias and the activation function.

A neural network [13] is a tuple $\mathcal{N} = (N, \to, w, \theta, f, I, O)$ with:

- the neurons $N = \{1, \ldots, n\}$
- the network structure $\to \subset N \times N$
- the weights $w = (w_{ij})_{i \to j} (w_{ij} \in \mathbb{R})$
- the thresholds/biases $\theta = (\theta_i)_{i \in \mathbb{N}} (\theta_i \in \mathbb{R})$
- the activation functions $f = (f_i : \mathbb{R} \to \mathbb{R})_{i \in \mathbb{N}}$
- the input neurons $I \subset N$
- the output neurons $O \subset N$

Within a neural network all input neurons I get an input. Their output is used as input for the neurons that are associated with them. The outputs of a neural network are the outputs of all output neurons O. To perform a clustering with the help of a neural network it has to be trained. The inputs for the training are the data points that need to be clustered. The output of the network is compared to a given result. Depending on this, the weights and biases of the neurons will be adapted. The training of a neural network by learning vector quantization (LVQ) is according to the following algorithm:

1. initialize a fixed number of prototypes w, assign each data point with its nearest prototype and move the prototypes into the balance point of the data points that are assigned to it
2. choose a data point x
3. adapt the nearest prototype w

   (a) if x is assigned to w, set $w := w + \eta(x - w)$
   (b) else, set $w := w - \eta(x - w)$

4. repeat from 2. As long as the changes are big enough

$\eta \in [0, 1]$ denotes the learning rate and can either be chosen constant or can be adjusted between the training. There are several extensions for LVQ that improve the behavior of the algorithm in specific situations, [14–16].

23.3 Clustering Regional-Specific Characteristics

23.3.1 Technical Characteristics of a Vehicle Concept

The vehicle concept is described as a structural design, which contains the fundamental development idea and verifies the reliability, regarding to the topological arrangement, so-called package. In this case, the vehicle concept includes the major elements and properties of the vehicle. Based on that, first concept solutions can be developed and a decision for or against a concept solution can be found. Based on an already known predecessor model the requirements can be defined (adapting construction). In this case, the general structural constitution of the vehicle concept and the associated requirements of a vehicle e.g. the determination of the vehicle shape or the choice of the drive concept remain unchanged. In order to structure requirements of a vehicle superior fields can be fixed, [17]. Depending on the degree of detail other categories can be subordinated to these fields. The following list shows a possible structure to describe technical requirements with superior fields:

- main dimensions
- car body
- basic design
- drive train design
- weights
- safety and ergonomics
- technique
- environment

To take the influences of the environment with defining the requirements into account, first it is sufficient to define the technical requirements in less detail. This is an approach, which is usual in the first steps of a vehicle developing process and makes the established structure easy to grasp. If the level of detail is increased, the complexity of the associated vehicle concept increases, too. As a rule, only the qualitative links of the approach are refined, [18].

23.3.2 Impact From the Environment on the Technical Characteristics of a Vehicle Concept

In the environment of a vehicle and throughout the production cycle, a large number of influencing factors can have an effect on a vehicle. These include obvious influences, like infrastructure and legislation, but also factors like the individuality of the vehicle's owner or the culture of a region, [1]. Some influencing factors are seldom taken into account in the conception of a vehicle, because they give only little information about the influence on technical

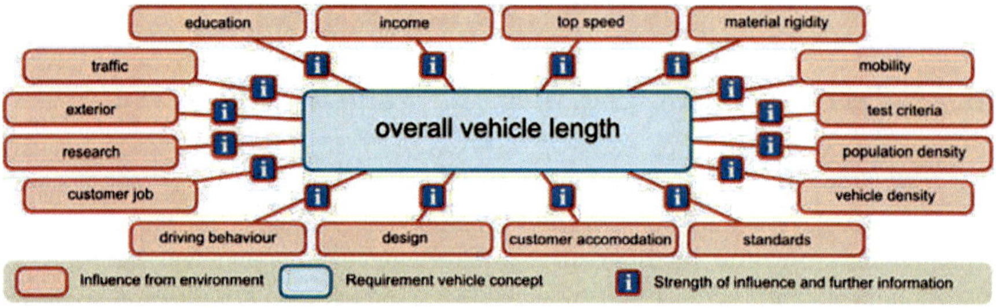

Fig. 23.1 Representation of the influencing factors out of the environment, which are affection the overall length of a vehicle

requirements. In order to develop a regional-specific vehicle, it is necessary to identify the influencing factors from the vehicle's environment, in the early steps of the developing process and show these effects on a vehicle concept. To identify the influence of a vehicle's environment on the concept, influencing factors have been identified first. Furthermore, the regional-specific differences have been analyzed. Afterwards a comparison in a relation system is made between the worked out influencing factors and the technical requirements. The particular influence is shown by edges. Because of the resulting model's complexity the relation system is documented by the editor "Artisan Studio" in the System Modeling Language (SysML). This enables a continuously expansion of the system with additional information. Moreover, the relational system can be analyzed specifically to the wishes of the engineer by the implementation of the software. For example, it is possible to show for one requirement all influencing factors, [19]. This is illustrated as an example in Fig. 23.1, for the overall vehicle length of a vehicle.

Figure 23.1 shows that obvious influencing factors, e.g. the volume of traffic in a region, have influences on the overall vehicle length. What makes the approach challenging is that difficult to consider factors, like the customer's driving characteristics, have also important influence. In addition to the influencing factors, which have an influence on the requirements, additional information about the influencing factors, e.g. the degree of influence or detailing explanations, can be put at disposal when designing a regional-specific vehicle. There is also the possibility to document, information about the regional-specific requirements of particular target regions, in the relation system. So regionally identified information about the volume of traffic can be allocated to the influencing factor traffic. In order to illustrate regional-specific influencing factors on a vehicle, Fig. 23.2 compares the best-selling vehicles in the countries Germany, India and the United States.

It can be seen in Fig. 23.2 that the best-selling vehicles in the Unites States are significantly longer than German ones. On the other hand, the German vehicles are longer than the best-selling Indian vehicles. If e.g. a vehicle concept has to be developed for all this three markets, is it to consider, which overall vehicle length appeals the largest target group. Because of the influences from the vehicle's

Germany: vehicle	length [in mm]	India: vehicle	length [in mm]	USA: vehicle	length [in mm]
1 VW Golf	4199	1 Maruti Alto	3620	1 Ford F-Serie	5415-6360
2 Audi A4	4703	2 Maruti WagonR	3595	2 Chevrolet Silverado	5222-6329

Fig. 23.2 Benchmark of the best-selling vehicles in the countries Germany, India and the United States in the year 2008

environment e.g. it can be concluded that a large number of Indian customers buy vehicles with a shorter length. Against this, customers in the United States have rather the possibility to drive vehicles with a long length, because of the environmental conditions. Taking a small number of variants of the overall vehicle length during the vehicle conception as an example, the overall length can be clustered, in order to get an economically acceptable solution in designing the vehicle concept for all three target markets.

23.3.3 Clustering Regional-Specific Requirements

Within the clustering of lengths of cars efforts are made to define the clusters in a way that they contain as most as possibly similar lengths. Every cluster correlates with one alternative that has to be developed. To be as economical as possible, the number of clusters should be rather small. The results of a prototype based clustering are prototypes that suggest values for the length of the cars. Now a clustering with two and three prototypes is executed with the values from Fig. 23.2. For this the LVQ approach is used. Because of the fact, that the Ford F-Series and the Chevrolet Silverado ranges of lengths are given, the smallest and the largest values are used. The result is a set $\begin{matrix} W = \{4,199,\ 4,703,\ 4,502,\ 3,620,\ 3,595,\ 3,760, \\ 5,415,\ 6,360,\ 5,222,\ 6,329, 4,805\} \end{matrix}$ of lengths.

23.3.3.1 Clustering with Two Prototypes

Both prototypes w_1, w_2 will be initialized with weights of 4,000 and 6,000. These weights stand for the lengths of the alternatives. Now all the data points will be assigned to the prototype nearest to them. The result is the following assignments (Table 23.1):

Table 23.1 Assignment between car length and two prototypes

Car [length (mm)]	Prototype (weight)
VW Golf (4,199)	w_1 (4,000)
Audi A4 (4,703)	w_1 (4,000)
Mercedes C220 (4,501)	w_1 (4,000)
Maruti Alto (3,620)	w_1 (4,000)
Maruti WagonR (3,595)	w_1 (4,000)
Maruti Swift (3,760)	w_1 (4,000)
Ford F-Series (min) (5,415)	w_2 (6,000)
Ford F-Series (max) (6,360)	w_2 (6,000)
Chevrolet Silverado (min) (5,222)	w_2 (6,000)
Chevrolet Silverado (max) (6,329)	w_2 (6,000)
Toyota Camry (4,805)	w_1 (4,000)

Now the LVQ algorithm chooses randomly data points and updates the weights of the prototypes. Here a fixed number of updates will be executed. The learning rate η is initialized with a value of 0, 1 and be increased by $1/10,000$ per step. Starting from these values the following prototypes were computed:

- $W_1 = 4181$
- $W_2 = 5840$

The clustering with two prototypes suggests the lengths 4,181 and 5,840 mm.

23.3.3.2 Clustering with Three Prototypes

The three prototypes w_1, w_2, w_3 will be initialized with weights of 3,500, 4,500 and 6,000. Now the following assignment occurs (Table 23.2):

The parameters were chosen identically to the case with two prototypes. After the executing of the LVQ algorithm the prototypes had the following values:

- W_1 = 3,770
- W_2 = 4,759
- W_3 = 6,229

The suggested lengths are 3,770, 4,759 and 6,229 mm.

These calculated values are proposals. The quality of them depends on several parameters. So they should not be seen as finally but can still be modified.

23.4 Conclusion

With the help of the developed system, for systematically identification of environmental influencing factors on a vehicle concept, regional-specific differences and the resulting requirements on the collective of requirements can be shown. This implies spreads in the collective of requirements. The consideration of these

Table 23.2 Assignments between car lengths and three prototypes

Car [length (mm)]	Prototype (weight)
VW Golf (4,199)	w_2 (4,500)
Audi A4 (4,703)	w_2 (4,500)
Mercedes C220 (4,501)	w_2 (4,500)
Maruti Alto (3,620)	w_1 (3,500)
Maruti WagonR (3,595)	w_1 (3,500)
Maruti Swift (3,760)	w_1 (3,500)
Ford F-Series (min) (5,415)	w_3 (6,000)
Ford F-Series (max) (6,360)	w_3 (6,000)
Chevrolet Silverado (min) (5,222)	w_2 (4,500)
Chevrolet Silverado (max) (6,329)	w_3 (6,000)
Toyota Camry (4,805)	w_2 (4,500)

spreads is especially necessary in the first steps of the developing process, because central requirements are fixed already at that time. At the same time, the depth of production cannot rise disproportionately, in order to develop economical vehicle variants. Therefore, spreads of requirements are clustered to prototypes with learning vector quantization.

The aim of the clustering is to consider the regional-specific requirements and simultaneously to keep the number of developing variants as low as possible. Furthermore, clustering of requirements allows taking several requirements simultaneously and the dependence between the characteristics into account. This allows to determine combinations of spreads which form the basis to support the engineer by identifying the vehicle concept.

References

1. Wallentowitz H, Freialdenhoven A, Olschewski I (2009) Strategien in der Automobilindustrie. Technologietrends und Marktentwicklungen. Vieweg + Teubner, Wiesbaden. ISBN 978-3-834807250
2. Pahl G, Beitz G, Feldhusen J, Grote KH (2007) Engineering design—a systematic approach, 3rd edn. Springer, Berlin. ISBN 978-1-84628-318-5
3. VDI (1993) Systematic approach to the development and design of technical systems and products. VDI-Gesellschaft Produkt- und Prozessgestaltung, Düsseldorf
4. Stechert C, Franke H-J (2009) Managing requirements as the core of multidisciplinary product development. CIRP J Manuf Sci Technol 1(3):153–158
5. VanBeek T-J, Tomiyama T (2008) Requirements for complex systems modeling. In: The 18th CIRP Design conference—design synthesis, p 32
6. Salustri F-A, Eng N-L, Weerasinghe J-S (2008) Visualizing information in the early stages of engineering design. Comp-Aided Des Appl 5(5):697–714
7. Ehrlenspiel K, Kiewert A, Lindemann U (2007) Cost-efficient design. Springer, Heidelberg. ISBN 3-540-34647-3
8. Renner I (2007) Methodische Unterstützung funktionsorientierter Baukastenentwicklung am Beispiel Automobil. Verlag Dr. Hut, Dissertation, Lehrstuhl für Produktentwicklung der Technischen Universität München, München. ISBN 978-3-89963-567-6

9. Schmitt J, Stechert C, Raatz A, Hesselbach J, Franke H-J, Vietor T (2009) Reconfigurable parallel kinematic structures for spreading requirements. In: Proceedings of IASTED—robotics and applications, Cambridge
10. Stechert C, Franke H-J (2009) Requirements models for modular products. In: Proceedings of international conference on research into design, Bangalore, India, pp 411–418
11. Ponn J (2011) Konzeptentwicklung und Gestaltung technischer Produkte - Systematisch von Anforderungen zu Konzepten und Gestaltlösungen, 2nd edn. Springer Verlag Berlin Heidelberg. ISBN 978-3-6422-0579-8
12. Kohonen T (1988) Self-organization and associative memory, 2nd edn. Springer, Berlin. ISBN 978-0387183145
13. Hammer B (2007) Neural computation, vorlesungsskript
14. Kohonen T (1990) The self-organizing map. In: Proceedings IEEE
15. Bojer T, Hammer B, Schunk D, Tluk K (2001) Relevance determination in learning vector quantization. In: Proceedings of European symposium of artificial neural networks (ESANN'01)
16. Hammer B, Villmann T (2002) Generalized relevance learning vector quantization. Neural Networks 15(8–9):1059–1068
17. Braess H–H (2007) Vieweg-Handbuch Kraftfahrzeugtechnik, 5th edn. Vieweg Verlag, Wiesbaden. ISBN 978-3-8348-0222-4
18. Prinz A (2011) Struktur und Ablaufmodell für das parametrische Entwerfen von Fahrzeugkonzepten, 1st edn. Logos Verlag, Berlin. ISBN 978-3-8325-2869-0
19. Nehuis F, Stechert C, Vietor T (2011) Werkzeugunterstützung bei der Ermittlung von Anforderungen für regionsspezifische Fahrzeugkonzepte. In: Proceedings of the 22nd DfX-symposium, München

Chapter 24
An Ontological Approach for the Integration of Life Cycle Assessment into Product Data Management Systems

H. Ostad-Ahmad-Ghorabi, T. Rahmani and D. Gerhard

Abstract The consideration of environmental aspects into product design is becoming more and more a key strategic feature of products. However, the inclusion of Life Cycle Assessment (LCA) and Eco design aspects in design is still far from being common practice. The complexity of the task, time-consumption and additional workload during design process may be some reasons therefore. There have been some efforts to integrate LCA in CAD systems and Product Data Management systems (PDM) to ease the handling of data and to assist the designer. However, these concepts suffer from inflexible data and database structures which are difficult to maintain and to update, in particular when complex products are designed in collaborative teams. This chapter presents an ontological approach to include environmental data in PDM systems. This approach reduces complexity of data structures, eases the use of environmental assessment methods and eases the sharing of product relevant information.

H. Ostad-Ahmad-Ghorabi (✉)
Magna Steyr Engineering Germany, Kolumbusstraße 1, 71063 Sindelfingen, Germany
e-mail: hesamedin@ostad.at

T. Rahmani · D. Gerhard
Vienna University of Technology, Institute for Engineering Design,
Getreidemarkt 9, 1060 Vienna, Austria
e-mail: touba.rahmani@tuwien.ac.at

D. Gerhard
e-mail: detlef.gerhard@tuwien.ac.at

H. Ostad-Ahmad-Ghorabi
SICENG GbR, Turmstraße 5, Holzgerlingen, 71088 Germany

A. Chakrabarti (ed.), *CIRP Design 2012*, DOI: 10.1007/978-1-4471-4507-3_24,

24.1 Introduction

PDM systems have become widely established in industry, in particular, when complex products have to be designed and design collaboration between different parties is required. To educate technically skilled employees for industry, technical and vocational schools play an important role in the Austrian market. To advance collaborative design projects among technical and vocational schools, Vienna University of Technology has created and maintains a PDM platform (within a project funded by Austrian federal government) [1]. This platform has been adapted to the specific needs of the technical schools to guarantee a smooth integration into the educational process of design. In a follow up project, the aspect of sustainable product development is further elaborated with the particular aim to integrate environmental considerations into the design education process and to raise an additional value of the established PDM platform.

To proceed in the improvement of the environmental performance of a product, it is necessary to evaluate its environmental performance first. This is however, a complex task, as it involves the handling of a huge amount of data, provoking additional workload in the design process. Design is already complex; including environmental consideration will make it even more complex. To integrate environmental considerations into the design process, Life Cycle Inventory Data (LCI) is added to the PDM system. This allows the students to focus on the environmental aspects of their design and to reduce the environmental impacts of their design over its entire life cycle. This enhanced PDM solution enables the technical schools to collaborate in design and to evaluate their design as to its environmental impacts. As a proof of concept, four different technical schools jointly designed an environmentally improved cordless drill driver. To ease the process of environmental evaluation, parameters needed for the evaluation are either retrieved from CAD files or manually defined in the PDM system. By using a reporting tool, data is linked to environmental impact indicators to calculate the occurring environmental impacts of a part, a component or a whole assembly. In order to improve the performance of the system, to reduce complexity, to be able to manage data structures, to repeat routines and to further ease the application of environmental evaluation methods, we propose an ontological approach. This chapter discusses how ontology was derived in the project and how it was linked to the CAD file structure in the PDM system and the environmental parameters.

24.2 State of the Art

The consideration of environmental aspects is becoming a key issue aspect in product development and product design. Many tools have been developed that help to track and evaluate the environmental impacts of a product over its entire life cycle. In any of the methods for reducing the environmental impacts of a product referred to as Eco design, Green Design, Design for Environment (DfE), Life Cycle Design (LCD) or Design for Sustainability (DfS), environmental information is necessary to start the environmental evaluation process. Life Cycle Assessment (LCA) is the most widespread assessment method for this purpose [2, 3]. However, there are some barriers that withhold LCA from being a common practice in industry, mainly because of data requirements [4], time-consumption or complexity. LCA requires much data that is constantly changing in the product development process, in particular in the early design stages. The early design stages would allow for the most efficient changes in the design, as they have the greatest potential for any improvement. But data in this stage may be unknown or fuzzy. Later in the development process, more data is available, but the potential for improvement is not so high, a phenomenon commonly known as the "design paradox". Databases such as the Eco invent database [5] shorten the time to gain first results of the environmental profile of a product. Ostad-Ahmad-Ghorabi et al. [6] parameterizes the product to extrapolate results for products that have similar features. Some other tools combine environmental databases with quick assessment tools, such as the Ecoindicator [7] or the tool Greenfly Online [8].

There have been some efforts to bring LCA into CAD systems [9]. Most of these approaches suffer from the lack of information in these early design stages. Furthermore, a CAD system is not the major source to gain comprehensive information about product structures and part information. Within the development process, more than one single CAD system may be used and products may contain more items than commonly modeled within CAD. Some approaches try to reduce the amount of parameters that have to be dealt with and try to deliver a good approximation of the pretended environmental impacts. Ostad-Ahmad-Ghorabi and Collado-Ruiz [10] developed a generic model to parameterize the LCA process out of parameters that are known during the initial definition of a product, specifically shown for knuckle-boom cranes. The parametric approach for knuckle-boom cranes enables the inclusion of LCA in early design stages. However, the development of the approach was faced with some drawbacks: first, the approach is much product specific. Second, the database used is difficult to maintain and to update. Third, the generic model cannot be easily transferred to another product. To face these problems, an ontological approach is proposed in this chapter.

By using ontology, a common understanding of different information structures can be achieved, knowledge can be reused and analyzed and general assumptions about a certain field of knowledge can be made [11]. Further, heterogeneous data can be integrated into ontology, which enables complex, semantic database queries. An ontology can therefore be regarded as an interface between databases and their environment.

Ontologies can be developed by using suitable software applications. The tool used in this project is Protégé. It allows the definition of classes and instances, interrelations can be defined and logical rules can be set [12]. Protégé can be used with any operating system.

24.3 Method

To develop an ontology to be integrated into a PDM system, the METHONTOLOGY approach, first developed by Fernandez et al. [13], was followed. The framework enables the development of ontologies at knowledge level. The method includes the identification of the ontology development process, a life cycle based on evolving prototypes, and particular techniques to carry out each activity [14]. The ontology life cycle considers the order in which the activities of the ontology development process should be performed. The METHONTOLOGY approach proposes the evolvement of prototypes for the development of ontologies. For the prototype, terms can be added, changed and removed throughout the evolution of the ontology.

The first step in the method is to define the specifications for the prototype. In parallel, knowledge acquisition is conducted. Once the first prototype has been specified, the construction of the conceptual model is built at the conceptualization phase. Then, the conceptualization, formalization and implementation of the knowledge are carried out. If some lack is detected in the ontology, the specification can be modified.

For the prototype in this chapter, the components of the ontology were developed by using the described METHONTOLOGY approach. Protégé is used to develop the ontology. The ontology is represented by using Web Ontology Language (OWL).

To proceed with Life Cycle Assessment, the basic approach would be to add environmental data to the CAD files stored in the PDM system. Some of these data can be derived from the CAD system; some have to be defined manually in the PDM interface. The additional data would then be stored somewhere in the databases of the system, which then can be accessed to retrieve an environmental profile.

For the case study of the gearbox, which is discussed in this chapter, the complexity of the conventional approach would already grow considerably since it consists of several parts and components, Although having a model at hand, the maintenance and update of this model is a time-consuming task; any change in the original data structure will require the establishment of a new, adapted model.

24.4 Development of a Prototype

To develop the ontology, a two-stage gearbox was taken as a case study. This gearbox has been designed by one of the cooperating vocational schools and is composed of several parts and components. The CAD files were stored in the PDM system.

To overcome the aforementioned obstacles, an ontology is developed for the gearbox, using the METHONTOLOGY approach and Protégé V3.2. The Protégé interface contains a series of Tabs that allow for the ontology to be applied to different areas. The OWL Viz Tab, included as a plug-in in Protégé, enables the class hierarchies in an OWL Ontology to be viewed and incrementally navigated, allowing comparison of the asserted class hierarchy and the inferred class hierarchy.

To include Life Cycle Assessment into the ontology, classes (concepts) were defined to be: *Life Cycle* (in order to assign life cycle stages to it), *General Information* (including all design parameters which can not be assigned directly to a particular life cycle of the gearbox), *Materials*, *Manufacture*, *Distribution*, *Use*, *End of Life*. Relations were defined by using terminologies such as: "consists of", "is used for determination", "determined by", "has effect on calculation", "has action", "influences" or "effected by". For example, the class Life Cycle *consists of* subclass Materials. Materials *has an action on* the design parameter Quality of Gear Stage (wheel and gear). Quality of Gear Stage *is effected by* Material Selection. Material Selection *has an effect* on Permissible Hertzian Stress which together with the Required Safety Factor *has an effect on calculation* of Existing Hertzian Stress. The Existing Hertzian Stress *is used for the determination* of the Module, which *is used for the determination of* the Dimension (of wheel and gear). Dimensions *influences* Weight. Weight in combination with the selected material can be directly linked to environmental data containing environmental indicator values for materials. This will already allow for the evaluation of the life cycle materials of the wheel and the gear.

The same logic is followed for all design parameters of the gearbox to retrieve those parameters, which according to the definition by Ostad-Ahmad-Ghorabi and Collado-Ruiz [10], are primary parameters. These are parameters that can be defined in the very early conceptual design stages and have a considerable influence on the environmental performance of the product. These parameters are shown in Table 24.1 for the case study of the gearbox.

Part of the final ontology is shown in Fig. 24.1.

24.5 Conclusion and Outlook

With the ontology derived in this chapter, environmental data can be integrated in early design stages and LCA is enabled. Design parameters can be retrieved from PDM systems and be interlinked with environmental indicators.

The ontology, however, has to be considered as a prototype with the intention to interlink technical design parameters known in early design stages with environmental data to establish an environmental profile of the product. Any change of a particular design parameter can easily be tracked as to its influence on the environmental profile. Of course, there are other approaches that may facilitate a similar output. The outcomes of this chapter can be understood as an intermediate step in a long-term research project. Although the ontology developed in this chapter is still specific to a particular product, it allows bringing together design and environmental considerations in a systematic way. To us, this approach is

Table 24.1 Primary parameters for the gearbox and their influence on other parameters in accordance to definition in [38]

Life cycle	Primary parameter	Has a link to/influences
General	Input power	E.g. Torque & weight of gear, weight of wheel, total weight of gearbox, etc...
	Revolution per minute input	E.g. Module
	Revolution per minute output	E.g. Forces on shaft
	Impact coefficient	E.g. Module
	Dynamic factor	E.g. Existing safety factor for wheel/gear
Materials	Quality of gear stage	E.g. Total weight of gearbox, Module, material selection, power dissipation, forces on shaft…
	Safety factor	E.g. Module
Manufacturing	Manufacturing sites	E.g. Consumption of resources for manufacturing, transportation distance, etc.…
Distribution	Type and weight of packaging	E.g. Total weight of component for distribution
	Transportation mode and distance	
Use	Amount of oil	E.g. Dimensions and total weight of gearbox
End of life	End of life processes	E.g. Resource consumption for end of life treatment

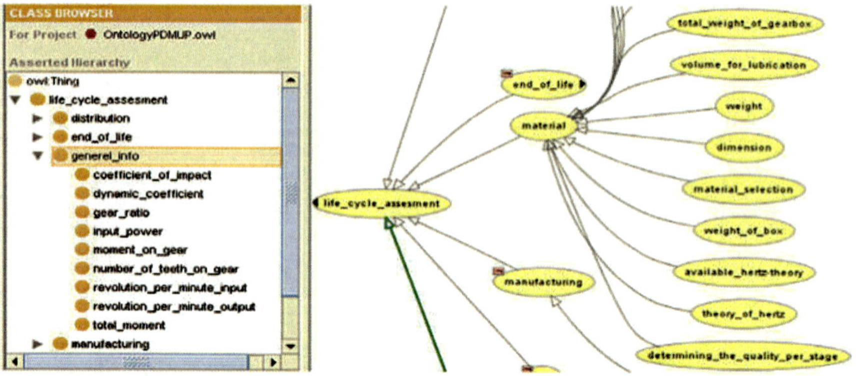

Fig. 24.1 Screenshot of part of the ontology derived with OWL Viz plug-in

considered to be an efficient one that brings LCA and PDM together in a user and administrator-friendly way.

The systematic approach will be used in our next research step to retrieve standardized ontology elements that can be merged together based on the characteristics of the product being analyzed. Collado-Ruiz and Ostad-Ahmad-Ghorabi have come up with definitions of similar or equivalent products in the scope of their LCP families [15]. In particular, they introduce a systematic approach to compare the environmental performances of products within the same family. A premise for their work is to have LCA results at hand, which may require an additional process of environmental evaluation. With the help of standardized ontology elements together with the LCP family approach, ontologies may be set up for similar products in a systematic and automatic way and environmental impacts of each life cycle stage of the product can be calculated out of this process.

This chapter showed that LCA and PDM can be brought together by using an ontological approach. Data from a PDM system was retrieved and different CAD systems were handled. The complexity of data structures was considerably reduced in comparison to conventional database relations. Also, it was possible to visualize results in a convenient way. All these aspects are important when it comes to successful implementation of environmental considerations in early design stages. Next steps in the research line include mapping of data in the PDM system with the ontology derived. This will help to derive a final list of independent and general (not product-specific) parameters and probably to optimize the amount of parameters involved. Further, the concept has to be optimized in order to allow starting the environmental evaluation process at any point in the ontology. The ontology shall automatically ask for all necessary information to calculate the environmental profile and give feedback to the engineering designer. Also, the acceptance of LCA with PDM systems in design education and also from an industrial point of view needs to be elaborated.

Acknowledgments The project discussed in this chapter has been funded by the Austrian Federal Ministry of Science and Research in the scope of the Sparkling Science research program.

References

1. Gerhard D, Grafinger M (2009) Integrative engineering design using product data management systems in education. In: Proceedings of the 11th international conference on engineering and product design education EPDE09, pp 79–84
2. Jeswiet J, Hauschild M (2005) Eco design and future environmental impacts. Mater Des (26)7:629–634
3. Ernzer M, Grüner C, Birkhofer H (2001) Implementation of DfE in the daily design work-an approach derived from surveys. In: Proceedings of 2001 ASME design engineering technical conference (DETC 2001)
4. Millet D, Bistagnino L, Lanzavecchia C, Camous R, Poldma T (2007) Does the potential of the use of LCA match the design team needs? J Cleaner Prod (15)4:335–346
5. Frichknecht R, Jungbluth N, Althaus HJ, G D, Dones R, Hecl T, Hellweg S, Hischier R, Nemecek T, Rebitzer G, Spielman N, Wernet G (2007) Eco invent overview and methodology. Data 2.0. Technical Report Swiss Centre for Life Cycle Inventories
6. Ostad Ahmad Ghorabi H, Bey N, Wimmer W (2008) Parametric eco design—An integrative approach for implementing eco design into decisive early design stages. In: Marjanovic D (ed) Proceedings of the design 2008—10th international design conference, pp 1327–1334
7. Goedkoop M, Spriensma R (2001) The eco-indicator 99: a damage oriented method for life cycle impact assessment. Methodology Report. PRi'e Consultants B.V

8. Horne R, Grant T, Verghese K (2009) Life cycle assessment: principles, practices and prospects. CSIRO Publishing, Australia
9. Ostad-Ahmad-Ghorabi H, Collado-Ruiz D, Wimmer W (2009) Towards integrating LCA in CAD. In: Norell M, Grimheden M, Skogstad P (eds) Proceedings of the international conference on engineering design. Stanford, vol 7. pp 301–310
10. Ostad-Ahmad-Ghorabi H, Collado-Ruiz D (2011) Tool for the environmental assessment of cranes based on parameterization. Int J Life Cycle Assess (16)5:392–400
11. O'Leary DE (1998) Using AI in knowledge management: knowledge bases and ontologies. IEEE Intell Syst 13:334–339
12. Musen M, NoyN, O'Connor M, Redmond T, Rubin D, Tu S, Tudorache T, Vendetti J (2008) The Protégé Ontology editor and knowledge acquisition system/stanford center for biomedical informatics research, stanford university. 2008.—Internetseite. http://protege.stanford.edu/. Accessed 20 Feb 2008
13. Fernández M, Gómez-Pérez A, Juristo N (1997) METHONTOLOGY: From ontological art towards ontological engineering. Symposium on ontological engineering of AAAI. Stanford (California)
14. Gómez-Pérez A, Rojas MD (1999) Ontological reengineering and reuse. International conference on knowledge engineering and knowledge management (EKAW), pp 139–156
15. Collado-Ruiz D, Ostad-Ahmad-Ghorabi H (2010) Comparing LCA results out of competing products: developing reference ranges from a product family approach. J Cleaner Prod 18:355–364

Chapter 25
Representation, Presentation and Visualization of Uncertainty

Reiner Anderl, Michael Maurer, Thomas Rollmann and André Sprenger

Abstract Uncertainty in the design process impacts performance parameters of products which are a major problem in the development of highly optimized load bearing systems. Typically, the information flow in the design process is a one way process from the design to subsequent product life cycle phases. Uncertainty about product properties leads to less economic and less safe products. In this paper an innovative approach to enable modeling and communication of uncertain properties along the product life cycle and the sophisticated presentation and visualization of these uncertainties is proposed. This approach is based on a three layer concept including a representation layer, a presentation layer and a visualization layer. The approach enables the visualization of a broad spectrum of process and product properties with respect to different specifications of each property e.g. single values, intervals, fuzziness and stochastic measures which may occur along the product life cycle.

R. Anderl (✉) · M. Maurer · T. Rollmann · A. Sprenger
Department of Computer Integrated Design (DiK), TU Darmstadt,
Petersenstr. 30, 64287 Darmstadt, Germany
e-mail: anderl@dik.tu-darmstadt.de

M. Maurer
e-mail: maurer@dik.tu-darmstadt.de

T. Rollmann
e-mail: rollmann@dik.tu-darmstadt.de

A. Sprenger
e-mail: sprenger@dik.tu-darmstadt.de

A. Chakrabarti (ed.), *CIRP Design 2012*, DOI: 10.1007/978-1-4471-4507-3_25,

25.1 Introduction

The development process of a load carrying structure is a knowledge intensive task. The geometric dimensioning of such a product requires engineering knowledge about product's stress and strength during product operation. In the product development process values for product properties are defined based on engineering methods and by using engineering tools. Product property values are, however, also determined by the manufacturing process and may even change during usage processes. These property values scatter due to process deviations which typically are known to the operator of the machine tool but not to development engineers and designers. Thus, available knowledge about processes and its exchange between stakeholders is essential for a successful product development.

Nowadays, lots of information is collected with different methods like e.g. quality methods, process capability or usage monitoring. They all deal with uncertainty, but they have different specific descriptions to capture uncertainty. Furthermore these uncertain product and process property values are related to each other within different life cycle phases [1]. There is a lack of understanding and controlling uncertainty in virtual product development today. Today's approaches focus only on geometric deviation and its inspection. In [2] an approach for the representation and visualization of geometric inspection data is presented. A holistic view on general properties, their deviations and relationships does not exist today.

There is a need for a uniform representation of uncertainty. Due to its complex nature, it is necessary to provide special presentation and visualization techniques. To represent knowledge about uncertainty an ontology-based approach for an information model is used and extended in for this purpose. This model allows the detailed representation of uncertain property values. It is also capable to describe relationships between properties. For the presentation of uncertainty sophisticated concepts are necessary in order to cope with the diversity of product and process properties. The differences in particular between geometric and non-geometric properties as well as differences between product and process properties require new presentation methods for the design of safe and economic products. For visualization new approaches are necessary to make engineers and designers understand uncertainty with respect to the product and process properties. Advanced design tools like Computer Aided Design (CAD) do not offer such capabilities. Other specialized tools like e.g. Computer Aided Tolerancing (CAT) focus mainly on geometric dimensioning, and do not provide a generalized approach for uncertainty representation presentation and visualization.

25.2 Uncertainty in Virtual Product Development

Virtual product development is based on the application of computer based methods and tools in product development processes.

While computer based methods and tools are advanced and powerful, there exists a fundamental issue on engineering data. This issue results from using typically numerically exact values in modeling, analysis, simulation and optimization as well as propagating these numerically exact values along successive processes (e.g. from design to manufacture). Due to this fundamental issue the German Research Community established a Collaborative Research Center (CRC 805) at Technische Universität Darmstadt entitled "Control of Uncertainties in Load-Carrying Structures in Mechanical Engineering". Within this Collaborative Research Center a new approach for representing, presenting and visualizing uncertainties has been developed.

25.3 The 3-Layer Approach

Uncertainty impacts virtual product development significantly but instead of some awareness by development engineers and designers computer based methods and tools only rarely take this issue into account.

To consider this issue a new approach has been developed consisting of a 3-layer architecture. The 3 layers are:

- The representation layer
- The presentation layer
- The visualization layer.

The representation layer is dedicated to the computer processable representation of uncertainty, integrated into the constructs of an appropriate information model. Thus a formal specification of uncertainty is provided and conceptually described within the ontology based information model. The presentation layer creates use-case defined objects which typically do not exist in a product or process model but which are necessary as an intermediate representation dedicated to the visualization of uncertainty. The visualization layer uses presentation object and maps these objects onto the functionality of computer graphics fulfilling the requirements for understanding and acceptance by development engineers and designers.

25.3.1 Representation Layer: Ontology Based Information Model

An ontology-based information model is developed for the representation of information about uncertainty and for embedding uncertainty into an integrated product and process model. This model must be able to provide information about future processes, their properties and the variation of appropriate properties.

An overview of the ontology-based information model is given in Fig. 25.1. It consists of partial models for product and process representation, uncertainty

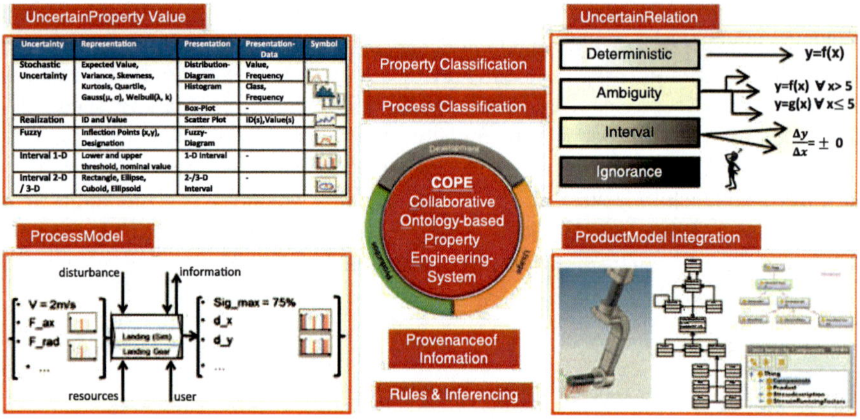

Fig. 25.1 Core elements of the ontology-based information model

properties, and uncertainty relations. The ontology is the connecting super model which classifies the elements, uses rules and inference to represent uncertainty. Thus, the ontology approach provides a naming space, cause and effect relationships as main extensions to product and process representation as available in the international standard ISO 10303 "Product Model Representation and Exchange" also known as STEP (Standard for the Exchange of Product Model Data) [3]. The developed information model further consists of a management level for uncertainty information as well as for the provenance of information.

The architecture of the ontology-based information model for uncertainty is presented in [4] and [5]. The approach takes into consideration to plan manufacturing and usage processes during the product development simultaneously and the integration of manufacturing and usage information into the product and process model which renders the possibility of capturing uncertainty related knowledge more comprehensively.

The information model is an ontology-based representation of the product and processes and integrates uncertainty to the integrated product and process model. The basic product and process model is presented in [5] and [6]. This part is used to aggregate information about uncertainty along product life phases and provide accessibility during product development.

The product is described through a 3D-CAD model. In the product model representation the 3D-CAD model is decomposed in certain elements for referencing uncertainty, allowing a precise localization of uncertainty. There are other approaches aiming at the representation of the product and assert knowledge [7–9], but these approaches do not offer the appropriate level of detail and the flexibility for the purpose of locating and managing uncertain property values. The elements of the model are mainly Features and BREP-elements.

By representing product elements, processes (and their uncertainty information) can be assigned to the product and identified within products' life. For the representation of this product and process model using ontology, a hierarchy for

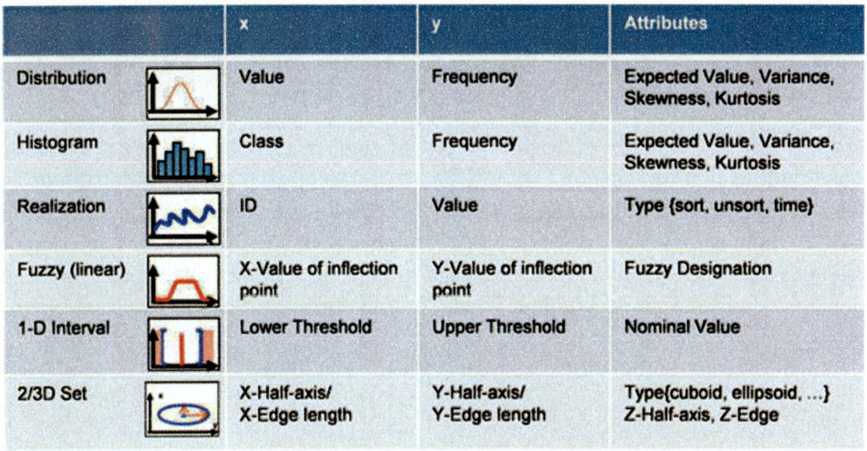

	x	y	Attributes
Distribution	Value	Frequency	Expected Value, Variance, Skewness, Kurtosis
Histogram	Class	Frequency	Expected Value, Variance, Skewness, Kurtosis
Realization	ID	Value	Type {sort, unsort, time}
Fuzzy (linear)	X-Value of inflection point	Y-Value of inflection point	Fuzzy Designation
1-D Interval	Lower Threshold	Upper Threshold	Nominal Value
2/3D Set	X-Half-axis/ X-Edge length	Y-Half-axis/ Y-Edge length	Type{cuboid, ellipsoid, ...} Z-Half-axis, Z-Edge

Fig. 25.2 Schematic overview of the uncertainty data type (UDT)

object mapping is created. This mapping assures that the hierarchic tree of the product and process model can be transformed in an ontology-based representation forth and back. This technique is implemented in a "CAD—OWL" connector which assures the consistency of the ontology-based information model in terms of the product and process representation [5].

Typically the values of properties within product and process models are affected by uncertainty which cannot be described with common data types in virtual product development environments yet. The UncertML approach [10] offers an Extensible Markup Language (XML) based model for the description of probabilistic values in the field of geospatial systems. This approach, however, is not able to represent estimated uncertainty like fuzziness or intervals and is focused on geospatial measurement systems. In order to describe uncertain values in the field of virtual product development a formalization for an appropriate representation has been developed. This formalization uses the uncertainty model of SFB 805 [11], but differentiates for data representation reasons. Detailed description refers to [6, 12] as well as its usage in [13]. An overview of the scheme is given in Fig. 25.2, depicting Uncertainty Data Types (UDT) as representations of stochastic uncertainty and estimated unknown uncertainty.

Stochastic uncertainty is represented as a distribution function, a histogram or a fuzzy function. Several stochastic key indicators like expected value, variance, skewness and kurtosis are used for an entire description of stochastic uncertainty. These key indicators represent the first four so called power moments in statistics. These parameters can also be used for a functional description of common distribution functions. The distribution function and the histogram itself can additionally be represented as a pair of values which are encoded in x and y columns. This is useful for the presentation of stochastic uncertainty for mechanical engineering without reprocessing raw data. The representation of (linear) fuzzy functions allows the definition of fuzzy intervals and is represented by vertices of linear functions which define different intervals. Estimated uncertainty is defined as a 1-D interval or 2/3D sets in a formalization. Intervals can be distinguished into bounded and unbounded intervals. The 2/3D sets are represented as rectangular or elliptic sets respectively cuboid or ellipsoid sets, in order to represent complex uncertainties in n-dimensional spaces. The representation is used for the storage of raw statistic data. With the application of methods like Monte Carlo Simulation (MCS) input samples are necessary to compute output samples. The in- and output samples are then analyzed in terms of their correlation which describes the dependency of the in- and output. This representation approach enables storage of these samples and further presentation and visualization in 2/3D scatter plots.

25.3.2 Presentation Layer

Product and process information tainted with uncertainty has to be presented to development engineers and designers during the design process. As described before uncertainty can occur in different product and process properties, so presentation and visualization techniques have to be provided. Uncertainty presentation comprises:

- Geometry and geometric tolerance related uncertainties,
- Load related uncertainties,
- Material related uncertainties,
- Process related uncertainties and
- Other uncertainties (e.g. economic, organizational,...)

In product design the geometry and geometric tolerance related uncertainties are crucial, in particular for Geometrical Design and Tolerancing purposes. Current 3D-CAD systems offer no specific presentation techniques for uncertainty in general and for geometry and geometric tolerance related uncertainty in particular. Geometry as part of the product model is built on mathematic models which imply numerically exact geometry. In order to integrate uncertainty into geometry it is necessary to model the uncertainty explicitly and to derive a presentation appropriate for visualization purposes.

uCloud—Presentation of uncertainty based on geometric tolerance

For the tangible presentation of geometric tolerance uncertainty the "uCloud" (Uncertainty Cloud) concept is developed. The uCloud concept creates a space which contains the probability distribution of body existence of the real part, e.g. for a part surface if the part has been produced many times.

> Definition: uCloud specifies the probability space where a real shape exists, based on representations of uncertainty which is defined by the appropriate processes, e.g. manufacturing.

As an example for cylindrical parts such as shafts, the uCloud is generated by set-theoretic operations (25.1) which comprise two bodies, each representing a maximum respectively minimum value of the geometric property, e.g. length and diameter of a cylindrical body.

V_{max_Body} : Volume of the maximum value of the maximum body
V_{min_Body} : Volume of the minimum value of the minimum body

$$V_{uCloud} = V_{max\_Body} \setminus V_{min\_Body} \tag{25.1}$$

The uCloud elements are available for cylindrical and prismatic parts, and are generated with respect to the specific product shape element. Since geometric tolerance deviations typically being small with respect to the dimensions, the technique of exaggeration known from FE simulations is used. The generated uCloud element is then subject to the application of geometric tolerance property visualization techniques.

25.3.3 Visualization Layer

The visualization layer is introduced to generate a visualization of the presentation objects (such as uClouds). Conceptually visualization is distinguished between three main visualization domains:

- Graphical visualization,
- Symbolic visualization,
- Structural visualization and
- Verbal visualization.

Graphical visualization uses the functionality of computer graphics [14] to generate an appropriate appearance of uncertainty [15, 16]. Such functionality comprises color, color intensity, transparency, colored patterns, etc. [17]. A cross reference table is necessary to attach the semantics of uncertainty to the graphical appearance of uncertainty.

Symbolic visualization refers presentation objects to pre-defined symbols as used e.g. in process visualization (e.g. activities, events, information flow, material flow) and enables the attachment of uncertainty information.

Structural visualization maps the presentation object onto structures such as lists, tables, tree-structures and graph-structures. Verbal visualization supports to express uncertainty lexically. The verbal visualization use the ontology approach as outlined in the ontology-based information model and creates textual output. Figure 25.3 depicts graphical visualization techniques for uncertainty with respect to the type of uncertain properties and its uncertain value description based on UDT. This approach for the visualization comprises geometric tolerance uncertainty visualization by

Visualisation techniques of:			
	Geometric Properties	Loading Properties	Other
Distribution	Colour density gradient in 3D-Model	Diagram based +Axis highlighting	Diagram based + Views
Histogram	Distinct colour density steps in 3D- Model	Diagram based +Axis highlighting	Diagram based + Views
Fuzzy (linear)	Colour density steps and gradient in 3D-Model	Diagram based +Axis highlighting	Diagram based + Views
1D Interval	Fixed hulls of upper and lower threshold	Diagram based +Axis highlighting	Diagram based + Views
2/3D Set	Combination of 2 or 3 hulls describing a body	Diagram based +Axis highlighting	NA
Realization	2/3D Scatterplot	2/3D Scatterplot	2/3D Scatterplot

Fig. 25.3 Visualization techniques for different types and descriptions of uncertainty

creating a cloud like space which contains the part surface of the real product. In order to visualize non-geometric properties structural visualization based on a diagram approach is used which, combined with an uncertainty information manager, allows filtering information.

Figure 25.4a depicts the uCloud for an interval. In this case an upper limit, a set value and a lower limit are graphically visualized. With respect to this information the real surface of the shaft is positioned in this transparent shaded space. This information state is specific for a chosen manufacturing process type. This technique is also available for the visualization of stochastic data by choosing a sigma level (c.f. Six Sigma) or choosing a confidence interval of the distribution function, depending on the available input data. In this case both sigma levels and the expected value would characterize three characteristic points, indicated by an additional specific symbol.

Figure 25.4b depicts the shaft with stochastic uncertainty information. In the case of a histogram given as input the color density range is mapped onto the min and max frequency and visualized by distinct elements which correspond to the classes of the histogram. The uCloud element color density corresponds to the probability which is given by the representation of the histogram.

Figure 25.4c depicts the case of stochastic uncertainty with a distribution function e.g. a Gaussian distribution. The uCloud element color density is mapped to the probability of the function. In this case, definite borders for the uCloud element needs to be integrated, thus this distribution function never equals zero. The borders could be derived through a chosen sigma level.

Each uCloud element has to be generated for each uncertainty according to the type of its value description. With an intersected view development engineers and designers are enabled to understand the quantity of uncertainty and are enabled to compare uncertain properties.

Through the visualization layer development engineers and designers are enabled to interpret different influences which occur in the product life cycle such as imperfect manufacturing, wear and corrosion. Development engineers and designers are also enabled to compare different uncertainties occurring in the same part or assembly.

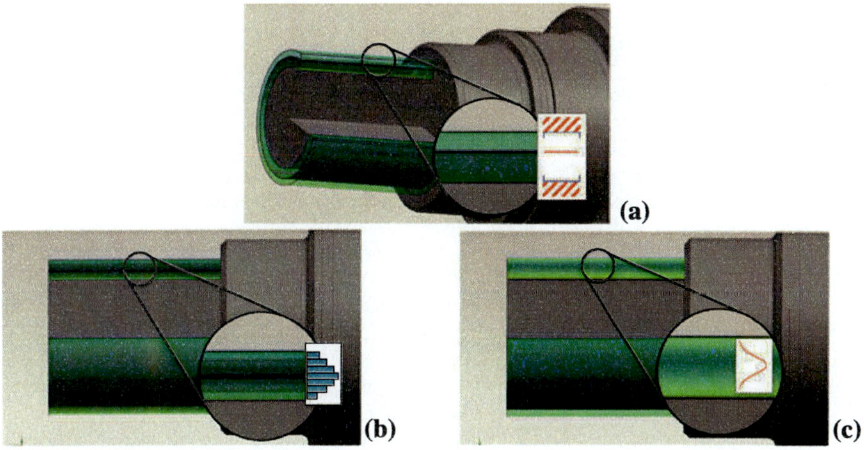

Fig. 25.4 Visualization: **a** Interval. **b** Histogram. **c** Distribution function

25.4 Outlook and Conclusion

The consideration of uncertainty in virtual product development becomes more and more important. Development and design strategies such as robust design indicate an increasing awareness about uncertainty. Capturing uncertainty requires the consideration of both, product and process properties to identity uncertainties and to take into account the impact of uncertainties during virtual product development.

Virtual product development comprises a set of computerized methods and tools which do not take into account uncertainties. Therefore a new approach is required to manage uncertainties throughout the phases of virtual product development. Such an approach is proposed by the three layer architecture including the representation layer using an ontology-based information model, a presentation layer and a visualization layer. This conceptual approach has been presented for cylindrical and prismatic parts focused on dimension tolerances. Presentation and visualization techniques for shape and positional tolerance are subject of current and future research.

Future research is dedicated to further areas of uncertainty such as load related uncertainties, material related uncertainties, process related uncertainties and other uncertainties, e.g. organizational and market uncertainties.

Acknowledgments We like to thank the German Research Foundation (Deutsche Forschungsgemeinschaft—DFG) for funding this project within the Collaborative Research Centre (CRC) 805.

References

1. Hanselka H, Platz R (2010) Ansätze und maßnahmen zur beherrschung von unsicherheit in lasttragenden systemen des maschinenbaus. In: Konstruktion 11/12
2. Wanderbäck F, Söderberg R, Carlson JS, Wahlborg P et al. (2009) Variation feedback and visualization of geometrical inspection data. Proceedings of the ASME 2009 international design engineering technical conferences and computers and information in engineering conference (IDETC/CIE), San Diego, DECT2009-86759
3. ISO-10303-1 (1994) Overview and fundamental principles, ISO
4. Mecke K, Sprenger A, Weitzmann O, Anderl R (2010) Ontology-based system for supporting the containment of uncertainty in the product life cycle. In: TMCE, faculty of industrial design engineering, Delft University of technology, eurotech-verlag, Ancona, pp. 1259–1272
5. Mecke K, Sprenger A, Mosch L, Anderl R (2010) Beherrschung von unsicherheit von lasttragenden systemen im maschinenbau durch ontologie-basierte informationsverarbeitung und visualisierungstechnik, KT 2010—Magdeburg
6. Mosch L, Sprenger A, Anderl R (2011) Consideration of uncertainty in virtual product design, proceedings of the ASME 2011 international design engineering technical conferences and computers and information in engineering conference (IDETC/CIE), Washington DC, DECT2011-47278
7. Ahmed S, Kim S, Wallace KM (2007) A methodology for creating ontologies for engineering design. J Comput Inf Sci Eng 2:132–140
8. Kim KY, Manley DG, Yang H (2006) Ontology-based assembly design and information sharing for collaborative product development. Comput-Aided Des 12:1233–1250
9. Patil L, Dutta D, Sriram R (2005) Ontology-based exchange of product data semantics. IEEE Trans on Autom Sci Eng 2(3):213–225
10. Williams M, Cornford D, Bastin L, Ingram B, R, (2008) UncertML: an XML schema for exchanging uncertainty. IN: GIS research UK 16th annual conference GISRUK 2008, Manchester (UK), ID: 11091
11. Engelhardt R, Koenen J, Enss G, Sichau A, Platz R, Kloberdanz H, Birkhofer H, Hanselka H (2010) A model to categorise uncertainty in load-carrying systems, MMEP 2010—international conference on modelling and management of engineering processes, Cambirdge, pp. 53–64
12. Sprenger A, Mosch L, Anderl R (2011) Representation of uncertainty in distributed product development, 17th European concurrent engineering conference, London, pp. 83–90
13. Sprenger A, Haydn M, Ondoua S, Mosch L, Anderl R (2011) Ontology-based information model for the exchange of uncertainty in load carrying structures. Proceedings of the ICUME2011, Darmstadt, pp. 55–66
14. Foley JD, Dam A van, Feiner S, Hughes J (1999) Computer graphics: principles and practice, 2nd edn. Addison-Wesley, reading, mass
15. Johnson CR, Sanderson AR (2003) A next step: visualization errors and uncertainty. In: Rhyne TM (ed) IEEE computer graphics and applications, September/October IEEE computer society
16. Tufte ER (2001) The visual display of quantitative information. 2nd edn. Graphics Press, CT USA
17. Ware C (2004) Information visualization: perception for design, 2nd edn. Morgan Kaufmann, San Francisco

Chapter 26
Implementation and Initial Validation of a Knowledge Acquisition System for Mechanical Assembly

N. Madhusudanan and Amaresh Chakrabarti

Abstract This paper describes the implementation and initial validation of a method for semi-automatically acquiring knowledge for an expert system to perform diagnosis in the domain of mechanical assembly. The approach developed in this research is a method of questioning experts for knowledge acquisition, starting with minimal domain knowledge, and no prior domain model. The software implementation of the knowledge acquisition method is a computer based tool titled ExKAV. The paper describes the layout of the software, its implementation details, and interfaces involved. An initial validation using two assembly situations is also presented, and, potential directions for extending this work are discussed.

26.1 Introduction—Mechanical Assembly

Assembly of parts and subassemblies that constitute a product is an important step towards the realization of the product. As an integrative process, assembly brings together not only the physical parts that are involved [1], but also a large number of people and technologies that must work together. Assembly planning involves deciding about the various aspects involved in the process of assembly, e.g. joining methods and materials, tooling, fixturing [2], feeding, motion paths of the components, assembly sequence, scheduling of assembly activities [3], etc.

N. Madhusudanan (✉) · A. Chakrabarti
Centre for Product Design and Manufacturing, Indian Institute of Science, Bangalore, 12, India
e-mail: madhu@cpdm.iisc.ernet.in

A. Chakrabarti
e-mail: ac123@cpdm.iisc.ernet.in

A. Chakrabarti (ed.), *CIRP Design 2012*, DOI: 10.1007/978-1-4471-4507-3_26,

Design problems discovered during production are expensive to resolve; hence, potential assembly difficulties encountered in manufacturing should be identified as soon as possible, preferably during the product design stage itself [4]. One way to do so is to follow Design for Assembly and Design for Manufacturing (DfA/DfM) guidelines [1], so that the resulting product is less difficult to manufacture and assemble. However, it is not clear how comprehensive traditional DfA/DfM guidelines are, which are often applied to certain kinds of assemblies only (e.g. automated assembly lines). There may be a wider array of domain-specific knowledge about how an assembly may be difficult to perform or what to do if a specific difficulty is faced, which is not captured in current DfA/DfM guidelines. Such knowledge often resides only with the experts in that domain.

26.1.1 Knowledge Based Systems

During the assembly planning stage, artificial intelligence systems can be utilized for capturing, storing and reusing expert knowledge about how an assembly can be made easier to assemble. Knowledge based systems (or expert systems) are examples of such systems. They have a wide base of knowledge within a restricted domain, and use complex inferential reasoning to perform tasks that are typically carried out by human experts [5]. The main components of a knowledge based system (KBS) are [6],

- A *knowledge base*, to store the knowledge acquired from experts.
- A *working memory* that contains information about the present state of the world, as seen by the expert system.
- An *inference engine* that performs the reasoning, based on the contents of the knowledge base and the present state of the world.

26.2 Knowledge Acquisition

Knowledge acquisition is defined as 'the process of eliciting and enumerating the knowledge of an expert in a particular field (domain) so that the expertise can be coded into an expert system' [7]. It consists of eliciting knowledge from the expert, interpreting it, and formalizing it using a representation [8]. Knowledge thus represented is stored in the knowledge base to be used later for inference. Knowledge acquisition has been a subject of study for quite some time, due to its importance and difficulty in the process of building an expert system [9].

26.2.1 Challenges in Knowledge Acquisition

Though procedures for manual knowledge acquisition have been developed [7], knowledge acquisition has been a bottleneck in the wider application of expert systems, e.g. Rank Xerox's LOOP, Nixdorf's TWAICE, and Teknowledge's S.1 [10]. This is due to the difficulty in encoding human knowledge into a machine. Also, the knowledge to be acquired may be explicit, implicit or tacit in nature. Tacit knowledge cannot always be made explicit, as in the case of many handling skills in various craft jobs [11, 12]. This makes its acquisition difficult—and addressing this problem becomes important since such knowledge is of greater value [13]. The knowledge to be acquired in this research is implicit or tacit in nature.

Another research challenge is the difficulty of automating the process of knowledge acquisition. Various reasons are found in literature as to why automation needs to be carried out—better productivity, development of better methodologies [14], quicker knowledge prototyping [14], more structured knowledge [15], as well as the difficulty and cost of manual knowledge acquisition [16].

26.3 Background—Questioning Procedure

Literature on automated knowledge acquisition reports various tools, methods and reasons for automating knowledge acquisition. A detailed analysis of the various systems built to automate knowledge acquisition is given in [17]. The discussion of literature concludes that, in the mechanical assembly domain, where there is no domain model available a priori, knowledge acquisition has not been automated. There is a lack of systems that can conduct a dialogue with experts for knowledge acquisition, as well as those that can acquire and apply knowledge in the context of specific situations, e.g. using them as examples [17].

We propose that knowledge can be acquired automatically using a method of questioning that resembles a natural dialogue between the computer and the expert; we call this the '*Questioning Procedure*'. The first set of questions in the procedure is used to identify any potential *issue* envisaged by the expert in a given assembly situation, and the cause and effect of this issue. The second set of questions identifies the *constraint* whose violation caused the issue, and the *parameters* that form this constraint. The third set of questions is to identify potential or realized *solutions* to the issue. Such *issues*, *constraints*, *parameters* and *solutions* have been identified as the knowledge types necessary for a computer to infer potential issues in a given assembly situation [17]. This questioning procedure is the basis for implementation of the proposed knowledge acquisition tool.

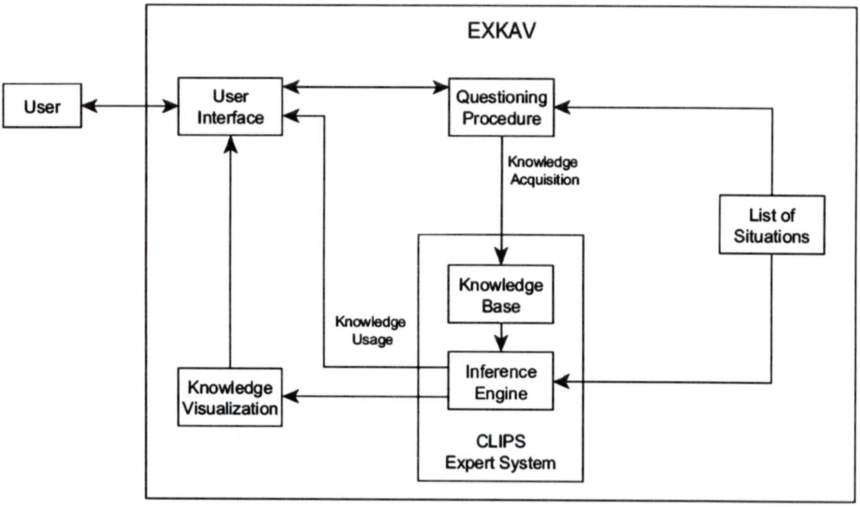

Fig. 26.1 Overview of ExKAV implementation

26.4 Implementation

The proposed questioning procedure discussed above has been implemented as a software tool titled ExKAV, an acronym for Expert Knowledge Acquisition and Validation. It has the following components, as shown in the Fig. 26.1:

- A CLIPS expert system, which has
  - a knowledge base to store expert knowledge
  - an inference engine that reasons using knowledge in the knowledge base to identify which issues might arise in a given assembly situation;
- A computer implementation of the questioning procedure;
- A graphical interface for interaction between a user and ExKAV. A user would be one of the following:
  - an assembly expert from whom knowledge is to be acquired, or
  - an assembly planner, who seeks the experts' advice;
- A list of situations, used for acquisition or usage of knowledge;
- A facility for visualizing the acquired knowledge in the form of a directed graph. However, this module will not be discussed in this paper, since it is not of direct relevance here.

The software is implemented in C++. The expert system component has been implemented using CLIPS expert system shell [18]. The interface is implemented using Microsoft Foundation Classes (MFC). 3D models used in presenting a situation are implemented using Dassault Systèmes' 3DVIA™ Player [19].

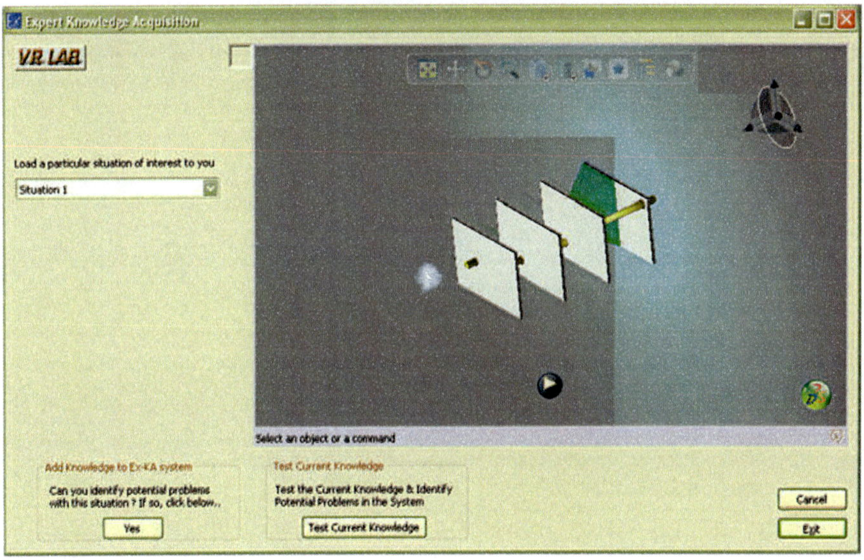

Fig. 26.2 Initial screen of the ExKAV system

26.4.1 User Interface for Knowledge Acquisition

In the knowledge acquisition mode, ExKAV presents an assembly situation to an expert, and acquires knowledge about potential issues and solutions. The initial screen of the system consists of a drop-down list to choose a situation amongst a list of assembly situations offered, see Fig. 26.2. There are two options thereafter—to either add knowledge about this situation, or test the knowledge that has been acquired. A display viewport is used to view the assembly situation in 3D, and to modify the view by scaling, translation or rotation. The assembly tree can be also be navigated in this viewport using the 3DVIA™ player. One example situation is shown in Fig. 26.2, which consists of four plates of equal dimensions with a hole in each of these. Assembly involves passing a cable bundle through these holes.

Once the expert chooses the option to add knowledge about an assembly situation, the next interface brought up is shown in Fig. 26.3. It contains questions about what potential issue in the assembly situation is identified by the expert, and the cause and effect of the issue. The expert types the answers for these questions into the forms provided in the interface. For the example situation, the issue has been described as '*The cable bundle may not be able to sustain the bend*'. The cause of this issue, ('*bend radius of the cable bundle is too less*') is then acquired.

The next interface (Fig. 26.4) consists of fields to dissect the cause of the issue into further pieces, such as the constraint and its parameters. The cause of the issue should be put in the form of '*X relation Y*', where *X* and *Y* are parameters. *X* and *Y* are then refined into the form 'an *Attribute of Object*'. This is done to reduce the

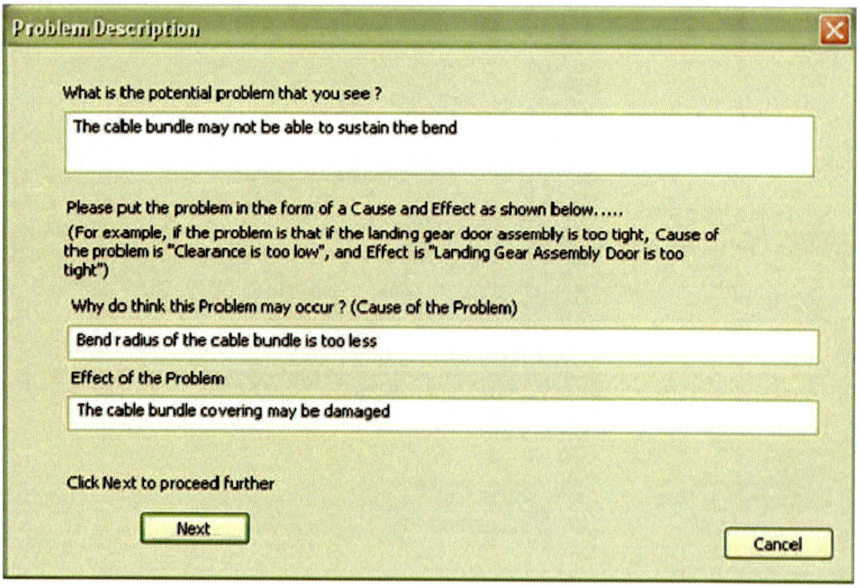

Fig. 26.3 The description of the potential issue spotted by the expert in the assembly situation

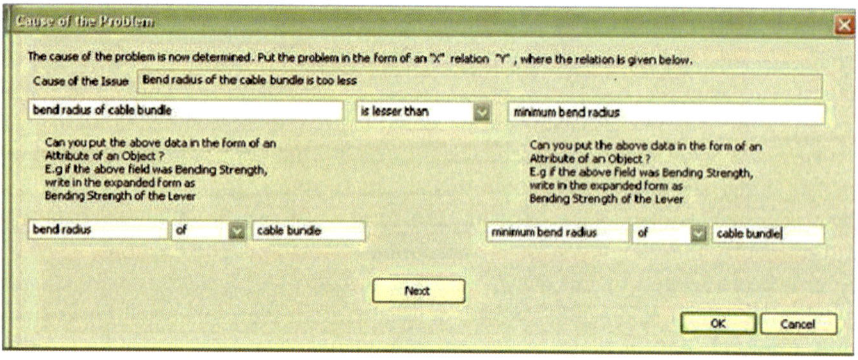

Fig. 26.4 Detailing of the cause of the issue

number of ways in which an expert might describe a parameter. Drop-down boxes help to choose from various connecting words and phrases e.g. relations such as '*is greater than*', '*is less than*', and connecting words for the parameters such as '*of*', '*in*', etc. For the example described above, the cause of the issue is '*bend radius of the cable bundle is too less*'. This is rewritten as '*bend radius of cable bundle*' -'*is less than*'- '*minimum bend radius of cable bundle*'.

The parameters are of two types—*basic* and *derived*, depending on how their values are to be obtained while applying the acquired knowledge. The values for *basic* parameters need to be supplied as necessary data, whereas those for derived

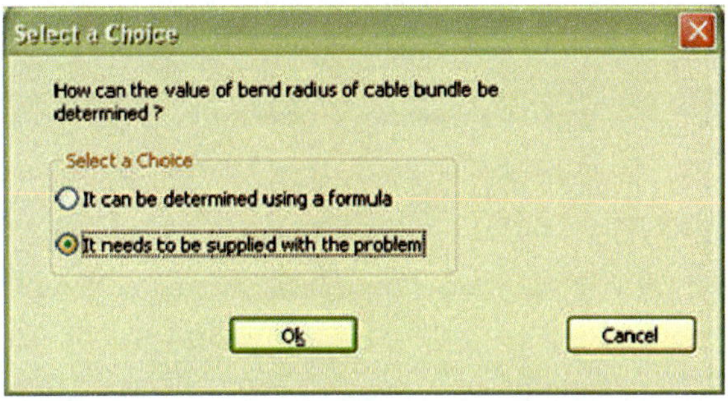

Fig. 26.5 Asking whether a parameter is a *basic* or *derived* parameter

parameters are calculated using relations among other, basic or derived parameters. The interfaces shown in Figs. 26.5, 26.6 are used to ask the user if a parameter is *basic* or *derived*. For example the *minimum bend radius* = *(1.84 * distance between plates)*, which means that minimum bend radius is a derived parameter.

The final interface window helps query for possible solutions to the issue. The possible solutions for the example issue are shown in Fig. 26.7.

Once the questioning procedure is completed, a set of CLIPS rules are automatically generated for every knowledge type, e.g. constraint, necessary data, and solutions. These rules are then written in a text-form into the knowledge base.

26.4.2 User Interface for Knowledge Application

In the knowledge application mode, the system utilizes the knowledge acquired in order to infer potential issues in an assembly situation. The user selects the option to test the acquired knowledge, using the window in Fig. 26.2. For the system to apply the knowledge stored in the knowledge base, it needs the values of the necessary parameters (discussed in the last section) that describe the current situation. A parameter value acquisition dialogue (Fig. 26.8) is used to acquire these values. The results of inference are then displayed using the interface shown in Fig. 26.9.

26.5 Initial Validation

An initial validation has been conducted on ExKAV to test whether ExKAV can engage in a dialogue with an expert, and whether potential issues and their solutions can be identified from the expert for an assembly situation. Also, the intention was to verify whether knowledge needed for identifying each issue can be detailed adequately to elicit the underlying constraints, parameters, and solutions. The knowledge acquired should be possible to be automatically represented by the system into a set of rules, and should be applicable to similar assembly situations.

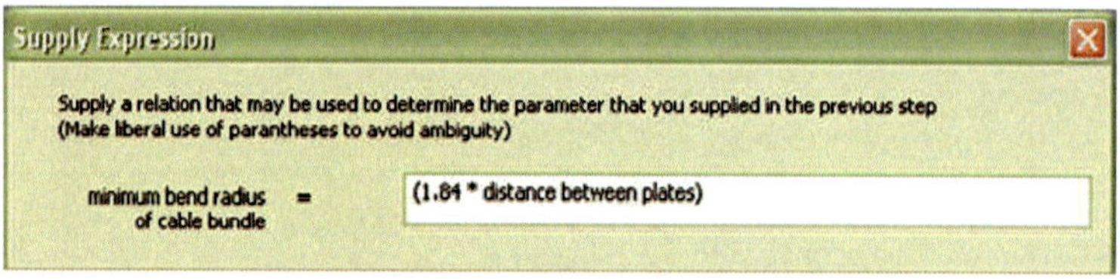

Fig. 26.6 Asking whether a parameter is a *basic* or *derived* parameter

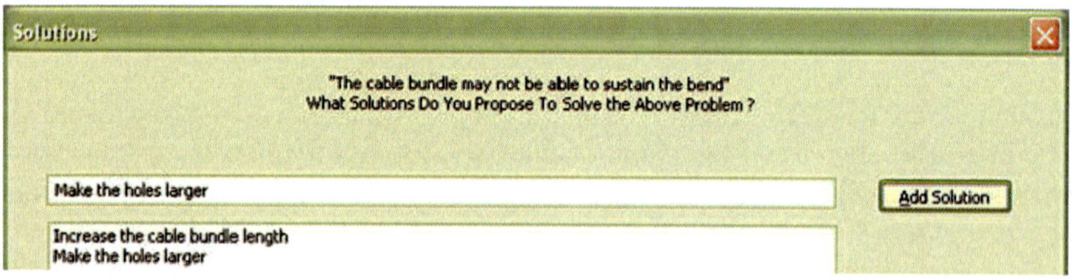

Fig. 26.7 The dialog window to input the resolutions to the potential issue

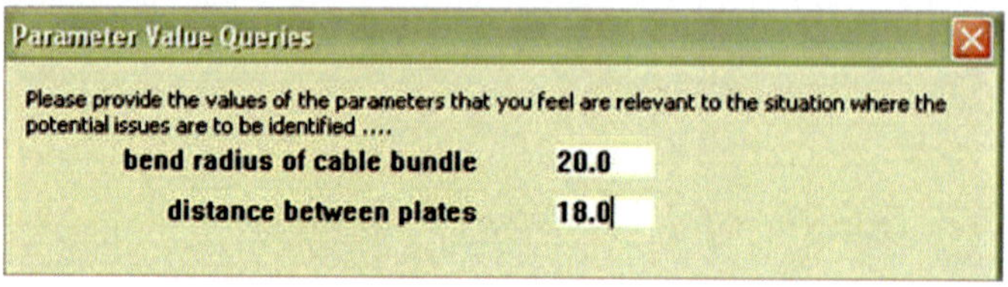

Fig. 26.8 Interface for acquiring parameter values

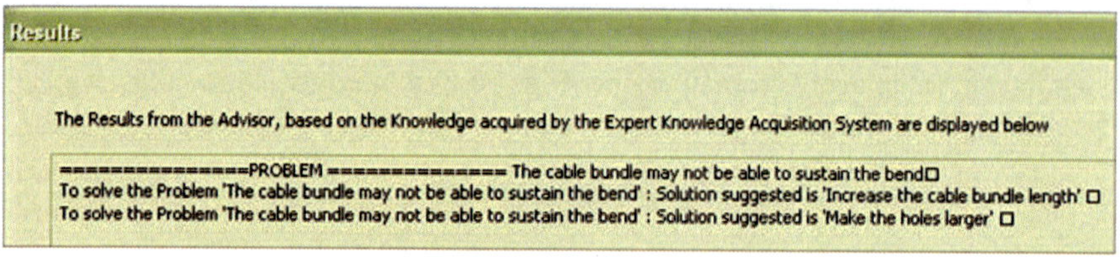

Fig. 26.9 The output after the inference procedure

26.5.1 Validation Method and Results

Two assembly situations were used for this initial validation. The first situation involves assembly of a cable bundle through four parallel plates, as described in Sect. 26.4.1. The second situation is the assembly of five parts shown in Fig. 26.10. After presenting the experts with each of these situations, they were taken through the questioning procedure. The presence of the researchers acting as an

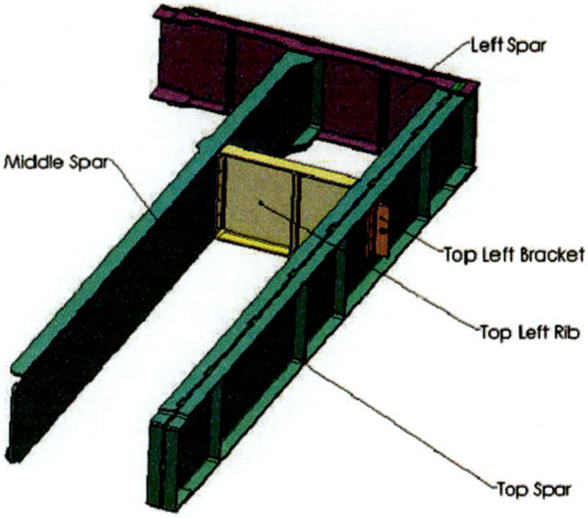

Fig. 26.10 Another situation used for the validation session

intermediary between the experts and the tool was needed to guide the experts through the questioning procedure during this initial evaluation, as time was not available to provide training to the experts in using the tool.

The validation sessions were conducted over a period of one year, due to limited access to experts in this domain. Three different sessions, each lasting for 40–90 min, were conducted. During each session, there were at-least two experts present. A total of 53 rules have been generated for seven issues identified by the experts in the three situations. The number of rules generated was six to seven for each issue identified. For each expert, the knowledge acquired was then used in the application mode, and all the issues and solutions generated during the corresponding knowledge acquisition session earlier were possible to be predicted subsequently by the rules generated in that session.

Amongst the potential issues identified, one issue was identified twice by the same expert in two sessions that were spread apart in time. However, the solutions suggested to the issue were different on each occasion. This illustrates that multiple iterations may lead to acquisition of more knowledge from the same situation.

26.6 Conclusions

This paper has discussed the implementation and initial validation of a knowledge acquisition system titled ExKAV. ExKAV presents an assembly situation to an expert and carries out a questioning procedure in the context of the situation. As found from the validation sessions, using ExKAV it has been possible to acquire knowledge from assembly experts using the proposed questioning procedure.

The questioning procedure and its implementation need to be expanded further to improve the knowledge acquisition process. For example, several aspects of an assembly situation, such as the processes and tools involved, are yet to be covered while presenting it to the expert. Self-organization and maintenance of the knowledge base is also a potential direction that needs to be explored.

The implementation can be extended further to include multiple constraints to represent the cause of an issue, rather than the current, single constraint. Also, greater automation is required in the acquisition of parameter values. Further, a more comprehensive validation needs to be conducted, using more experts and more situations, to evaluate the tool's capabilities in an extensive manner.

Acknowledgments The authors acknowledge the contribution of various experts from Indian Institute of Science in this research. The Boeing Company has provided financial support under contract PC36016 at SID, IISc; Dr. Al Salour and his colleagues from The Boeing Company, USA have been helpful in providing valuable technical support.

References

1. Whitney DE (2004) Mechanical Assemblies. Oxford University Press, Oxford
2. Dewar R, Carpenter I, Ritchie J, Simmons J (1997) Assembly planning in a virtual environment. In: innovation in technology management—the key to global leadership. PICMET'97
3. van Holland W, Bronsvoort W (2000) Assembly features in modeling and planning. Rob Comput Integr Manuf 16:277–294
4. Choi ACK, Chan DSK, Yuen AMF (2002). Application of virtual assembly, tools for improving product design. Components 377–383
5. Hart A (1986) Knowledge Acquisition for Expert Systems. Kogan Page, London
6. Sasikumar M, Ramani S, Muthu Raman S, Anjaneyaulu K, Chandrarasekhar R (1993) Rule Based Expert Systems. Narosa Publishing, New Delhi
7. Br′ule A, Blunt JF (1990) Knowledge acquisition. Computing. McGraw-Hill, NY
8. Schreiber G, Akkermans H, Anjewierder A, de Hoog R, Shadbolt N, de Velde WV, Wielinga B (2000) Knowledge engineering and management, the common KADS methodology. MIT Press, Cambridge
9. Wielinga BJ, Schreiber AT, Breuker JA (1992). Kads: a modelling approach to knowledge engineering. Knowl Acquis 4:5–53, ISSN 1042-8143
10. Savory S (1988) Some views on the state of the art in AI, Nixdorf computer AG. Ellis Harwood Books, Paderborn
11. Pomerol J, Brezillon P (2001) About some relationships between knowledge and context. Context 1:461–464
12. Gabbiadini S, Facoetti G, Colombo G, Rizzi C (2009). Knowledge management and customised 3D modelling to improve prosthesis design. In: Proceedings of the ASME 2009 international design engineering technical conferences and computers and information in engineering conference IDETC/CIE, San Diego (California), USA
13. Inkpen A (1998) Learning, knowledge acquisition, and strategic alliances. Eur Manage J 16(2):223–229
14. Gruber TR (1989) The acquisition of strategic knowledge, perspectives in artificial intelligence, vol 4. Academic, London

15. Vlaanderen MJ (1990) Automated knowledge acquisition for expert systems. Master's thesis, Erasmus University, Rotterdam
16. Greenwell M (1988) Knowledge engineering for expert systems. Ellis Harwood Books, Chichester
17. Madhusudanan N, Amaresh Chakrabarti (2011), An interactive questioning based method to acquire knowledge for knowledge-based systems, International conference on trends in product life cycle, modeling, simulation and synthesis- PLMSS-2011
18. CLIPS Homepage (2011) Homepage : http://clipsrules.sourceforge.net/
19. 3D via player (2011) URL: http://dl.3dvia.com/software/3dvia-player/

Part V
Design For X (Safety, Manufacture, Assembly, Cost, Risk, Reliability, Modularity, etc.)

Chapter 27
Robust Adaptable Design Considering Changes of Parameter Values in Product Operation Stage

Jian Zhang, Deyi Xue and Peihua Gu

Abstract Increasing competition in the global marketplace demands products be adaptable to the changes of functional requirements, operation environments, and technology advancement. Adaptable design can be used for developing adaptable products that can be easily changed, such as reconfigured or upgraded, in the product operation stage. In addition, since the uncertainties in product parameters and environment parameters usually lead to the variations of functional performance, robust design has to be carried out to identify the design whose functional performance is the least sensitive to the variations of product and environment parameters. In this research, a robust adaptable design method is developed to identify the product whose parameter values can be changed in the product operation stage. First, a mathematical model is established to describe the relationships among product functional performances, parameters, and variations of these parameters. Then, an optimization-based design model is developed to achieve the robust adaptable design. A case study has also been developed to demonstrate the effectiveness of the newly developed robust adaptable design method.

J. Zhang · D. Xue
Department of Mechanical and Manufacturing Engineering, University of Calgary, Calgary, AB T2N 1N4, Canada
e-mail: zjia@ucalgary.ca

D. Xue
e-mail: dxue@ucalgary.ca

P. Gu (✉)
Department of Mechatronics Engineering, Shantou University, Shantou, 515063, Guangdong, China
e-mail: peihuagu@stu.edu.cn

A. Chakrabarti (ed.), *CIRP Design 2012*, DOI: 10.1007/978-1-4471-4507-3_27,

27.1 Introduction

The competition in the global marketplace requires products be adaptable to the various changes in functional requirements, operation environments, and technology advancement in their operation/use stage. Adaptable design is a new design methodology to design adaptable products that can be easily modified for satisfying the various changes [1]. Significant research efforts have been devoted in adaptable design to create adaptable products that can be reconfigured or upgraded in the operation stages [2–5].

Any product is designed to perform its expected functions. During the operation of a product, the uncertainties in both product and environment parameters can lead to the variations of functional performance. Many robust design methods have been widely studies and employed to improve product robustness through minimizing the sensitivity of functional performance to uncertainties in parameters [6–9]. For an adaptable product, since its functional performance should also be robust considering the uncertainties in parameters, robust design has to be carried out to identify the design whose functional performance is insensitive to the uncertainties in parameters.

The robustness of the adaptable product, however, has never been considered in the present adaptable design methods. In addition, the existing robust design methods cannot be used directly to design an adaptable product. In order to design an adaptable product whose functional performance is insensitive to uncertainties in parameters, a robust adaptable design method has to be developed. An adaptable product can be changed to satisfy the various changes through modifications of product structure and parameters. In this research, only the changes of parameter values in the operation stage of a product are considered.

27.2 Modeling of Functional Performances and Parameters

27.2.1 Modeling of Functional Performances

Suppose m is the number of the required functional performances, all the functional performances of a product, $\boldsymbol{F}$, can be described by: $\boldsymbol{F} = \{F_1, F_2, \ldots, F_m\}$. In the operation stage of a product, the different required values of the i-th functional performance measure, F_i, is described by:

$$F_i = \begin{cases} z_{i1}, & p_{i1} \\ z_{i2}, & p_{i2} \\ \vdots & \vdots \\ z_{ih_i}, & p_{ih_i} \end{cases}, \quad i = 1, 2, \ldots, m \tag{27.1}$$

where h_i is an integer representing the number of possible values of $F_i, P_{ij}(i = 1, 2, \ldots, m, j = 1, 2, \ldots, h)$ represent the probabilities that the functional performance F_i should be selected as z_{ij}.

27.2.2 Modeling of Parameters and Parameter Variations

In this research, the parameters are classified into four categories: design parameters, changeable parameters, adaptable parameters, and the other parameters.

The design parameters are those to be achieved in adaptable design. Values of design parameters cannot be modified in the operation stage. If l represents the number of design parameters of a product, the collection of the design parameters, $\boldsymbol{P}^{(D)}$, can be described by: $\boldsymbol{P}^{(D)} = \left\{P_1^{(D)}, P_2^{(D)}, \ldots, P_l^{(D)}\right\}$.

The changeable parameters are those whose values can change independently in the operation stage. For example, the wind speed and temperature for a wind turbine can change in the operation stage. If n represents the number of the changeable parameters of a product, the vector of changeable parameters, $\boldsymbol{P}^{(C)}$, can be described by: $\boldsymbol{P}^{(C)} = \left\{P_1^{(C)}, P_2^{(C)}, \ldots, P_n^{(C)}\right\}$. In the operation stage of a product, the changeable parameters can change to different values with different probabilities. To simplify the discussion, changes of changeable parameter values to only discrete values are considered. The values of the i-th changeable parameter, $P_i^{(C)}$, are described by:

$$P_i^{(C)} = \begin{cases} y_{i1}, & q_{i1} \\ y_{i2}, & q_{i2} \\ \vdots & \vdots \\ y_{ik_i}, & q_{ik_i} \end{cases}, \quad i = 1, 2, \ldots, n \tag{27.2}$$

where k_i is an integer representing the number of possible values of $P_i^{(C)}, q_{ij}(i = 1, 2, \ldots, n, j = 1, 2, \ldots, k_i)$ represent the probabilities of the changeable parameter values y_{ij}.

The adaptable parameters are those modified by the user or computer in the operation stage of a product to satisfy the various changes of product functional performances and changeable parameters. The air flow rate of an air conditioner is changed, for example, when the temperature in a room is changed. The air flow rate is considered as an adaptable parameter. If r represents the number of adaptable parameters of a product, the collection of adaptable parameters, $\boldsymbol{P}^{(A)}$, can be described by: $\boldsymbol{P}^{(A)} = \left\{P_1^{(A)}, P_2^{(A)}, \ldots, P_r^{(A)}\right\}$.

In the product operation stage, the parameter values can fluctuate from their targets due to uncertainties. These parameter variations are usually difficult to measure, and these parameter variations can lead to the variations in functional performance. For a robust adaptable product, the functional performances should be insensitive to the parameter variations. If t represents the number of the parameter variations, the vector of parameter variations, δ, can be described by: $\boldsymbol{\delta} = \{\delta_1, \delta_2, \ldots, \delta_t\}$.

In engineering design, variations due to uncertainties can be observed for design parameters, changeable parameters and adaptable parameters. In addition, variations can also be observed for the parameters whose target values are selected as constants. In this work, these parameters are called the other parameters. If s represents the number of the other parameters, the vector of the other parameters, $\boldsymbol{P}^{(O)}$, can be described by: $\boldsymbol{P}^{(O)} = \left\{P_1^{(O)}, P_2^{(O)}, \ldots, P_s^{(O)}\right\}$.

27.3 Robustness of a Design to Evaluate an Adaptable Product

With robust adaptable design, the changes of functional performances and changeable parameter values should be adapted by changing the values of adaptable parameters. In addition, the values of design parameters should be optimized to reduce the sensitivity of functional performance to parameter variations caused by uncertainties in parameters.

In this work, it is assumed that the functional performance of a product can be calculated from design parameters, changeable parameters, adaptable parameters, and the other parameters:

$$F_i = f_i\left(\boldsymbol{P}^{(D)}, \boldsymbol{P}^{(C)}, \boldsymbol{P}^{(A)}, \boldsymbol{P}^{(O)}\right),\ i = 1, 2, \ldots, m \tag{27.3}$$

where F_i is the functional performance, and $\boldsymbol{P}^{(D)}$, $\boldsymbol{P}^{(C)}$, $\boldsymbol{P}^{(A)}$ and $\boldsymbol{P}^{(O)}$ are collections of design parameters, changeable parameters, adaptable parameters and other parameters, respectively.

Under the influence of parameter variations caused by uncertainties, the functional performance measures can deviate from their target values. If δ represents a set of parameter variations, the deviation of functional performance measure can be calculated by:

$$\begin{aligned} \Delta F_i = & f_i\left(\boldsymbol{P}^{(D)} + \Delta \boldsymbol{P}^{(D)}, \boldsymbol{P}^{(C)} + \Delta \boldsymbol{P}^{(C)}, \boldsymbol{P}^{(A)} + \Delta \boldsymbol{P}^{(A)}, \boldsymbol{P}^{(O)} + \Delta \boldsymbol{P}^{(O)}\right) \\ & - f_i\left(\boldsymbol{P}^{(D)}, \boldsymbol{P}^{(C)}, \boldsymbol{P}^{(A)}, \boldsymbol{P}^{(O)}\right), i = 1, 2, \ldots, m \end{aligned} \tag{27.4}$$

When multiple functional performance measures are modeled by different units, these performance measures, $F_i(1, 2, \ldots, m)$, should be converted into comparable evaluation indices, I_i, between 0 and 1 using:

$$I_i = \Psi_i(F_i), i = 1, 2, \ldots, m \tag{27.5}$$

where $\Psi_i()$ is the non-linear relationship between the F_i and I_i. In this work, satisfaction levels are used as the evaluation indices.

According to Eqs. (27.3)–(27.5), the deviation of satisfaction index, I_i, can be calculated by:

$$\Delta I_i = \Psi_i(F_i + \Delta F_i) - \Psi_i(F_i), i = 1, 2, \ldots, m \tag{27.6}$$

In this research, the variance of the ΔI_i is used to evaluate the robustness of functional performance measure F_i.

Suppose $V_{F_{ij}}(y_{1i_1}, y_{2i_2}, \ldots, y_{ni_n})$ represents the variance of the ΔI_i on condition that the changeable parameters are selected as $P_1^{(C)} = y_{1i_1}, P_2^{(C)} = y_{2i_2}, \ldots, P_n^{(C)} = y_{ni_n}$ and the target value of functional performance measure F_i is selected as z_{ij}. Suppose $q(y_{1i_1}, y_{2i_2}, \ldots, y_{ni_n})$ represents the probability of the combination of changeable parameters $P_1^{(C)} = y_{1i_1}, P_2^{(C)} = y_{2i_2}, \ldots, P_n^{(C)} = y_{ni_n}$, according to Eq. (27.2), the probability can be calculated by $q(y_{1i_1}, y_{2i_2}, \ldots, y_{ni_n}) = q_{1i_1} q_{2i_2} \cdots q_{ni_n}$. If $V_{F_{ij}}$ represents the overall variance of the ΔI_i on condition that the target value of functional performance measure F_i is equal to z_{ij}, the $V_{F_{ij}}$ can be calculated by:

$$V_{F_{ij}} = \sum_{i_1=1}^{k_1} \sum_{i_2=1}^{k_2} \cdots \sum_{i_n=1}^{k_n} \left[q(y_{1i_1}, y_{2i_2}, \ldots, y_{ni_n}) \cdot V_{F_{ij}}(y_{1i_1}, y_{2i_2}, \ldots, y_{ni_n}) \right] \tag{27.7}$$

If V_{F_i} represents the overall variance of the ΔI_i considering the functional performance measure F_i, the V_{F_i} can be calculated by:

$$V_{F_i} = \sum_{j=1}^{h_i} \left(p_{ij} \cdot V_{F_{ij}} \right), i = 1, 2, \ldots, m \tag{27.8}$$

In this research, the overall variance of the functional performance indices of an adaptable product, V_F, can be calculated by:

$$V_F = \sum_{i=1}^{m} (\omega_i \cdot V_{F_i}) \tag{27.9}$$

where $\omega_i (i = 1, 2, \ldots, m)$ are the weighting factors between 0 and 1 representing importance of functional performances F_i.

27.4 Robust Adaptable Design Based on Optimization

For a robust adaptable design, adaptable parameters should be changed based on changes of functional performances and/or changeable parameters. In addition, the product functional performances should also be insensitive to the variations of parameters. In this work, optimization is employed to identify the robust adaptable design:

Find: the design paremeters, $\boldsymbol{P}^{(D)} = \left\{ P_1^{(D)}, P_2^{(D)}, \ldots, P_l^{(D)} \right\}$

$$\begin{array}{lll} \text{Minimize}: & ; V_F & \\ \text{Subjectto}: & ; F_i = z_{ij}, \quad i = 1, 2, \ldots, & j = 1, 2, \ldots \\ & ; g_u(\boldsymbol{P}^{(D)}, \boldsymbol{P}^{(C)}, \boldsymbol{P}^{(A)}) = 0, & u = 1, 2, \ldots \\ & ; k_v(\boldsymbol{P}^{(D)}, \boldsymbol{P}^{(C)}, \boldsymbol{P}^{(A)}) \leq 0, & v - 1, 2, \ldots \end{array} \tag{27.10}$$

where z_{ij} represents the target value of the required functional performance F_i.

In the operation stage, the changeable parameters can have different values, and a functional performance can have different target values. The values of adaptable parameters have to be modified to adapt to the changes of required functional performances and/or the changeable parameters. In this research, it is assumed that the values of adaptable parameters can be calculated from changeable parameters and functional performances.

According to the optimization model given in Eq. (27.10), the algorithm for the robust adaptable design has been developed as follows:

STEP 1:

Assign a set of initial values of design parameters, $\boldsymbol{P}^{(D)}$. Let the required value of functional performance measure F_i be z_{ij} $(i = 1,\ j = 1)$ and the changeable parameters be $P_1^{(C)} = y_{1k_1}, P_2^{(C)} = y_{2k_2}, \ldots, P_n^{(C)} = y_{nk_n}$ $(k_1 = k_2 = \ldots = k_n = 1)$.

STEP 2:

According to the design parameter values, the changeable parameter values and the target value of functional performance, use the relationship given in Eq. (27.3) and/or heuristics to calculate the corresponding values of adaptable parameters, $\boldsymbol{P}^{(A)}$.

STEP 3:

Calculate variance $V_{F_{ij}}(y_{1i_1}, y_{2i_2}, \ldots, y_{ni_n})$ used in Eq. (27.7). If the variances for all the combinations of changeable parameter values have been calculated, the $V_{F_{ij}}$ can be obtained using Eq. (27.7). Otherwise, change the values of changeable parameters, and go to STEP 2.

STEP 4:

If $j = h_i$ where h_i represents the number of possible values of the functional performance measure F_i, calculate the variance V_{F_i} using Eq. (27.8). Otherwise, increase the value of j by 1 (i.e., $j = j + 1$), and go to STEP 2.

STEP 5:

If $i = m$ where m represents the number of the required functional performances, calculate the variance V_F using Eq. (27.9). Otherwise, increase the value of i by 1 (i.e., $i = i + 1$), and go to STEP 2.

Fig. 27.1 Pictures from the manufacturing shop. **a** The coating machine, and **b** the drying oven (Courtesy of Huaying Soft-packing Equipment Plant Ltd.)

STEP 6:

Evaluate whether the overall variance of the functional performance indices of an adaptable product, V_F, has been minimized considering the design parameters, $\boldsymbol{P}^{(D)}$. If the V_F has not been minimized, modify the values of design parameters, and go to STEP 2. Otherwise, the current values of design parameters are selected as the optimal design.

27.5 A Case Study

Coating machine, as shown in Fig. 27.1a, is a typical equipment in light industry to apply a thin liquid film on a substrate, such as paper, to improve surface properties. The drying oven, shown in Fig. 27.1b, is a key module in the coating machine to dry the coating liquid on the substrate [10]. In this research, design of the drying oven was selected as the case study example to demonstrate the newly developed robust adaptable design method.

In the operation process of the drying oven, the liquid film on the substrate can be dried when the substrate goes through the drying oven. In this case study, the solvent content after drying, M (%), was selected as the functional requirement to be satisfied. According to the equations introduced by Cohen et al. [10] and Jin [11], the solvent content after drying, M (%), can be calculated by:

$$M = 1 - S - \frac{50E \cdot L}{3V \cdot G} \tag{27.11}$$

E the drying rate of the oven (kg/(h·m^2))
L the length of the drying oven (m)
V the speed of the substrate in the drying oven (m/min)
G the unit coating weight (g/m^2)
S the solid content of the coating (%)

In this case study, it is assumed that the required target values of the solvent content after drying should be $z_{11} = 3\ \%$, $z_{12} = 5\ \%$, and $z_{13} = 8\ \%$, and the probabilities of the required target values are equal to 1/3. The relationships between the satisfaction index, I_1, and functional performance measure, M, is modeled by:

$$I_1 = \begin{cases} 1 - (M - 3\%)^2, & \text{for the required value of } M \text{ is } 3\,\% \\ 1 - (M - 5\%)^2, & \text{for the required value of } M \text{ is } 5\,\% \\ 1 - (M - 8\%)^2, & \text{for the required value of } M \text{ is } 8\,\% \end{cases} \tag{27.12}$$

In this case study, the length of the drying oven, L, and the speed of the substrate in the drying oven, V, were selected as design parameters. The drying rate of the oven, E, was selected as the adaptable parameter which can be modified by the user in the product operation stage. The unit coating weight on the substrate, G, and the solid content of the coating, S, were selected as the changeable parameters. In the operation stage, the values of the unit coating weight on the substrate can be $y_{11} = 70$ (g/m^2) and $y_{12} = 90$ (g/m^2). The values of solid content of the coating can be $y_{21} = 70\ \%$, $y_{22} = 75\ \%$ and $y_{23} = 80\ \%$. In this case study, it is assumed that the values of changeable parameters follow the discrete uniform probability distributions.

Under the influence of uncertainties in parameters, the values of the unit coating weight, solid content and the speed of substrate can fluctuate from their target values. In this case study, it is assumed that the variances of the unit coating weight, the solid content, and the substrate speed are 1 (g^2/m^4), 0.3 % (−) and 0.01 (m^2/min^2), respectively.

According to Eq. (27.7), if $V_{F_{1i}}$ represents the overall variance of the ΔI_1 on the condition that the target value of functional performance measure M is equal to z_{1i}, the $V_{F_{1i}}$ can be calculated by:

$$V_{F_{1i}} = \sum_{j=1}^{2} \sum_{k=1}^{3} \left[\frac{V_{F_{1i}}\left(y_{1j}, y_{2k}\right)}{6} \right] \tag{27.13}$$

where $V_{F_{1i}}\left(y_{1j}, y_{2k}\right)$ represents the variance of the ΔI_1 on the condition that the changeable parameters are selected as $G = y_{1j}$ and $S = y_{2k}$ ($j = 1, 2$, $k = 1, 2, 3$), and the target value of functional performance measure M is equal to z_{1I} ($i = 1, 2, 3$).

According to Eqs. (27.8), (27.9) and (27.13), the overall variance of the functional performance indices of the drying oven can be calculated by:

$$V_F = \sum_{i=1}^{3} \left\{ \sum_{j=1}^{2} \sum_{k=1}^{3} \left[\frac{V_{F_{1i}}\left(y_{1j}, y_{2k}\right)}{18} \right] \right\} \tag{27.14}$$

According to Eqs. (27.10)–(27.14), the optimization model of the robust adaptable design can be establish as follows:

Find: The length of the drying oven, L, and the speed of the substrate in the drying oven, V

Table 27.1 The optimization results of the robust adaptable design of the drying oven

Cases	Case 1	Case 2	···	Case 18
Length of the drying oven, L (m)	6	6	···	6
Speed of substrate in the drying oven, V (m/min)	25	25	···	25
Solvent content after drying, M (−)	3 %	5 %	···	8 %
Unit coating weight on the substrate, G (g/m^2)	70	70	···	90
Solid content of the coating, S (−)	70 %	75 %	···	80 %
Drying rate of the oven, E (kg/(h·m^2))	4.725	3.5	···	2.7
The variance $V_{F_{1i}}(S, G)$, $i = 1, 2, 3$	0.003	0.012	···	1.48×10^{-32}
The overall variance, V_F	0.004			

$$\begin{aligned} \text{Minimize} \ : \quad & V_F \\ \text{Subject to} \ : \quad & M \in \{3\,\%, 5\,\%, 8\,\%\} \\ & S \in \{70\,\%, 75\,\%, 80\,\%\} \\ & G \in \{70, 90\} \\ & 0 \le E \le 10 \\ & 5 \le L \le 12 \\ & 15 \le V \le 120 \end{aligned} \tag{27.15}$$

In the optimization process, the adaptable parameter, E, can be calculated by:

$$E = \frac{3G \cdot V \cdot (1 - S - M)}{50L} \tag{27.16}$$

Based on the possible values of the required functional performance and the changeable parameters, 18 possible cases in the product operation stage can be created considering different combinations of these values. Using the algorithm introduced in Sect. 27.4, the optimization results can be obtained as shown in Table 27.1.

Form Table 27.1, we know that the values of design parameters, i.e., the length of the drying oven and the speed of substrate in the drying oven, are equal to 6 (m) and 25 (m/min), respectively. According to Eq. (27.16), the values of adaptable parameter (the Drying rate of the oven) under different cases during the operation stage of the drying oven can also be obtained. By using the optimized design parameter values and the adaptable parameter values, the overall variance V_F considering the 18 possible cases of the adaptable design has been minimized.

27.6 Conclusions

A robust adaptable design method has been developed to identify the adaptable product whose parameter values can be changed in the product operation stage. The relationships among product functional performances, parameters, and variations of parameters have been described by a mathematical model. An optimization-based design model has been developed to achieve the robust adaptable design. The newly developed robust adaptable design method has the following advantages.

- Compared with the current adaptable design methods, the newly developed robust adaptable design method can be used to identify an adaptable product whose functional performances are the least sensitive to uncertainties of parameters.
- Compared with the current robust design methods, the newly developed robust adaptable design method can be used to identify a product whose parameter values and functional performances can be changed in the operation stage.

Despite the progress, a number of issues need to be further addressed to improve the current robust adaptable design method. The following two aspects will be considered in our future research: (1) Development of the robust adaptable design method considering the changes of structure/configuration in the operation stage of a product. (2) Development of a generic robust adaptable design method considering the changes of product as the functions of the product lifecycle time parameter.

Acknowledgments The authors wish to thank the National Science Foundation of Guangdong Province, China (No: 8351503101000001), the Natural Sciences and Engineering Research Council (NSERC) of Canada, and the National Natural Science Foundation of China (No: 51175315) for providing financial supports to this research.

References

1. Gu P, Hashemian M, Nee AYC (2004) Adaptable design. CIRP Annu—Manuf Technol 53(2):539–557
2. Sand J, Gu P (2006) AdaptEx: extending product lifecycles through strategic product upgrades. In: ElMaraghy HA, ElMaraghy WH (eds) Advances in design. , , pp 111–119
3. Li Y, Xue D, Gu P (2008) Design for product adaptability. Concurrent Eng: Res Appl 16(3):221–232
4. Gu P, Xue D, Nee AYC (2009) Adaptable design: concepts, methods and applications. Proc Inst Mech Eng, Part B, J Eng Manuf 223(11):1367–1387
5. Xue D, Hua G, Mehrad V, Gu P (2012) Optimal adaptable design for creating the changeable product based on changeable requirements considering the whole product lifecycle. J Manuf Syst 31(1):59–68
6. Parkinson A, Sorensen C, Pourhassan N (1993) A general approach for robust optimal design. J Mech Des, Trans ASME 115(1):74–80
7. Taguchi G, Chowdhury S, Taguchi S (2000) Robust engineering. McGraw-Hill, USA
8. Du X, Chen W (2000) Towards a better understanding of modeling feasibility robustness in engineering design. J Mec Des, Trans ASME 122(4):385–394
9. Beyer HG, Sendhoff B (2007) Robust optimization—a comprehensive survey. Comput Methods Appl Mech Eng 196(33):3190–3218
10. Cohen E, Brookling G, Edgar B (1992) Modern coating and drying technology. Wiley, USA
11. Jin RH (2007) A design of the oven for coating. Hangzhou Chem Ind 37(3):37–40

Chapter 28
A Method to Compute Early Design Risk Using Customer Importance and Function-Flow Failure Rates

Bryan M. O'Halloran, Robert B. Stone and Irem Y. Tumer

Abstract The general method for using customer importance to validate a design misses the important opportunity to influence how that design is created. The intent of this research is to capture risk during the conceptual design stage. Risk is calculated using function-flow failure rates and customer importance. This allows the designer to effectively identify what functionality should be given additional importance during the generation and selection of design concepts. Functional risk using customer importance has not yet been investigated. In general, risk is implemented later in the design process. A generic process to calculate the risk is presented, then applied to an example where a subset of function-flows have been identified as generating 75 % of the risk.

28.1 Introduction

In general, customer importance gives designers evaluation criteria and metrics for use later in design. Using this information in this form offers designers a basic *design validation* scheme. However, this information becomes more useful when it is used during the design process to create the final solution. In a similar respect, reliability is often used during the design process to make design changes.

B. M. O'Halloran (✉) · R. B. Stone · I. Y. Tumer
Oregon State University, Corvallis, OR 97330, United States of America
e-mail: ohallorb@onid.oregonstate.edu

R. B. Stone
e-mail: rob.stone@oregonstate.edu

I. Y. Tumer
e-mail: irem.tumer@oregonstate.edu

A. Chakrabarti (ed.), *CIRP Design 2012*, DOI: 10.1007/978-1-4471-4507-3_28,

Mitigating unreliable components to improve the health of a design is good practice. Strictly using reliability will miss fixing the problems that are most important to the customer. Merging customer importance with reliability allows the designer to effectively identify what should be mitigated in the design. This need is addressed here by transitioning customer importance to consequence and reliability to the probability of failure. These values are used to calculate the risk of losing functionality in the design.

28.2 Background

28.2.1 Methods for Early Design Risk and Reliability

Research in reliability engineering has focused on moving analyses earlier in the design process. The Function Failure Design Method (FFDM) is used in the early stage of design and provides the relationship between functions and failure modes. FFDM is a structured formulation of the function-failure analysis method introduced by Tumer and Stone [1]. FFDM has several advantages including reduced user workload, using an archived failure knowledge base, being usable during early design, using the Functional Basis, component taxonomy, and failure mode taxonomy as a formalized failure language, and being practical for electrical and mechanical systems [1]. This method also aids the designer by using a function-based concept generator approach helping to streamline the design process [2]. FFDM utilizes knowledge bases that link product function to failure modes. Data in the knowledge base is archived in a function-component matrix and a component-failure mode matrix, which reduces the need for a designer to have a large intellectual knowledge base. The function-failure mode matrix is produced by multiplying the function-component and component-failure mode matrices together.

Two independent advances have been made to improve FFDM. The Risk in Early Design (RED) method has prioritized failures using severity and occurrence [3, 4]. An interactive tool has been developed to aid designers in creating reliable designs (http://idecms.srv.mst.edu/ide/). The Function Failure Rate Design Method (FFRDM) has prioritized failures using only occurrence [5]. This method was constructed by applying failure rates of components to the knowledge base. The designer can use a functional model to determine which failure modes are the most probable to occur.

System Failure Modes and Effects Analysis (System FMEA) is a bottom up approach to risk analysis used to provide design recommendations to reduce the risk of failure, or risk of losing functionality [6–8]. The Risk Priority Number (RPN), used to determine which functions to provide recommendation for, is determined by the designer and does not use any information from the customer. The designer makes an assumption that each function holds an equal importance to the customer, and in the event of a failure, the function that fails is irrelevant. These analyses provide the designer information before components have been selected to facilitate making customer-oriented design decisions.

Quantitative Risk Assessment (QRA), also referred to as Probabilistic Risk Assessment (PRA) in the nuclear power industry and Performance Assessment within waste repositories, is a top-down design risk methodology. Fundamentally, this method answers three question; what can go wrong?, how likely is this to occur?, and what are the consequences [9]? A typical QRA relies on executing five independent steps. The first step is to list all undesirable events. Next the initiating events (IE) are found, accounting for disturbances to normal operation. The first two steps allow for specific analyses to be conducted such as Fault Tree and Event Tree, answering the first question. Step four uses all evidence to evaluate probabilities for each scenario, answering question two. Finally, the list of scenarios is rank ordered by risk value. The benefits of QRA include improved exploration over other commonly used methods of potential risks, improved likelihood of finding risk associated with complex interactions within the system, and it focuses on uncertainty quantification to facilitate decision making [10]. QRA is a comprehensive failure analysis and requires a large workload to perform. The method provided in this research is simple to perform and produces high-level results specifically tied to the customer requirements.

Previous research has determined a method to calculate system reliability during functional design [11]. This work presents a process to compute constant function-flow failure rates using component failure rates and historical data relating components to functions. The historical data from the Design Repository (http://designengineeringlab.org/delabsite/repository.html) leverages information from existing designs to more accurately model the function-flow failure rates. A five-step process is used to calculate the reliability. Examples show the usefulness of the methodology and compare the results to the traditional reliability block diagram. This research uses the function-flow failure rates to generate a probability of failure for the element risk calculations.

28.2.2 Functional Modeling

Functional modeling is a standard part of many engineering design methodologies and is used to describe a design at an abstract level. Generating a functional model is done early in the design process before components have been chosen in an original design problem or before reviewing existing component choices in a redesign problem. The design process, in general, follows five steps; *project definition and planning*, *specification definition*, *conceptual design*, *product development*, and *product support* [12]. The functional design method is used in the first stage of conceptual design.

The format of functional models consists of functions connected by flows. The three types of flow include material, energy, and signal. Stone [13] standardized

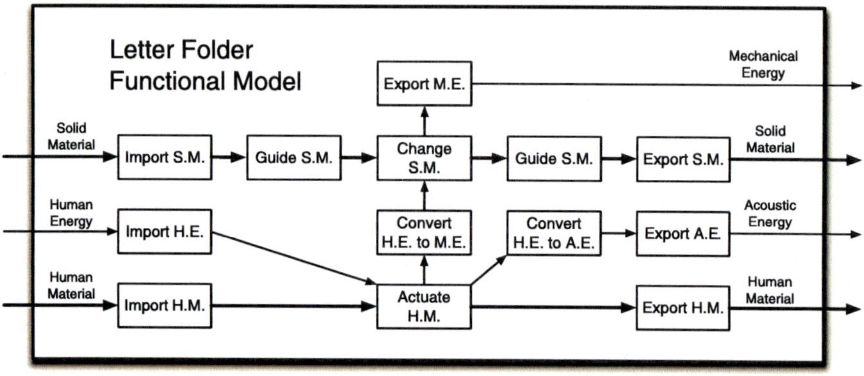

Fig. 28.1 Functional model for the letter folder design

functional modeling by creating a common Functional Basis, which provided a set number of functions and flows to describe the entire design space. The Functional Basis provides consistency across functional models of different designs. This research uses failure rates for the terms in the functional basis. The functional basis is also used as the functional language to generate the functional model in Fig. 28.1.

28.3 Research Approach

This section presents a process to calculate function-flow risk. The intent is to provide a step-by-step process that the designer can reproduce on other designs. The requirements to complete this process include customer requirements for the intended design and a functional model.

Step 1: Gather Failure Rate and Customer Importance Data

First, failure rate data is needed to calculate the probability of failure. To gather function-flow failure rate data, tables in [11] can be used. These tables provide minimum, maximum, and weighted average failure rates for each function-flow in the functional basis. Prior work presented a method to calculate these failure rates for other functional taxonomies or function-to-component relationships. The *weighted average* failure rates are recorded for each function-flow in the functional model. The weighted average is used since this value observes all occurrences where a component has solved a function, then methodically uses these relationships with component failure rates to calculate the average failure rate.

Second, customer importance is required to determine consequence data. Consequence data is typically set on a scale from 1 to 10 where 1 is the least important and 10 is the most important. There are a variety of ways to obtain customer importance data. It is encouraged in this step to use an ethnographic survey where the customer is presented with a list of requirements to provide

Table 28.1 Functional-flow failure rate data for the letter folder design [11]

Function(flow)	Failure/rate
Import'solid	7.57
Guide'solid	17.01
Change'solid	113.18
Guide'solid	17.01
Export'solid	9.69
Import'human'energy	3.12
Import'human'material	3.14
Actuate'human'energy	0.18
Actuate'human'material	0.18
Export'human'energy	1.39
Export'human'material	3.65
Convert'human'energy'to' acoustic	X
Export'acoustic	9.26
Convert'human'energy'to' mechanical	5.31
Export'mechanical	4.89

feedback. In general, ethnographic surveys use a 1–5 scale. This can be transitioned to a 1–10 scale by multiplying by 2. When multiple customers are involved, an average value for each customer requirement should be calculated. An example of the result to step 1 is shown in Table 28.1 and Table 28.2.

Step 2: Generate Consequence Data

Transitioning from customer importance data to function-flow importance depends on the mapping between customer needs and flows in the functional model. During the generation of the functional model, customer requirements are correlated to flows. The importance value for each customer requirement is directly translated across this mapping to arrive at an importance for a flow. In many cases, this will result in a flow listed several times with varying importance values. These values should be averaged to determine a final flow importance value. Any function that exists along a flow naturally receives the importance rating of that flow (i.e., all functions along a flow have the same importance value).

Risk calculations use the term consequence, not importance. This research transitions the terminology of customer importance to consequence by noting that the two are equal. The higher importance a specific function has, the higher the consequence is to the customer for losing that functionality. Similarly, lower importance results in a lower consequence of loss. An example of this step can be seen in Table 28.4.

Step 3: Element Risk Calculations

Risk is defined using the following equation:

$$K(t) = P_{fail}(t) * C \qquad (28.1)$$

Table 28.2 Ethnographic survey for importance of requirements based on individual customers

Evaluation[1–5]	Customer requirements	Customer 1	Customer 2	Customer 3	Avg customer
OPERATION	Quiet operation	3	2	3	2.67
	Simple to use	5	5	5	5.00
	Lowest functioning individuals can operate	5	5	5	5.00
	Requires minimal dexterity	5	4	4	4.33
	Involves user into the process	3	2	3	2.67
	Ergonomic	5	5	5	5.00
CAPABILITIES	Bi-folds paper	3	3	3	3.00
	Tri-folds paper	5	5	5	5.00
	Folds different paper sizes	5	5	5	5.00
	Different patterns or styles	3	2	3	2.67
GENERAL	Portable	3	1	2	2.00

where K(t) is the risk as a function of time, P_{fail}(t) is the probability of failure as a function of time, and C is the consequence. The probability of failure is defined using the following equation:

$$P_{fail}(t) = 1 - R(t) \tag{28.2}$$

where R(t) is the reliability as a function of time. Function-flow failure rates are used to calculate the probability of failure and ultimately the risk associated to each function-flow.

The function-flow consequence is used as an indicator to the designer to show where the customer places the most importance in the design. Combining this with the probability of failure presents a true risk value to the designer. The risk values for each function-flow quantitatively represent the impact of failure. For example, function-flows with a low probability of failure and a relatively high consequence only raise a moderate to low risk level to the customer. The customer has identified not wanting to lose the functionality, although the low probability of failure assures the customer that the likelihood this loss will occur is low. On the other hand, the risk level is raised when the probability of failure and the consequence are both high values. The product of two high values generates a high risk for lost functionality. An example of this step is summed up in Fig. 28.2.

28.4 Example

An example showing how to calculate function-flow risk is presented here. This example uses the design of a letter folder, which is currently employed at a rehabilitation facility. The intent of this letter folder is to be used by low-functioning

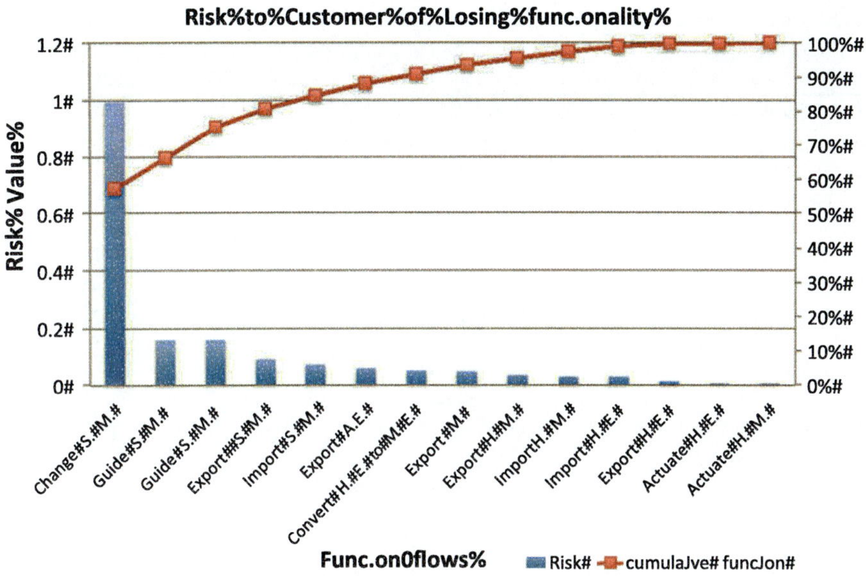

Fig. 28.2 Risk indicators for losing functionality

individuals to tri-fold a standard 8.5 × 11 inch piece of paper. Further details about the design are shown throughout this section as needed.

The functional model in Fig. 28.1 was constructed using the functional basis and according to the procedures outlined by Stone [13]. The high-level, or black-box, function of the letter folder is to fold paper. Using language from the functional basis, this functionality is described by the verb-noun pair *change solid material*. This functional language was chosen specifically to match with the failure rates used in step 1 to gather the failure rate data. In order to achieve the customer requirements, additional functionality has been included in the model.

Step 1: Gather Failure Rate and Customer Importance Data

Using the tables generated in prior work [11], failure rates are recorded for each function-flow in the functional model. The function *convert human energy to acoustic* did not have any values so a risk calculation cannot be performed. This information is shown in Table 28.1.

An ethnographic survey, seen in Table 28.2, was used to generate how important each requirement was for each customer. The values under each customer, on a scale with increasing importance from 1 to 5, indicate how important each requirement is to a specific customer. Actual customer names were not included for confidentiality. The final column was added after the survey was finished to display the average customer importance for each customer requirement.

Step 2: Generate Consequence Data

Each customer requirement was then correlated to a flow in the functional model (e.g., Correlate CR's to FLOW). These values are multiplied by 2 to convert

Table 28.3 Average customer importance for each flow

Customer requirements	Correlate CR's to FLOW	Avg customer importance(X2)
Quiet operation	Acoustic energy	5.33
Simple to use	Human material and energy	10
Lowest functioning individuals can operate	Human material and energy	10
Requires minimal dexterity	Human material and energy	8.67
Involves user into the process	Human material and energy	5.33
Ergonomic	Human material and energy	10
Bi-folds paper	Mechanical energy and solid material	6
Tri-folds paper	Mechanical energy and solid material	10
Folds different paper sizes	Solid material	10
Different patterns or styles	Solid material	5.33
Portable	Human material and energy	4

to a 1–10 scale. Table 28.3 displays the average customer importance for each flow(s).

Table 28.3 shows the same flow listed several times in the second column. These flows evolved from different customer requirements, but each only exists once in the functional model. This means that each flow enters and exits the functional model only one time each. During this final step, each value in the *Avg customer importance (X2)* column of Table 28.3 is averaged for a specific flow. For example, the values 10.00, 10.00, 8.67, 5.33, 10.00, and 4.00 average to a value of 8.00, which becomes the final customer importance value for Human Material and Human Energy. The term customer importance was employed in the original ethnographic study. A natural change in terminology takes place where customer importance is called consequence. Table 28.4 lists the final consequence values for each flow in the functional model.

A final transition must be made from flow consequence to function-flow consequence. Each function along a flow naturally inherits the consequence value of that flow. For example, *import, guide, change, guide, and export solid material* all individually have a consequence of 7.83. Functions cannot be distinguished specifically from flow to determine consequence values using the method described in this research. However, the probability of failure is specific to each function-flow, which presents unique risk values.

Step 3: Element Risk Calculations

Equations 28.1 and 28.2 were used to arrive at the results in Fig. 28.2. The Pareto chart displays the risk value using the bar graph and the cumulative risk function using the line graph. The Pareto principal states that approximately 80 % of the effects come from 20 % of the causes. It was identified that 75 % of the risk is a result of 21 % of the functionality; *guide-, change-, and guide-solid material.* This result shows that three of the functions along *solid material* flow can be

Table 28.4 Final consequence value for each flow in the functional model

Flow	Consequence
Acoustic energy	5.33
Human material and energy	8.00
Mechanical energy	8.00
Solid material	7.83

grouped together and identified as the risk indicators for the letter folder design. Specific risk values include *guide solid* (9 %), *change solid* (57 %), *and guide solid* (9 %). These three function-flows should be carefully solved and evaluated later in the design process.

28.5 Discussion

It is difficult to justify changing a functional model, because doing so tells the customer they are getting something different than what they originally requested. Computing risk values of function-flows does *not* constitute reasoning to change the functional model even though risk is calculated during functional design. Instead, these values are used to indicate the risk to the designer of losing functionality based on what the customer has identified as important. The designer should use the risk values to identify which components can carefully solve the function-flows to achieve maximum reliability and reduce the likelihood of lost functionality. Consequence values, used in the risk calculation, are set values that come straight from the customer. Reducing *component reliability* is an optimal mitigation strategy for high-risk function-flows. Efforts to select highly reliable components should be made during concept generation and detailed design based on functions with high consequence.

The highest risk function-flow for the letter folder, used as an example in this paper, is *change solid* with approximately 57 % of the total risk. In this design, *change solid* is the least desirable function-flow to have a high-risk value since this is also the black-box function. However, this is critical information for the designer to have during the early stage of design. When design concepts are created and evaluated, solutions to *change solid* should hold more weight than solutions to other function-flows in the design.

28.6 Conclusion

In this research, risk combines the consequence of losing functionality with the probability that functionality will be lost. The general method of using customer importance to validate a design misses the important opportunity to influence how that design will be solved. The intent of this research is to capture risk during the conceptual stage of design. This information is used to indicate to the designer which function-flows require more effort during the generation and selection of design concepts. A generic process to arrive at the functional risk is shown in detail and includes an example. This process can be employed using different techniques to generate customer importance and function-flow failure rates.

28.7 Future Work

The future goal of this research is to provide an integrated risk and reliability methodology to be used in early design. Reliability will be used to meet design requirements while risk will inform the designer of important areas of the design that require specific attention for generating design concepts.

Acknowledgments This research was funded in part by DARPA (Subaward to FA8650-10-C-7079 with Palo Alto Research Center). The opinions, findings, conclusions, and recommendations expressed are those of the authors.

References

1. Tumer IY, Stone RB, Van Wie Michael (2004) The function-failure design method. Mech Des 127:397–407
2. Stone MESRB, Tumer IY (2005) Linking product functionality to historic failures to improve failure analysis in design. Res Eng Des 16:96–108
3. Lough KG, Stone RB, Tumer IY (2006) The risk in early design (RED) method: likelihood and consequence formulations. In: ASME international design engineering technical conferences, Philadelphia
4. Lough KG, Stone RB, Tumer IY (2007) The risk in early design method (RED). J Eng Des, vol 18
5. O'Halloran BM, Stone RB, Tumer IY (2011) Link between function-flow failure rates and failure modes for early design stage reliability analysis. In: international mechanical engineering conference and exposition, Denver
6. Stamatis DH (2003) Failure mode and effect analysis: FMEA from theory to execution. Milwaukee: ASQ quality press
7. Modarres M, Kaminskiy M, Krivtsov V (2010) Reliability engineering and risk analysis a practical guide: CRC press
8. Blischk WR, Murthy DNP (2000) Reliability modeling, prediction, and optimization. Wiley NY
9. Kaplan S, Garrick J (1981) On the quantitative definition of risk. Risk Anal 1:11–27
10. Apostolakis GE (2003) How useful is quantitative risk assessment? In: massachusetts institute of technology engineering systems division working paper series: ESD-WP-2003-05
11. O'Halloran BM, Stone RB, Tumer IY (2011) Early design stage reliability analysis using function-flow failure rates. In: Design, theory, and methodology Washington
12. Ullman DG (2003) The mechanical design process, 3rd edn. McGraw-Hill, New York
13. Stone R, Wood K (2000) Development of a functional basis for design. J Mech Des 122: 359–370

Chapter 29
Integrating Systematic Innovation, Interaction Design, Usability Evaluation and Trends of Evolution

S. Filippi and D. Barattin

Abstract Day by day, more R&D divisions of modern industries start adopting inventive design tools and methods. Creativity needed in designing complex products cannot be left to subjective behavior; it must be helped and exploited by a systematic approach. The research described in this paper aims at developing a design framework focused on interaction issues, by exploiting the systematic approach of the theory of inventing problem solving TRIZ. The final result should integrate design, evaluation, and evolution issues. For this reason, the starting point consists in three tools already developed by the authors' research group: the interaction design guidelines—IDGL, the usability evaluation multi-methods—UEMM, and ITRE, a gatherer of interaction trends of evolution. All of them contain generic elements both of the TRIZ theory and the interaction design field; for this reason the proposed integrated approach could be exploited in completely different contexts. A first prototype of the framework has been developed as a Microsoft Access database. Its validation has started with two experiences in the field. Results are reported and discussed in the last section of the paper.

29.1 Introduction

Marketing laws force shifting the focus from technologies to user needs and this requires new design methods. They should implement the usability concept as a major goal. At the same time, products must answer to rigorous innovation

S. Filippi (✉) · D. Barattin
DIEGM Department, University of Udine, via Delle Scienze, 208, 33100 UD, Italy
e-mail: filippi@uniud.it

D. Barattin
e-mail: daniela.barattin@uniud.it

A. Chakrabarti (ed.), *CIRP Design 2012*, DOI: 10.1007/978-1-4471-4507-3_29,

requirements. All of this should come by minimizing time and resources. Our research group has already studied and developed three methods related to interaction issues: the interaction design guidelines—IDGL, the usability evaluation multi-method—UEMM—and the interaction trends of evolution—ITRE. They belong to interaction design, evaluation and evolution respectively. These three methods have already demonstrated their effectiveness thanks to early Microsoft Access prototypes. Up to now they have been far from each other, and this implies wasting time in transferring data among them.

The main goal of this research is integrating these three methods in a new one, to verify if their synergic exploitation generates improved results. IDGL, UEMM and ITRE are quite similar to other methods already present in literature, regarding architecture and components. This is important because it gives the research a wider scope, in order to be applied in different environments, where other tools and methods rather than IDGL, UEMM and ITRE could exist.

The paper develops as follows. The background section offers an overview of the state of the art regarding interaction design, evaluation, and trends of evolution. After that, IDGL, UEMM and ITRE are described. The next section is the core of the research. First, the elements that could act as building blocks for relating the three methods are highlighted; then, the generation of the relationships takes place. Finally, the validity and effectiveness of the results start to be checked thanks to two experiences in the field, focused on designing a refrigerator and an energy delivery system. Some discussion and hints for future research close the paper.

29.2 Background

The systematic innovation is the most important fundamental of this research. Two out of the three domains considered here—design and evolution—heavily exploit it. Moreover, the three domains present some tools and methods well documented in literature, as summarized in the following paragraphs

29.2.1 Systematic Innovation

Systematic innovation has become important in the last thirty years mainly thanks to the TRIZ theory. It aims at systematizing and generalizing the design process, in order to boost the generation of innovative ideas, concepts, and products, as much as possible [1–3]. TRIZ provides tools to manage many aspects of design, mainly the mechanical one, as the forty inventive principles, the contradiction matrix, the features and the trends of evolution (http://www.triz40.com, [4–8]).

Recently TRIZ has been deeply revised and updated, because modern design problems are much more complex than the ones Altshuller dealt with. For these reasons TRIZ has evolved towards a more articulated theory named OTSM. This is the Russian acronym for the general theory of powerful thinking [9, 10]. OTSM revises the bases of TRIZ in a formal way, while keeping contradictions as central element. The classic TRIZ tools have been conveniently modified to manage more than one contradiction simultaneously.

29.2.2 Interaction Design

Interaction design deals with the dialogue between users and products, based on the usability concept [11, 12]. It is quite young as a discipline, but there already exist methods and tools that effectively help the interaction design process. For example, the interdisciplinary approach with a user-system focus [13] describes a complete design process. It collects input data about product usage and users' behavior and impressions thanks to a software package named dynamic product usage information system—D'PUIS. Process mining—ProM—techniques allow analyzing these data, while the compromise decision-support-problem—cDSP—tool is devoted to generate the interaction design guidelines [14].

Some other elements can be used supporting the interaction design process. Some examples are the ISO 9241 standard (http://www.usabilitynet.org), the Shneiderman's eight golden rules of interface design (http://www.usabilitynet.org) and the ten heuristics developed by Nielsen [15]. They are accounted for as checking tools during the research activities.

29.2.3 Usability Evaluation

Usability evaluation was born before interaction design. Its main goal is evaluating the dialogue between users and products. Many evaluation methods are described in literature [16, 17]. They belong to two main categories, requiring real user involvement or not. Moreover, sometimes very different skill and knowledge are needed to adopt one evaluation method instead of another.

Some studies focus on highlighting the best criteria to characterize the usability evaluation methods [18, 19], in order to introduce some automatisms in selecting the best methods given the characteristics of the evaluation field. The most used criteria are the costs, time and resources available for the evaluation, the accuracy and objectiveness of the results, the development stage of the product under evaluation, etc. [19]. These criteria are exploited by algorithms like the analytic hierarchy process—AHP [20].

29.2.4 Trends of Evolution

Trends of evolution are postulates consisting in the states through which a product evolves. These states are snapshots and do not contain any indication about the way a product passes from a state to the next ones [3]. This specific issue is in charge of the evolution laws. Altshuller highlighted eight trends in TRIZ, from system combination to segmentation and coordination among different actions. Some software packages implementing TRIZ theory, as TriSolver (http://www.trisolver.eu), consider and exploit the trends of evolution. Many researchers have exploited TRIZ trends of evolution and customized them in order to make them compatible with their own application domains, as in manufacturing [21], or biology [8].

29.3 Methods and Tools

29.3.1 IDGL

The interaction design guidelines—IDGL—is a method that suggests products easy to use, ready to go, needless of heavy modifications or redesign [22]. The design procedure runs as follows. First, the product under development is characterized. Then, a questionnaire is automatically generated, in order to collect users' needs and expectations. The house of interaction—HOI, a data structure derived from the classic house of quality, collects and analyses the answers, highlighting meaningful interaction requirements that will be exploited for the development of a list of solution concepts. This happens thanks to two TRIZ tools, the forty inventive principles and the contradiction matrix, revised to make them compatible with the interaction design domain. The result consists in forty-seven interaction principles and in the relationship matrix, an improved release of the contradiction one. Thanks to this matrix, the IDGL selects the best interaction principles related to users' needs and organizes some examples as design suggestions in the guidelines.

29.3.2 UEMM

The usability evaluation multi-method—UEMM—suggests the activities to perform in order to verify the goodness and correctness of a product from the usability point of view. The classic evaluation methods are the building blocks of the UEMM. The goal is selecting the best evaluation methods, given the peculiarities of the adoption field. The UEMM exploits the analytic hierarchy process—AHP—to select the evaluation methods [20, 23]. Microsoft Access has been used again to implement the UEMM. The UEMM adoption activities start with the input of data describing the application fields, as available design resources, strategies, and product features. Thanks to all these pieces of information, together with an importance weighting algorithm, the UEMM generates an ordered list of the best methods to use during usability evaluation activities.

29.3.3 ITRE

The interaction trends of evolution—ITRE—collects pieces of information that describe the evolution of several aspects of the user-product interaction. Its goal is to help foreseeing the direction changes of the development of a specific product. The theoretical background comes again from the TRIZ theory, and the process exploited in highlighting trends in the interaction field is the same. The results consist in nine interaction trends, each of them showing examples that help in comprehending their meaning and their possible exploitation as design tools. An Access database implements ITRE. Thanks to the input of the product features, ITRE locates it in the evolution scenario, by showing where it is from each trend point of view. Then, ITRE suggests possible improvements automatically.

29.4 Activities

The first paragraph of this section focuses on highlighting the elements that could allow the integration of IDGL, UEMM and ITRE. Next, possible relationships among them are discovered, together with a classification of them.

29.4.1 Highlighting Integration Elements

The search of pieces of information representing possible contact points requires some criteria as the following ones: the elements must be complete and independent, they must have an active role with clear goals and they should be generic as much as possible. Table 29.1 shows the elements highlighted thanks to this investigation, classified by domain instead of by method to emphasize once again the effort in dealing with these topics from a point of view as general as possible.

Table 29.1 Elements that could be used for the integration

Interaction design	Usability evaluation	Trends of evolution
ID1 Final design solutions	UE1 Context forms	TE1 Trend examples
ID2 Interaction principles	UE2 Pairwise comparison matrix	TE2 Evolution trends
ID3 Interaction requirements	UE3 Decision matrix	TE3 Patterns
ID4 Relationship matrix	UE4 Usability evaluation multi-methods	TE4 Product evolution forms
ID5 Questionnaires	UE5 Usability evaluation methods	
ID6 Product forms		
ID7 House of interaction		

29.4.2 Building the Relationships

The elements of Table 29.1 are now collected in a vocabulary to be used as the starting point to define some sort of grammar rules, the relationships among the methods. Again, some criteria are needed. The one used for the generation of the element set has been simply to avoid duplicates. In this case, no one was found, so all the elements have been considered. Figure 29.1 (left) shows the result of these activities. The larger circle represents the boundary of the interaction domain, regarding design, evaluation, and evolution. The smaller one contains all the vocabulary elements used to generate the integrated method. Open triangles represents the three methods IDGL, UEMM and ITRE, with their own elements.

An association between two elements could be set if some similarity exists about structure or behavior, or if an element can positively influence another one without losing its identity and its native functionalities.

Concerning the first criterion, there are pieces of information that show similar structure and/or behavior. For this reason they can be classified in two categories and represented by common structures as follows.

- Input category. This set contains the product forms (ID6), the context forms (UE1) and the product evolution forms (TE4). All of them collect information required to start the activities scheduled by the methods, and focused on the specific product and its adoption domain. The resulting new structure, named input form, contains all the pieces of information collected by the previous ones.
- Output category. Here there are the final design solutions (ID1), the usability evaluation multi-methods (UE4) and the trend examples (TE1). All of them are expected results of the three methods. The new output form structure contains all the outcomes of the method application in the field: design solutions, evaluation multi-methods and examples of trend adoption.

An important remark concerning the method integration is that the elements of the output category can be somehow considered also as inputs. For example, ID1 can be thought about as a sort of early prototype and used as input for the evaluation inside the UEMM. Moreover, the UE4 outcomes, expressed as evaluation

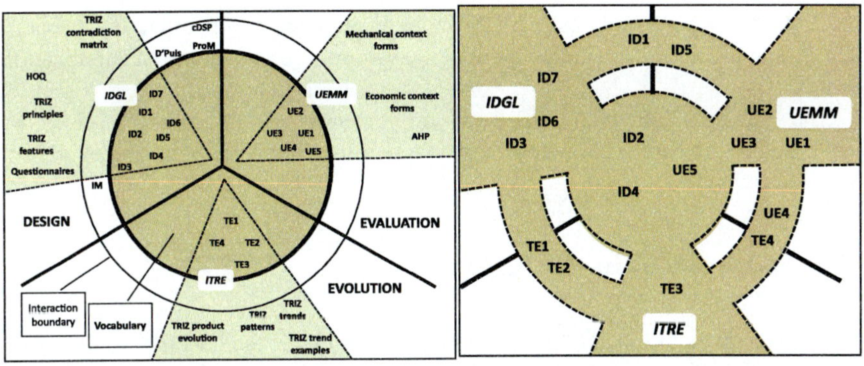

Fig. 29.1 Highlighting the vocabulary elements (*left*) and representation of the relationships among the three methods (*right*)

notes and remarks, could be investigated to study the evolution of interaction in ITRE. TE1 could represent the starting product data for the IDGL, and so on.

The other elements in Table 29.1 are too different each other, so the second criterion takes place. In order to satisfy it, elements must be active. We intend an element as active here if it acts as a control or a mechanism, as meant in the IDEF formalism [24]. A control is something that sets the requirements/rules for a specific function to execute its job properly; a mechanism supports function execution as a helping tool. The items of Table 29.1 can be considered controls and mechanisms as follows.

- The evolution trends (TE2) and the usability evaluation methods (UE5) can be controls for the generation of design solutions, because they allow checking if these solutions follow the correct evolution path.
- The interaction principles (ID2) are controls for ITRE, because each time a new state is added in a trend, they allow checking its correctness.
- The interaction requirements (ID3) are mechanisms for UEMM. They are the basic elements the usability evaluation focuses on, because users have voted them as the most important issues to consider.
- The interaction requirements (ID3) and the usability evaluation methods (UE5) can be mechanisms for ITRE. Given that the UE5 are needed for evaluating the product, they can be used as well in the following activities where product evolution is evaluated. Once again, the ID3 represent the starting point for a correct evaluation during the insertion of a new state in an evolution trend.

Figure 29.1 (right) shows the result of the second step of the integration procedure. The larger circle represents the vocabulary boundary. The inner circle shows the common elements of the three methods, while the elements allowing the dialogue between pairs of methods are in the three sectors around it. The highlighted collection of controls and mechanisms shows that they are not linked anymore to the method they belong to, but to the moment they are needed and exploited during the design process. This represents the shifting from specific products towards general ones, and this goes again towards generalization, because it allows applying the integrated method in a wider set of situations and application contexts.

The integrated method has been implemented in Microsoft Access, by exploiting many parts of previous implementations of the three separated methods.

29.5 Experiences in the Field

Two experiences in the field have been carried on to test the benefits coming from the adoption of the integrated method. Four designers made two groups and applied the separate methods and the integrated one respectively. Some metrics have been defined to allow highlighting both quantitative and qualitative differences. These metrics consist in the following four indicators: M1—number of solution concepts; M2—time required by the design process; M3—number of modifications suggested by the evaluation phase; M4—provenience variety of the pieces of information in the solution concepts.

29.5.1 Design of a Refrigerator

The goal of this experience is evaluating the contribution of the integrated method in designing a refrigerator as usable as possible. First design activities are common to the separated vs. the integrated approach. Fifteen users fill the questionnaire generated automatically by the software package and their answers are inserted in the HOI. The analysis of these data allows highlighting 24 interaction requirements.

Now the procedures diverge. The adoption of the three methods separately comes first. The IDGL alone generates forty-eight solution concepts. The multi-method proposed by the UEMM to evaluate the solution concepts contains seven evaluation methods. Sixteen remarks are highlighted. The application of the corrections generates a final amount of solution concepts equal to fifty-one. ITRE allows checking the validity of the solution concepts. Some problems are highlighted again and, after their correction, designers have fifty-three solution concepts available, all of this in two weeks.

The procedure based on the integrated method starts with the availability of evolution trends and interaction principles together. The trends are mainly used to highlight directions to follow and the principles to develop solution concepts. The number of resulting solution concepts is equal to seventy-two. They are automatically stored in the output form. Then the generation of the evaluation multi-method takes place. Six usability evaluation methods are present in it. Seven revisions constitute the outcome of this activity. Once the solution concepts are corrected, they can be easily checked by considering the principles and trends that

Table 29.2 Indicator values of the two experiences in the field

Approach	Refrigerator				Energy delivery system			
	M1	*M2*	*M3*	*M4*	*M1*	*M2*	*M3*	*M4*
Three separated methods	53	2 weeks	16	Scarce	71	5 weeks	21	Scarce
Integrated method	75	1 week	7	High	94	2 weeks	7	High

the software package makes available. At the end of this reviewing activity, designers have seventy-five solution concepts ready to be exploited, all in a week. The results are summarized in Table 29.2.

29.5.2 Design of an Energy Delivery System for Cars

The second experience in the field deals with the design of an energy delivery system for cars. The procedure is the same as in the previous case; both the separated and integrated approaches are performed again. In order to highlight the effectiveness of the integrated one, one of the resulting solution concepts is "Cars connect autonomously to energy supply system when it is low on battery and this can happen virtually everywhere". Table 29.2 reports the values of the indicators also for this second experience.

29.6 Discussion

From the qualitative point of view, the solution concepts generated using the integrated approach contain explicit links to interaction principles and trends of evolution, while the ones coming from the adoption of the separated methods contain direct links only to principles. Moreover, the reviewing phase highlights less criticisms and this bears the good quality of the solution concepts. More elements must be considered when dealing with quantitative concerns. First, the integrated approach generates more solution concepts than the separated one. This is not a mere numerical improvement, but there is the synergic exploitation of several pieces of information in them. Time required by the two processes is different, and in the integrated approach it is always shorter, more or less equal to one half. An important consideration focuses on the tools derived by the TRIZ theory. They have been enhanced here, by contextualizing them thanks to interaction-related contents and integrating them with new functionalities like the exploitation of the positive relationships in the relationship matrix, etc.

29.7 Conclusions

The research described in this paper aimed at generating an interaction design method to take advantage from the synergic integration of three existing methods, the IDGL for interaction design, the UEMM for usability evaluation and the ITRE for managing trends of evolution. This integration started by highlighting common elements, as generic as possible, in order to make the procedure applicable in other contexts, presenting similar design methods and tools. These common elements allowed developing relationships among the three methods, both from the data structure and the algorithmic point of view. Many pieces of information have been merged and this implied a strong reduction of redundancy. Two experiences in the field started to demonstrate the effectiveness and the added value of the integrated design approach. The amount of solution concepts rose up and the quality of them seems to be better than before. Moreover, the integrated approach requires fewer resources to be applied.

Some open problems still remain, like making the selection of the meaningful trends automatic, defining a roadmap for the adoption of the usability evaluation multi-methods (now this activity is completely left to designers), and searching for other tools to enhance the relationships among the three methods.

References

1. Altshuller G, Rodman S (1999) The innovation algorithm: TRIZ, systematic innovation and technical creativity. Technical Innovation Center Inc, Worcester
2. Zlotin B, Zusman A (1999) The ideation approach to the search, development, and utilization of innovation knowledge. J Altshuller Inst TRIZ Stud 21–29
3. Lin CC, Luh DB (2009) A vision-oriented approach for innovative product design. Adv Eng Inform 23:191–200
4. Pin SC, Haron F, Sarmady S, Talib AZ, Khader AT (2011) Applying TRIZ principles in crowd management. Saf Sci 49:286–291
5. Zhang J, Shang J (2010) Research on developing environmental protection industry based on TRIZ theory. Proced Environ Sci 2:1326–1334
6. Domb E (2011) The 39 features of Altshuller's contradiction matrix. Available on http://www.triz-Journal.com/archives/1998/11/d/index.htm. Accessed 20 may 2011
7. Craig S, Dekoninck E, Harrison D, Cripps A (2006) An analysis of the technological development of the transparent building envelope using the TRIZ trends of evolution. Proceedings of the 13th CIRP international conference on life cycle engineering. Leuven, Belgium
8. Bogatyrev NR, Bogatyreva OA (2009) TRIZ evolution trends in biological and technological design strategies. Proceedings of the 19th CIRP design conference-competitive design. Cranfield University, pp 293–299
9. Cavallucci D, Khomenko N (2007) From TRIZ to OTSM-TRIZ: addressing complexity challenges in inventive design. Int J Prod Dev 4:4–21
10. Becattini N, Cascini G, Rotini F (2011) Correlations between the evolution of contradictions and the law of identity increase. Proc Eng 9:236–250
11. Hertzum M (2010) Images of usability. Int J Hum-Comput Interact 26(6):567–600

12. ISO (2007) ISO 20282-3: Ease of operation of everyday products—part 3: test method for consumer products. ISO 20282. International Organization for Standardization Geneva
13. Koca A, Funk M, Karapanos E, Rozinat A, Martens J, Brombacher A (2008) A soft reliability: an interdisciplinary approach with a user-system focus. Qual and Reliab Eng Int 25(1):3–20. Available on http://www.softreliability.org/DokuWiki/publications. Accessed 04 Oct 2011
14. Koca A, Brombacher A, Panchal J, Mistree F (2009) Engineering soft reliability in product realization. Proceeding of the ASME 2009: international design engineering technical conferences, pp 739–751
15. Kamper RJ (2002) Extending the usability of heuristics for design and evaluation: lead, follow get out of the way. Int J Hum-Comput Interact 14(3):447–462
16. Koutsabasis P, Spyrou T, Darzentas J (2007) Evaluating usability evaluation methods: criteria, method and a case study. Proceeding of the 12th international conference on human-computer interaction: interaction design and usability pp 569–578
17. Freiberg M, Baumeister J (2008) A survey on usability evaluation techniques and an analysis of their actual application. Institute of Computer Science, University of Wurzburg, Germany
18. Hartson HR, Andre TS, Williges RC (2001) Criteria for evaluating usability evaluation methods. Int J Hum-Comput Interact 13(4):373–410
19. Doroodchi M, Nikmehr N (2008) Overview and evaluation of usability evaluation methods for E-learning system. Proceedings of the 2008 international conference on E-learning, E-business, enterprise information system and E-government, EEE. Las Vegas, pp 123–128
20. Saaty TL (2008) Decision marking with the analytic hierarchy process. Int J Serv Sci 1:83–98
21. Roderburg A, Klocke F, Koshy P (2011) Principles of technology evolutions for manufacturing process design. Proced Eng 9:294–310
22. Filippi S, Barattin D (2011) Exploiting TRIZ tools in interaction design. Proceeding of the TRIZ future 2011, ETRIA 2011. Dublin, Ireland
23. Kurttila M, Pesonen M, Kangas J, Kajanus M (2000) Utilizing the analytic hierarchy process (AHP) in SWOT analysis. Policy Econ 1:41–52
24. Tsironis L (2008) Empowerment the IDEF0 modeling language. Int J Bus Manag 3:109–118

Chapter 30
Robust Design of a Dynamic Mechanical System Based on Component Modal Synthesis

Y. Chen, J. Pang, J. Zhang, D. Xue and P. Gu

Abstract The dynamic performances of a mechanical product under uncertainty have significant influence on the quality of the product. In this research, the relationships between dynamic performance and design parameters are derived based on component modal synthesis (CMS). The first natural frequency is used in this work as the dynamic performance measure. The accuracy of the dynamic model is verified by FEA method. An optimal robust design model is introduced to minimize the dynamic performance variation. An industrial application is given to illustrate the effectiveness of the robust design approach considering dynamic performance.

30.1 Introduction

The product quality is influenced by many uncertainty factors in design and environment parameters such as design dimensions and physical and mechanical characteristics of materials (e.g., damping coefficient, thermal conductivity, friction coefficient and strength limit). The actually manufactured dimensions are usually different from the expected design dimensions. The material properties at different

---

Y. Chen · J. Pang
Department of Mechanical Engineering, Tianjin University, 300072 Tianjin, China
e-mail: chenyongliang@tju.edu.cn

J. Zhang · D. Xue
Department of Mechanical and Manufacturing Engineering, University of Calgary, Calgary T2N 1N4, Canada
e-mail: dxue@ucalgary.ca

P. Gu (✉)
School of Engineering, Shantou University, 515063 Shantou, China
e-mail: peihuagu@stu.edu.cn

A. Chakrabarti (ed.), *CIRP Design 2012*, DOI: 10.1007/978-1-4471-4507-3_30,

locations of a component may also be different. To reduce the impact of uncertainty factors, robust design is often used. Robust design is an approach to identify the design parameters that provide the minimum impact on system performance [7]. The analytical robust design method introduced by Gu et al. [3] is a new method to establish the relationship among design parameters, performance measures, and their variations. Compared with the robust design based on design experiment where design parameters are only assigned with small numbers of values, the analytical robust design method can be used to identify the optimal design parameters accurately.

In analytical robust design, however, the relationships among functional performance, design parameters, and variations of performance measures and design parameters need to be obtained first. Accurate and efficient robustness assessment is essential for obtaining a robust solution. Huang and Du [5] investigated features of model-based methods for robustness assessment. Zhang et al. [11, 12] used Taylor series expansion to calculate the variations of functional performances considering explicit mathematical functions. Du and Zhang [2] proposed a semi-second-order Taylor expansion method to evaluate robustness. Sun et al. [8] developed a method to calculate the variations of functional performance using engineering models described by implicit mathematical functions.

Most applications of robust design have been concerned with static performance in mechanical engineering and process systems, and applications in structural dynamics are rare. Zang et al. [10] studied the robust design of a vibration absorber with mass and stiffness uncertainty based on Taguchi approaches, which shows that robust design methods have great potential for application in structural dynamics to deal with uncertain structures. Lu and Li [6] proposed a stability-based robust design approach which incorporated system eigenvalues that are less sensitive to model uncertainty. In dynamic analysis of mechanical structures, Rayleigh method is often used to derive the relationships between the first natural frequency and design parameters [1]. For complex structures with several components, component modal synthesis (CMS) offers an excellent tool for the dynamic analysis of the mechanical systems with uncertainties [4]. CMS is a substructure coupling analysis method frequently employed in studying structural dynamics. CMS allows the derivation of the behavior of the entire assembly from its components. CMS replaces the system matrices by a smaller set of interface DOFs between substructures and truncated sets of normal mode generalized coordinates. Despite the advantages of CMS, modeling of the relationships between dynamic performance and design parameters by explicit mathematical functions is not a trivial task in industrial applications. To address the challenges, the symbolic mathematical tool, Matlab, is used in this research.

This paper introduces a robust design method where relationships between dynamic performance and design parameters are derived by free interface CMS. The natural frequency of the structure is used to determine the nontrivial solutions of the generalized Eigenvalue equation. From these implicit functions, the dynamic performance variations can be calculated. An optimal robust design method is developed to minimize the dynamic performance variation. An industrial application is developed to minimize the variations of first natural frequencies for a storage roller in a mechanical system.

30.2 Modeling of the Relationships Between Dynamic Performances and Design Parameters Using Free Interface CMS

Component Modal Synthesis (CMS) was introduced by Hurty in 1960 [9]. Since then, many methods and applications have been developed based on this approach [9]. In CMS, the dynamic characteristics of a complex structure can be modeled effectively and accurately by a few modes with modal coordinates. The CMS can reduce the number of degrees of freedom to simplify the computation process.

For an undamped system, each CMS substructure is defined by stiffness and mass matrices. By using the substructure motion equations, the system's motion equations can be obtained. The matrix equation of the motion for the system is described by:

$$\overline{M}\ddot{p} + \overline{K}p = 0 \tag{1}$$

where $\overline{M}$ is the generalized mass matrix of the structure, $\boldsymbol{p}$ is the generalized modal coordinate, and $\overline{\mathbf{K}}$ is the stiff matrix.

The connection points in the whole structure should satisfy the constraint equations. Among the generalized modal coordinates, since they have to satisfy the constraints, only some coordinates are independent. These independent coordinates are described as $\boldsymbol{q}$.

The transformation from $\boldsymbol{q}$ to $\boldsymbol{p}$ is conducted by:

$$p = Sq \tag{2}$$

For the free interface method, the transformation matrix S has the form:

$$S = \begin{bmatrix} -B_d^{-1} & B_1 \\ 1 & 0 \end{bmatrix} \tag{3}$$

where B is the matrix defined in the constraint equations. B is divided into two submatrices B_d and B_I in the same way as the independent coordinates.

After replacing the relevant terms in Eq. (1) using Eqs. (2) and (3), the equation of motion in the reduced space is obtained. The reduced stiffness and mass matrices of the CMS substructure are described by:

$$M = S^T \overline{M} S \tag{4}$$

$$K = S^T \overline{K} S \tag{5}$$

In the reduced system, master DOFs are used to couple the CMS super-element to other elements and/or CMS super-elements.

The following equation can be obtained:

$$M\ddot{q} + Kq = 0 \tag{6}$$

The mode shape and the corresponding natural frequency can be obtained using the following equation.

$$\left(K - \omega_r^2 M\right)\varphi_r = 0 \tag{7}$$

where ω_r and φ_r are the r-th mode shape and the corresponding natural frequency of the vibration.

The natural frequency is solved using the following equation:

$$\det\left(K - \omega_r^2 M\right) = 0 \tag{8}$$

As $\boldsymbol{K}$ and $\boldsymbol{M}$ are the functions of the design parameters for the structure, the relationship between the natural frequency and design parameters is defined by these implicit functions in Eq. (7) and (8). For some structures, the explicit mathematical functions can also be obtained from the implicit functions using the symbolic mathematical tools.

30.3 Calculation of the Dynamic Performance Variations by Taylor Series Expansion

In dynamic design of a mechanical system, uncontrollable factors often cause variations of design parameters, leading to deviations of dynamic performances from their target values. In this research, the deviations of design parameter values from their expected values are modeled by variations of the uncontrollable factors. Suppose F_i ($i = 1,\ldots, n$) represent the n dynamic performance measures, D_j ($j = 1,\ldots, m$) represent the m design parameters, μ_j represent the m mean values of design parameters D_j, and $\Delta D_j = D_j - \mu_j$ represent the m deviations of design parameters D_j from their expected values, the relationships between dynamic performance measures and design parameters can be described using the following implicit mathematical functions:

$$\begin{cases} f_1(F_1,\cdots,F_n,\ D_1,\cdots,D_m) = 0 \\ \quad\vdots \\ f_l(F_1,\cdots,F_n,\ D_1,\cdots,D_m) = 0 \end{cases} ;\quad l \geq 1 \tag{9}$$

Suppose F_i ($i = 1, \ldots, n$) follow the distributions (R_i, σ_{Fi}^2), where R_i represent the desired functional performance measures, and σ_{Fi}^2 (i = 1, ..., n) represent the variances of dynamic performance measures. In the optimal robust design, the σ_{Fi}^2 need to be minimized. In this work, the D_j ($j = 1, \ldots, m$) are assumed to be mutually independent and follow normal distributions $N(\mu_j, \sigma_{Dj}^2)$. The nominal values of design parameters μ_j need to be determined in robust design, while the variances σ_{Di}^2 are obtained from industrial practices.

According to the Taylor series expansion, the following relationships can be obtained ([11], [8]):

$$\sigma_{Fi}^2 = \sum_{k=1}^{m}\left[\sum_{j=1}^{l}\left(c_{ij}b_{jk}\right)^2\sigma_{Dk}^2\right] \tag{10}$$

where $b_{ik} = \partial f_i/\partial D_k$ ($i = 1, \ldots, l; j = 1, \ldots, n; k = 1, \ldots, m$), and c_{jk} ($j = 1, \ldots, n$; $k = 1, \ldots, l$) are the elements of C. The C is the left inverse of A (i.e., $CA = I$) where $a_{ij} = \partial f_i/\partial F_j$.

30.4 An Optimal Robust Design Model

In robust design, the dynamic performance should satisfy the required functional requirement. By taking the expected values of both sides of Eq. (9), the following relationships can be obtained:

$$f_i(R_1, \cdots, R_n, \mu_1, \cdots, \mu_m) = 0, \quad i = 1, \ldots, l \tag{11}$$

In this study, variations of dynamic performances are modeled by the variances of these performance measures. When robustness of a product is determined by several functional performance measures, the impact of different individual functional performance measures on the product robustness can be defined by different multipliers. Suppose σ_{F}^2 is the variance of the overall performance, σ_{F}^2 can be defined by $\sum w_i \sigma_{F_i^2}$, where σ_{Fi}^2 (i = 1,…, n) are the variances of Fr_i, and w_i are the weighting factors representing the importance of the relevant performance measures.

In robust design of a mechanical product, dynamic performances should satisfy the required functional requirements, and the variations of dynamic performances under the influence of uncontrollable factors should be minimized. According to Eqs. (10) and (11), an optimal robust design model can be defined by:

$$\text{Minimize} : \sigma_{\mathrm{F}}^2 \tag{12}$$

$$\text{Subject to } f_i(R_1, \cdots, R_n, \mu_1, \cdots, \mu_m) = 0, \quad i = 1, \ldots, l$$

30.5 A Case Study

Paper storage roller is a key component in zero-speed automatic splicer, which is a key auxiliary module in printing equipment especially in web printing press. Many uncertain factors in the paper storage roller affect its dynamic performance. These uncertainties are caused by design dimensions, material parameters, production processes and operation environment. For the design of the storage roller, in addition to satisfy the design requirements, the variations of performance should be minimized. Robust design is a way to effectively reduce the sensitivity of product performance to uncertain factors.

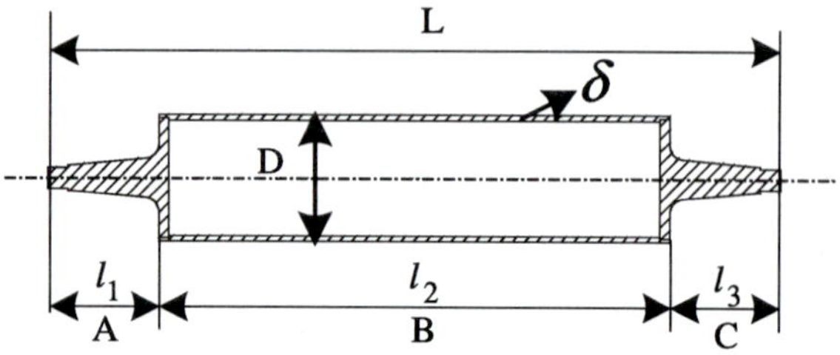

Fig. 30.1 Structure, substructures and parameters of the paper storage roller

30.5.1 Calculation of the Natural Frequency of the Roller

According to different manufacturing processes, many types of roller structures, including monobloc casting, assembling and sleeve, can be used. In this case study, the monobloc casting type model with symmetrical structure was selected. The section drawings are shown in Fig. 30.1. Carbon structural steel was selected as the material.

The storage roller is divided into three sections, as shown in Fig. 30.1. The whole roller is composed of three substructures, which are respectively numbered as A, B and C. Since section A and section C have the same geometric and material properties, their equivalent elastic modulus is E, length is l_1 (i.e., $l_3 = l_1$), and linear mass density is m_1 (i.e., $m_3 = m_1$). For section B, equivalent elastic modulus is E, length is l_2, and linear mass density is m_2.

Based on the free interface CMS method, substructures A and C are cantilever beams, and substructure B is free beam on both ends.

Using symbolic operation toolbox in Matlab, the first natural frequency of the whole storage roller is obtained from Eq. (8). In the obtained formula, the first natural frequency is described as a function of the design parameters.

To verify the accuracy of the derived formula, the finite element analysis (FEA) model in ANSYS was built.

In this work, 6 groups rollers with different wall-thickness δ were selected, and the results calculated using the CMS method and the finite element method are shown in Table 30.1.

Since the maximum relative error between the CMS result and the finite element method result is 1.9 %, the formula derived from CMS can effectively satisfy the engineering requirements.

30.5.2 Optimal Robust Design of the Storage Roller

When the values of l_1 and l_2 are determined, the first natural frequency f_1 of the storage roller is mainly determined by d, d_1, D and E where D is the outer diameter of section B, and d is the inner diameter of section B, and d_1 is the diameters of section A or C. The design objective is to satisfy the equation $f_1 = 293$ Hz while minimizing

Table 30.1 Comparision between the CMS method and the finite element method

Wall-thickness δ (mm)	8	7	6	5	4	3
CMS method f1 (Hz)	235.37	247.80	262.97	282.15	307.66	344.41
Finite element method f1 (Hz)	234.60	247.60	262.40	281.09	305.28	337.79
Relative error (%)	0.33	0.08	0.22	0.37	0.77	1.92

Table 30.2 Comparison of the initial design and the dynamic robust design

	Design variable		Performance and its variance	
	d_I (mm)	d (mm)	μ_{f_1} (Hz)	$\sigma^2_{f_1}$
Initial design	42.4	172.6	289.8	3.68
Optimal robust design	47.03	168.99	293	3.62

the variance of f_I. In this work, d and d_I are selected as design variables, and D and E are selected as noise variables. According to the design requirement, the initial values were selected as $D = 180$ mm, and $E = 2.1 \times 10^5$ N/mm^2.

According to the above discussion, the optimal robust design of the paper storage roll can be modeled by:

$$\begin{cases} \quad \min(\sigma^2_{f_1}) \\ s.t. \quad \mu_{f_1} = 293, \\ \quad y \leq 19, \\ \quad m \leq 14, \\ \quad 40 \leq d_1 \leq 50, \\ \quad 164 \leq d \leq 174 \end{cases} \tag{13}$$

where σ^2_{fI} is the variance of first natural frequency f_I, μ_{fI} is the mean value of the first natural frequency, y is deflection, and m is mass.

In this work, the variances of d_I, d, D and E were selected as 0.1, 0.06, 0.01, and 6,000, respectively.

From the above discussion, the mathematical model described in Equation (13) is a nonlinear programming problem with single-objective and multi-constraints. The SQP (Sequential Quadratic Programming) method was used to solve this problem. By comparing the optimal robust design and the initial design as shown in Table 30.2, we can see the standard deviation of the performance can be reduced by the optimal robust design method.

30.6 Conclusions

Advantages of the developed optimal robust design method considering dynamic performances are summarized as follows.

- By using the component modal synthesis and symbolic mathematical tool, the relationships between dynamic performance and design parameters for a complex mechanical system can be derived and modeled by the analytical mathematic functions.
- The dynamic performance variances can be calculated using the Taylor series expansion, and the optimal robust design considering dynamic performances can be achieved by minimizing the dynamic performance variances.
- The industrial application has demonstrated that the optimal robust design method is effective for improving robustness of the mechanical systems.

Acknowledgments The authors wish to thank the Natural Science Foundation of Guangdong Province, China (No: 8351503101000001), the Key Technologies R&D Program of Guangdong Province, China (Nos: 2009A080202010 and 2010A080202010), and Natural Sciences and Engineering Research Council (NSERC), Canada for providing financial supports to this research.

References

1. Chen YL, Liu S, Wang HT, Gu PH (2010) Analytical robust design of a gear milling machine headstock. Appl Mech Mater 37–38:162–165
2. Du X, Zhang Y (2010) An approximation approach to general robustness assessment for multidisciplinary systems. J Comput Inf Sci Eng 10(1):011003 (9 pages) doi:10.1115/1.3290765
3. Gu PH, Lu B, Spiewak S (2004) A new approach for robust design of mechanical systems. CIRP Ann Manuf Technol 53(1):129–133
4. Hinke L, Dohnal F, Mace BR, Waters TP, Ferguson NS (2009) Component mode synthesis as a framework for uncertainty analysis. J Sound Vib 324:161–178
5. Huang BQ, Du X (2007) Analytical robustness assessment for robust design. Struct Multi Optim 34(2):123–137
6. Lu XJ, Li HX (2011) Robust design for dynamic system under model uncertainty. J Mech Des 133(2)021006 (7 pages) doi:10.1115/1.4003342
7. Park GJ, Lee TH, Lee KH, Wang KH (2006) Robust design: an overview. AIAA J 44(1):181–191
8. Sun GZ, Hong B, Zhang J, Xue DY, Gu PH (2011) Robust design with engineering models described by implicit mathematical functions. J Mater Sci Forum 697–698:661–664
9. Yang X, Tang HL, Liao BY (1983) Machine tool dynamics. Machine Industry Press, Beijing
10. Zang C, Friswell M, Mottershead J (2005) A review of robust optimal design and its application in dynamics. Comput Struct 83(4–5):315–326
11. Zhang J, Li SP, Bao NS, Zhang GJ, Xue DY, Gu PH (2010) A robust design approach to determination of tolerances of mechanical products. CIRP Ann Manuf Technol 59:195–198
12. Zhang J, Bao NS, Zhang GJ, Gu PH (2009) Analytical approach to robust design of nonlinear mechanical systems. Front Mech Eng China 4(2):203–214

Chapter 31
Adaptronic Solution Principles: Potential to Flexible Design

David Inkermann, Carsten Stechert and Thomas Vietor

Abstract This contribution introduces the concept of adaptronic solution principles (ALP) to support the development of products for changing requirements. Based on an approved understanding of flexible products, three aspects are proposed to characterize ALP. Since adaptronic systems are object of current research, a set of solutions is analyzed concerning the aspects design flexibility, product flexibility and realization concept. This analysis points out major characteristics of ALP and their potential to close conflicting requirements and affect product properties within the use phase automatically. Finally, the introduced ALP are applied for the development of an adaptronic coupler within a new drive concept for serial kinematic manipulators.

31.1 Introduction

The development of sustainable products requires an entire consideration of the product and its interactions with the joining systems and consumers. Since the demand for individual products has increased rapidly new development and production strategies are needed to deal with the ever-growing number of product variants. Furthermore, foreseen and unforeseen changes of environments and use cases of a product result in challenges which have to be handled within the product development process. Figure 31.1 introduces four basic design approaches to

D. Inkermann (✉) · C. Stechert · T. Vietor
Technische Universität Braunschweig, Institute for Engineering Design,
Langer Kamp 8, 28106 Braunschweig, Germany
e-mail: d.inkermann@tu-braunschweig.de

A. Chakrabarti (ed.), *CIRP Design 2012*, DOI: 10.1007/978-1-4471-4507-3_31,

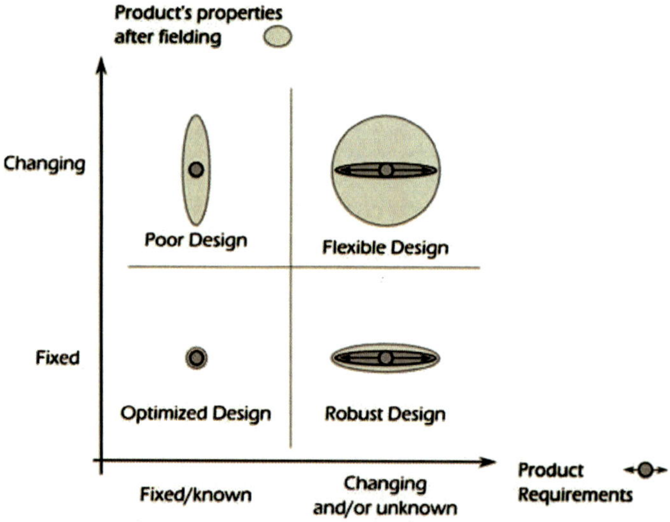

Fig. 31.1 Flexibility, robustness and interrelation to products properties and requirements (based on [1])

fulfill product requirements and provide adequate properties during the use of the product.

In case of well-known and fixed product properties within use phase, an optimized design (e.g. light weight road bike frame for top athletes) is reasonable. When the requirements are unknown or change during the development, product properties can be fixed in order to cover a wide range of these properties in the use phase (e.g. use of reliability factors for the dimension of a bike frame).

Contrary to this robust design approach a flexible design intends to fulfill changing requirements in an optimal way after launching a product, for instance by means of an adjustable saddle height for bikes. While robust design normally results in oversizing and therefore increased system weight, flexible design allows for optimal system weight and enhanced performance. Following the concept of flexible design, changes during the development process as well as changes of environments and customer needs after the product launch can be handled more easily. In order to conquer the fuzziness of the term flexibility and border the scope of this contribution, the following definition of flexibility according to [1] will further be used:

> The flexibility of a design is the property of a system or a product that allows it to respond to changes in its initial objectives and requirements—both in terms of capabilities and attributes—occurring after the system or product has been fielded, i.e., is in operation, in a timely and cost-effective way.

This definition also covers the understanding of adaptive products proposed e.g. by [2]. With regard to reduced duration for changes not only manual adaption within the use phase is suitable but automatic adaptions have to be offered. Providing automatic adaption often results in mechatronic systems expanding the function of a mechanical system by the spatial and/or functional integration of sensors and actuators and the use of a control system to guarantee functionality [3]. In this contribution the concept of adaptronic solution principles (ALP) as a special class of mechatronics is introduced. Based on the assumption that effective adaption is often limited by the functionality of the mechanical system, rather than control algorithms, the focus of this contribution is on integration of multifunctional elements. Furthermore, the application of ALP within the design process is discussed and the influence of both product and process flexibility is highlighted.

31.2 Adaptronic Solution Principles

As pointed out before, adaptronic systems constitute a specific class of mechatronic systems. While mechatronics consist of discrete constructional elements, an adaptronic solution intends to integrate functionality into constructional elements. According to the introduced structure of an adaptronic systems [4] adaptronic solution principles (ALP) are hereinafter referred to principles that are based on the application of multifunctional materials. The right hand side of Fig. 31.2 illustrates the major elements of an ALP related to the given structure of an adaptronic system. The basis for this class of solution principles are physical effects allowing modification of certain functional variables. With regard to the underlying physical effect the multifunctional element is integrated into the mechanical structure defining number, arrangement, and shape of the working surfaces. The impact of the ALP to systems properties is described by a mathematical model, which is also used to design the control of the system. In this way system properties are adapted

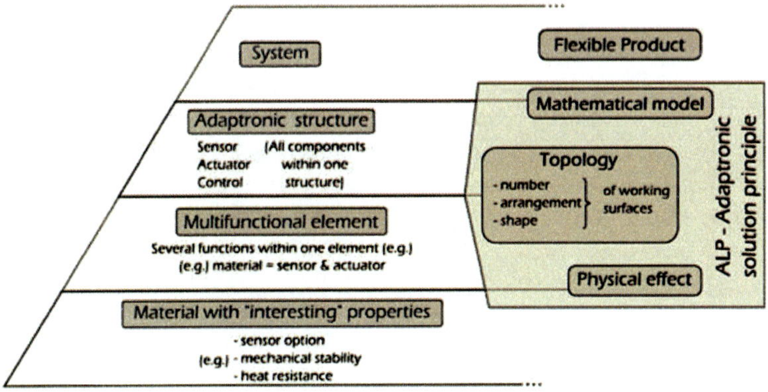

Fig. 31.2 Structure of adaptronic systems (based on [4]) and elements of an ALP

to changing environments which leads to a flexible product according to the aforementioned definition.

The introduced elements of an ALP allow for a more generic view of adaptronic systems suitable for earlier phases of the design process. In order to assure reasonable applications of ALP, two questions have to be addressed:

1. How can ALP be used to increase flexibility of a product as well as flexibility within the design process. What kind of functional extensions are provided by an ALP and how can they be used to extend the useability of a product?
2. Which restrictions (functional and geometrical) have to be attended to allow optimal spatial integration of sensors and actuators?

To gain a deeper insight to the use of ALP in engineering design three aspects for the description of adaptronic solution principles are proposed. Those will be presented in the following section.

31.3 Aspects to Describe Adaptronic Solution Principles

In order to derive ALP for more flexible designs numerous adaptronic systems have been analysed. Aside from known classifications covering specific applications such as active noise and vibration reduction e.g. [5] or specific groups of multifunctional elements [6] the presented analysis intends to address the aforementioned questions. Hence, it contains three aspects:

- *Design Flexibility* characterizes the possibility to achieve new degree of freedom within the design process when applying an ALP. These degrees of freedom are identified by analyzing the demanded properties [7] of the overall system. For instance reducing system weight while providing high stiffness at the same time. The aspect design flexibility therefore supports the top-down view starting from the requirements and objectives defined for the whole system.
- *Product Flexibility* characterizes the ability of an ALP to affect product properties and functionality in order to cope with changing requirements after fielding the products. Based on the elements of the ALP (physical effect), the influence on system properties and objectives is analyzed. Contrary to the design flexibility this represents a bottom-up analysis. With regard to adaption strategies, only automatic runtime adaption is considered.
- *Realization Concept* characterizes kinematic (placement of actuators), working principle, and topology (arrangement, number and shape of working surfaces) of an ALP [8].

In the following section these aspects are described in more detail.

31.3.1 Design Flexibility

During the design process numerous decisions have to be made by the designer. In order to match defined product requirements each decision is based on information gathered within the task clarification phase [9]. Since this information is rather vague, early decisions can endanger the success of a product development. Moreover, several requirements get into conflict with one another when choosing a specific working principle. In order to provide more flexibility within the design process ALP should provide the potential to close conflicting requirements. This potential is pointed out using the 39 technical parameters proposed by [14]. First, conflicts addressed on system level have to be stated. Breaking these conflicts down to component level more specific conflicts can be formulated. Within the presented analysis (cf. Table 31.1) both levels are addressed.

31.3.2 Product Flexibility

While the aspect of design flexibility focuses on the design phase, product flexibility addresses the use phase of a product. With regard to a reasonable application of ALP the analysis captures product properties enabling the coverage of changing requirements while using the product. Furthermore, the possibility to adjust adaption strategy and range is analyzed. For instance, the adaption strategy of the start-up-clutch [10] is based on the characteristic of the deployed multifunctional material e.g. shape memory alloy. Thus, the adjustment of the adaption requires either a replacement of the material or the adjustment of the pre-stressing. An automatic adjustment e.g. change of control algorithm is not possible. Moreover, additional functions provided for the user are considered. For instance an additional function could be the monitoring of an important product property e.g. monitoring of the bearing condition or the capturing of dynamic process forces within a milling process.

31.3.3 Realization Concept

Third aspect of the analysis is the realization concept of an ALP. While the first two aspects provide information for the selection of an adequate ALP the following points characterize the implementation and show up restrictions to implement the chosen ALP in the further development process. The realization concept contains kinematic aspects as well as the working principle.

Kinematic. In order to affect several product properties additional degrees of freedom have to be provided. With regard to the kinematic the following approaches are differed ([11], [12]):

- *Serial actuator placement.* Placed between two structural elements or at the end of the passive structure the actuator serves an additional degree of freedom. For instance the compensation-mechanism introduced by Puzik et al. [15] constitutes a serial placement of actuators at the end of the passive structure.
- *Parallel actuator placement.* With regard to a parallel placement of the actuators two cases have to be distinguished [12]: either the actuator is placed between a basis and the endpoint of the passive structure (*fully parallel*) or the actuator element is arranged parallel to one or more elements of the passive structure (*semi parallel*). In both cases the actuator affects inherent degrees of freedom of the passive structure resulting in elastic deformations.

Working principle. The working principle describes the physical effect used to affect functional properties, e.g. adjustment of friction via variation of normal force [10]. Based on the selected effect an adequate multifunctional element as well as a specific topology is required [8]. Therefore, the kind of multifunctional elements and the structural shape of the applied actuator e.g. circular stack actuator or rectangular patch are part of the aspect working principle.

31.4 Examples for Adaptronic Solution Principles

With regard to the introduced aspects, numerous adaptronic systems have been analyzed in order to derive generic system characteristics for the application of ALP. In Table 31.1 four systems are listed, representing a wide range of applications: machine tools, robotic systems, and automotive. Since the detailed analysis and description of the working principles has been part of earlier work [8], focus of this analysis are the aspects: design flexibility, product flexibility, kinematic, and physical effect of the underlying ALP.

As can be seen in Table 31.1 each ALP provides specific properties to increase design as well as product flexibility. In order to highlight the benefit of the proposed aspects to characterize ALP on an abstract level, in the following section the application of ALP for the development of machine elements for high dynamic power transmission will be introduced.

[1] Joint project of the Institute for Robotic and Process Information (iRP) and the Institute for Engineering Design (IK) at the TU Braunschweig.

Table 31.1 Representatives of adaptronic systems and derived system characteristics related to the aspects design flexibility, product flexibility, and realization concept

No.	Adaptronic solution	Characteristics of ALP
1.	3D-piezo-compensation-mechanism for online-error-compensation of serial robotic systems  [15]	**Design flexibility** (addressed design conflicts) 1. High productivity versus high manufacturing precision 2. Low weight of moving object versus high strength **Product flexibility** • Affect stiffness of the robotic system • Automatic adjustment of adaption characteristic • Enable detection of process forces **Realization concept** (kinematic and physical effect) • Serial placement of three additional axis • Compensation of elastic deformations due to contrary movement
2.	Start-up clutch with NiTi shape memory actuators 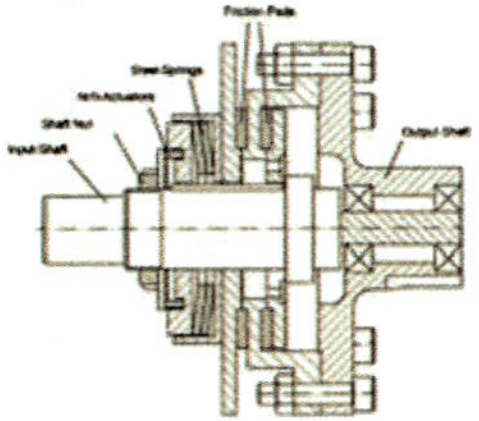 [10]	**Design flexibility** (addressed design conflicts) 1. Low weight of moving objects versus high adaptability/universality 2. Low volume of moving objects versus high adaptability/universality **Product flexibility** • Covering spreading torque demands • No automatic adjustment of adaption characteristic • Automatic compensation of abrasive wear **Realization concept** (kinematic and physical effect) • Serial placement of actuators • Adjustment of friction via normal force

(continued)

Table 31.1 (continued)

No.	Adaptronic solution	Characteristics of ALP
3.	Active modules for increased stiffness of a grinding machine 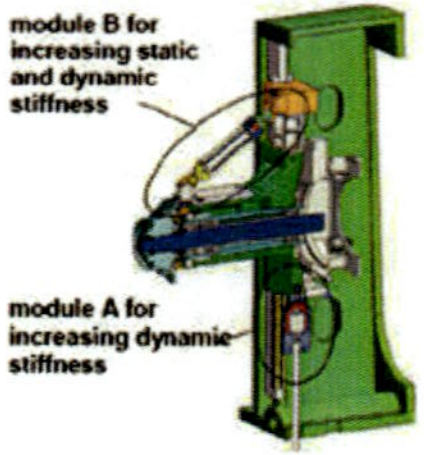 [16]	**Design flexibility** (addressed design conflicts) 1. High productivity versus high manufacturing precision 2. High strength versus high adaptability/universality **Product flexibility** • Affect stiffness of the grinding machine • Automatic adjustment of adaption characteristic • Enable detection of process forces **Realization concept** (kinematic and physical effect) • Semi parallel placement of actuators • Compensation of elastic deformations by contrary movement
4.	Revolute joint with quasi-static clearance adjustment for parallel kinematic robots 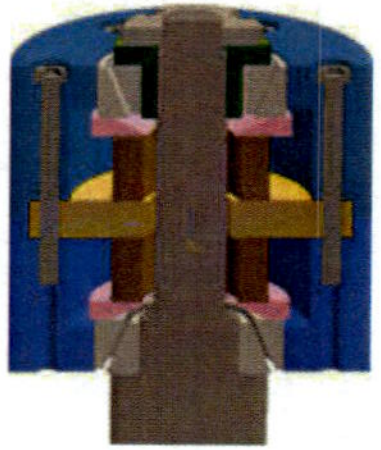 [17], [18]	**Design flexibility** (addressed design conflicts) 1. High productivity versus high manufacturing precision 2. Low loss of energy versus high strength **Product flexibility** • Affects friction and stiffness • Automatic adjustment of adaption characteristic • Joint locking for calibration and reconfiguration of the kinematic structure **Realization concept** (kinematic and physical effect) • Serial placement of actuators • Automatic adjustment of friction via normal force • Compensation of bearing wear

31.5 Applying ALP for High Dynamic Power Transmission

In a joint project[1] current work focuses on the application of ALP for high dynamic variable torque transmission in serial kinematic manipulators. Starting point of research is a new drive concept proposed to enhance the dynamics of serial robots and reduce dimensions of the manipulator's joints. Instead of employing an electric motor in each joint, specific machine elements, called adaptronic couplers, are used to transmit drive energy to each joint directly. Figure 31.3 (left) depicts an abstract view of the drive concept with adaptronic couplers. The power of a central drive is transmitted through the entire kinematic using shafts, and traction mechanism drives [13].

The proposed drive concept requires machine elements capable of transmitting variable torques from an input shaft in a highly dynamic manner thus enabling torque, velocity, and position control of the joint. Moreover, the torque-to-mass ratio of this machine element has to reach a maximum to allow reducing the structural mass. These requirements result in two design conflicts which have to be addressed by an ALP:

- low weight of moving objects *vs.* high energy, and
- low volume of moving objects *vs.* high adaptability/universality.

Based on the introduced ALP (cf. Table 31.1) a prototype of an adaptronic coupler has been developed in order to validate the proposed drive concept. In Fig. 31.3 (right) a CAD-model of the current prototype is given. The torque transmission is based on the friction force generated at the bush-shaft contact by

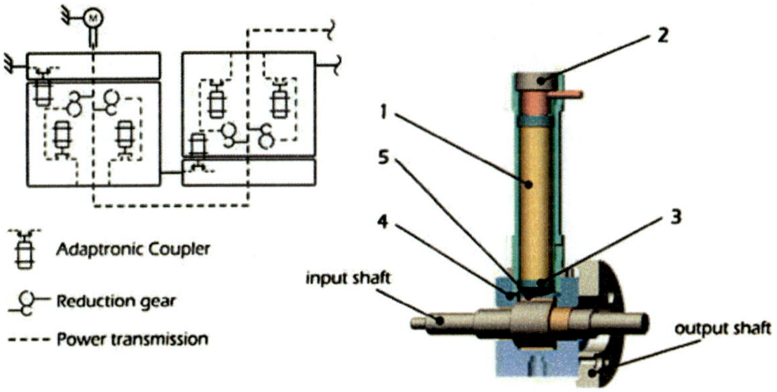

Fig. 31.3 *left:* Schematic illustration of a serial manipulator based on adaptronic couplers. A central drive at the robot base provides energy which is transmitted mechanically through the kinematic structure. Adaptronic couplers within each joint allow rotation in both directions. *right:* CAD-model of the current prototype (based on [13])

[1] Joint project of the Institute for Robotic and Process Information (iRP) and the Institute for Engineering Design (IK) at the TU Braunschweig.

means of a piezoelectric stack actuator (1) (semi parallel placement). The initial pre-stress applied by the threaded cap (2) and the expansion of the actuator results in a displacement of the lower washer (3) and a deformation of the upper part of the housing (4), which accommodates the sliding bush (5). Due to this friction variation a continuous adaption of the torque transmitted from the rotating input shaft to output side is permitted. Since only one actuator is employed, the transmitted torque can only cause a rotation of the output element against gravity. First experimental results show promising position and force/torque control performance of the adaptronic coupler [13].

Since the stated design conflicts are addressed and enhanced functionality is provided, the introduced application highlights the advantages of ALP. Covering a wide range of torque transmission the introduced concept of the adaptronic coupler allows nearly completely decoupling of the drive train and therefore soft actuation of the kinematic structure. This enhanced functionality can facilitate new applications of the robotic system (e.g. collaborative work of robot and humans).

31.6 Conclusion and Perspectives

In this contribution ALP were introduced as a mean to support the development of products capable for changing requirements during product use. Based on an approved definition of the term flexibility and the discussion of different development approach three aspects to characterize ALP were proposed. Analyzing numerous adaptronic solutions from different fields of applications the potential of ALP to address both design and product flexibility was highlighted. The advantage of ALP was further illustrated using the example of a specific machine element for high dynamic power transmission in serial kinematic manipulators.

As stated out before the application of ALP represents an effective approach to flexible the design process as well as to develop products with flexible properties. However, the potential to close conflicting requirements and provide enhanced functionalities (cf. Table 31.1) is evident the realization of adaptronic solutions results in challenging design tasks caused by multidisciplinary work. The proposed aspects to describe ALP support designer's decision to make use of these principles capable to close several design conflicts while development and providing flexibility during the use-phase of a product and therefore achieve competitive design solutions.

Further research will focus on extensive analysis of adaptronic solutions and the computer aided allocation of derived design relevant information using e.g. the system modeling language SysML. Moreover, the potentials of ALP to affect product properties will be pointed out introducing detailed models. In order to improve the assessment of ALP the use of functional requirements as well as the systematic detection of undesired effects will be part of further work. Apart from this methodical work further research is intended to determine the practical use of ALP in product development. Therefore, work will focus on new concepts to realize adaptronic couplers as well as the investigation of prototypes.

References

1. Saleh JH, Hasting DE, Newman DJ (2001) Extracting the Essence of flexibility in system design. The third NASA/DoD workshop on envolvable hardware, pp. 59–72
2. Chmarra MK, Arts L, Tomiyama T (2008) Towards design-time and runtime adaptability. EDIProD engineering design in integrated product development conference, pp. 33–40
3. VDI 2206 Design methodology for mechatronic systems. Beuth-Verlag GmbH, Berlin
4. Janocha H (ed) (2007) Adaptronics and smart structures: basics, materials, design, and applications. Springer, Berlin
5. Matthias M, Bös J, Hanselka H (2010) Active noise and vibration reduction: classification scheme and application examples. 17th international congress on sound and vibration ICSV17
6. Langbein S, Sadek T (2010) Design for functional integrated shape memory alloy systems. 11th international design conference DESIGN 2010, pp 835–844
7. Weber C (2005) CPM/PDD—An extended theoretical approach to modelling products and product development processes. In: Bley H, Jansen H, Krause F-L, Shpitalni M (eds.) Proc. of the 2. German-Israeli symposium for design and manufacture Advances in methods and systems for products and processes. Berlin/Germany, pp. 159–179
8. Inkermann D, Stechert C, Vietor T (2011) Einsatz multifunktionaler werkstoffe für die entwicklung einer adaptiven gelenkachse. In: Proc. of 8. Paderborner Workshop Entwurf mechatronischer Systeme. HNI-Verlagsschriftenreihe Paderborn, pp. 231–244
9. Franke HJ Krusche T (1999) Design decisions derived from product requirements. 9th CIRP design conference—integration of process knowledge into design, pp. 371–382
10. Predki W, Bauer B (2010) Concept of a start-up clutch with nickel-titanium shape memory alloys. Forschung im Ingenieurwesen 74(1), pp. 41–47
11. Preumont A (2002) Vibration control of active structures: an introduction. Springer, New York
12. Simnofske M (2009) Adaptronische Versteifung von Werkzeugmaschinen durch strukturintegrierte aktive Module. Dissertation, TU Braunschweig
13. Kubus D, Inkermann D, Claassen P, Tacke C, Vietor T, Wahl FM (2010) Adaptronic couplers—an alternative drive principle. IEEE/ASME international conference on advanced intelligent mechatronics, pp. 1207–1214
14. Altshuller G (1984) Creativity as an exact science. Gordon and Braech, New York
15. Puzik A, Meyer C, Verl A (2010) Robot machining with additional 3-D-piezo-actuation-mechanism for error compensation. ISR/ROBOTIK 2010, pp. 416–421
16. Simnofske M, Hesselbach J (2006) The increase of the dynamic and static stiffness of a grinding machine. International design engineering technical conference and computers and information in engineering conference—IDETC/CIE, pp. 23–30
17. Pavlovic N, Keimer R, Franke HJ (2008) Adaptronic revolute joints for parallel robots based on simultaneous quasi-statical axial and radial clearance adjustment. International design engineering technical conference and computers and information in engineering conference IDETC/CIE
18. Pavlovic N, Otremba R, Inkermann D, Franke HJ, Vietor T (2010) Passive and adaptive joints for parallel robots. In: Schütz D, Wahl FM (eds.) Robotic systems for handling and assembly. STAR, Nr. 67, Springer, Berlin, pp. 429–444

Chapter 32
Effect of Cell Shape on Stress Strain Behavior of Aluminium Foam

C. Mahesh, A. Deb, S. V. Kailas, C. Uma Shankar, T. R. G. Kutty and K. N. Mahule

Abstract The basic objective in the present study is to show the effect of cell shape on the stress strain behavior of aluminum foam using finite element analysis. Five different shapes are considered in the development of a representative unit cell that is capable of describing the complex geometry of closed-cell metallic foams. Stress–Strain behaviors for different models are compared and desirable candidate for numerical representation of aluminum foam is suggested.

32.1 Introduction

Cellular structures such as closed cell aluminum foam which are highly porous are known to have high stiffness combined with a very low specific weight and high energy absorption capacity, making them an excellent material candidate for the improvement of automobiles crashworthiness. It is necessary to understand the deformation mechanism of the foams for the development of suitable material constitutive equations. For this it is essential to obtain detailed knowledge of the dependence of the mechanical behavior on the foam structure. In order to come up with a general conclusion of cell shape, we are trying to capture the cell micro-structure in an average sense which can be in general representative of various

C. Mahesh · A. Deb (✉) · S. V. Kailas · C. U. Shankar · T.R. G. Kutty · K. N. Mahule
Center for Product Design and Manufacturing, Indian Institute of Science, Bangalore, India
e-mail: adeb@cpdm.iisc.ernet.in

C. Mahesh
e-mail: mahesh@cpdm.iisc.ernet.in

S. V. Kailas
e-mail: satish.kailas@gmail.com

A. Chakrabarti (ed.), *CIRP Design 2012*, DOI: 10.1007/978-1-4471-4507-3_32,

Fig. 32.1 Closed cell metallic foam

foam samples. The morphology of closed cell metallic foam is very intricate (Fig. 32.1). It consists of a spectrum of irregular cells. And each foam sample will be different from other sample from same block. If we have to represent the randomness, that will be only applicable to that foam sample. We are trying to come up with one cell representation which will reflect the behavior of randomness. For which average cell size, cell wall thicknesses are considered.

The modeling procedures are mainly based on repeated unit cell constructed from idealized cellular structures. Gibson and Ashby [1] are among the first to develop unit-cell-based models. They presented a skeletal cubic unit-cell to model open-cell foams and implemented dimensional arguments to relate the strength and elastic constants of the unit cell to its relative density. They also identified bending deformation of the cell trusses to be the dominant mode of deformation. Santosa and Wierzbicki (SS&TW) [2] model idealized closed-cell aluminium foam structure as an assembly of closely packed truncated cube and pyramids. This model is referred as cruciform-pyramidal foam model and the repeating unit-cell of this model is shown in Fig. 32.6a This model is supported with analytical formulations. The relative density (Eq. 32.1)

$$\frac{\rho^*}{\rho_s} = 3\frac{t}{b} + (4\sqrt{3} - 6)\left(\frac{c}{b}\right)^2 \frac{t}{b} \qquad (32.1)$$

$\rho*$ density of foam (kg/m^3)
ρ_s density of base material (kg/m^3)
t cell-wall thickness (mm)
b cube-width (mm)
c half-diameter of the pyramid (mm)

In this paper we present a comparison of different three dimensional unit cell to model closed cell aluminium foam using a multi-cell approach to come up with best numerical approximation for aluminium foam.

32.2 Finite Element Models and Analysis

The finite element calculations are performed by using the original commercial finite element code LS-DYNA_971. Cell walls are modelled as piecewise linear plastic material (MAT24 of LS-DYNA_971 material model), Belytschko-Tsay shell elements with Young's Modulus of 69 GPa, a Poisson ratio of 0.3, yield stress of 86.9 MPa and ultimate strength 171 MPa [2, 3]. The input files for LS-DYNA_971 are produced using Altair Hypermesh v10.0. Cell wall thickness is maintained uniform for a particular unit cell models. Density of 0.27 g/cc is maintained constant for all the shaped used by varying cell wall thickness.

In current study we have considered 5 different basic cell configurations

1. Cube
2. Cube with curved cell walls
3. Tetrakaidecahedron
4. Tetrakaidecahedron with curved cell walls
5. Cell configuration proposed by Wierzbicki

32.2.1 Foam Represented With Uniform Cubical Cells

Foam represented with uniform cubical cells as shown in Fig. 32.2.

32.2.2 Foam Represented With Cubes With Faces Curved Out of Plane

Foam represented with cubes with faces curved out of plane as shown in Fig. 32.3.

32.2.3 Foam Represented With Uniform Polyhedral Cells (Called as Tetrakaidecahedra)

Foam represented with uniform polyhedral cells (called as tetrakaidecahedra) as shown in Fig. 32.4.

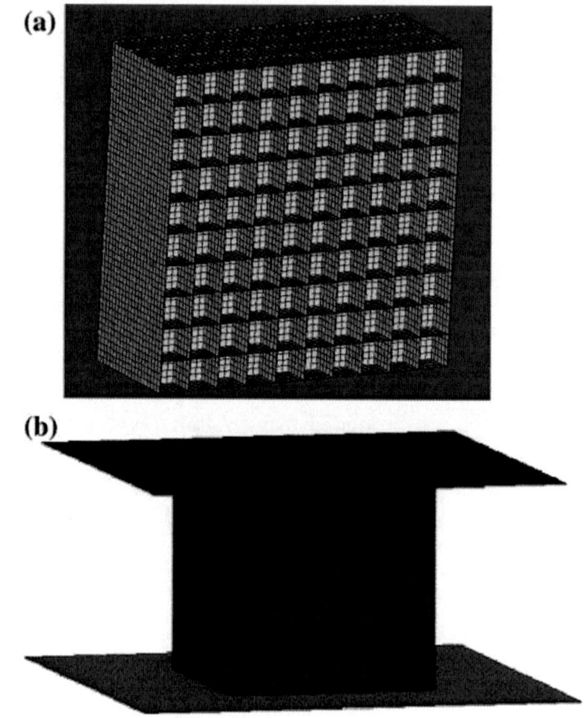

Fig. 32.2 **a** BLT shell-based finite element model of foam with uniform cubical cells (one face is hidden from view so that multi-cellular structure can be seen). **b** Foam model of Fig. 32.2 **a** supported on a fixed plate and compressed by assigning a fixed downward velocity to a *top rigid plate* (density: 0.27 gm/cc)

32.2.4 Foam Represented With Uniform Polyhedral Cells With Curved Faces

Foam represented with uniform polyhedral cells with curved faces as shown in Fig. 32.5.

32.2.5 Foam Represented With Cell Configuration Proposed by Wierzbicki

Figure 32.6 shows Santosa-Wierzbicki representation of foam and current finite element model of a single cell. In multi-cell finite element model developed from unit cell shapes suggested each face is shared between two cells. Multi-cell models are constrained to move in negative z direction by providing a rigid wall at base and a uniform velocity is applied on the top platen, which in turn crushes the specimen. And motion of top platen is constrained in x and y direction.

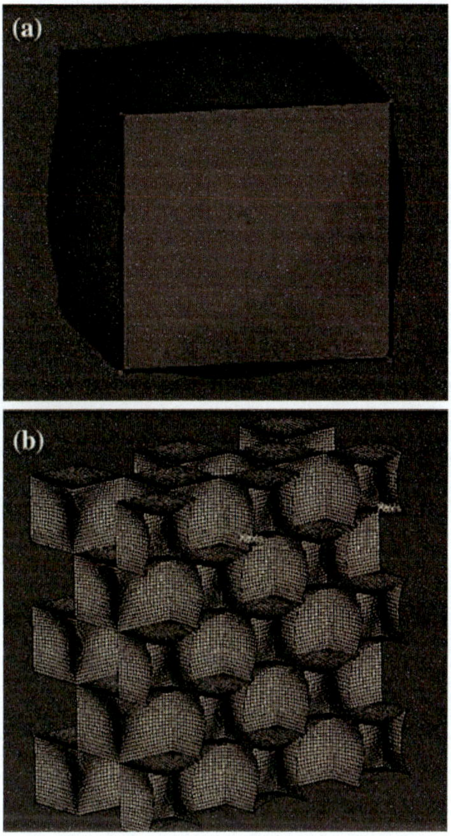

Fig. 32.3 **a** Geometry of a single cell of the shape of a *cube* with curved faces; **b** multi-cell finite element model based on *cubes* meshed with BLT elements (density: 0.27 gm/cc)

32.3 Results

FEA-based predictions of stress–strain response for all five models are compared (Fig. 32.7). Variation of cell wall thickness with different shape is as shown in Table 32.1.

Amongst the responses given in Fig. 32.7, it appears that the modeling of foam with tetrakaidecahedral cells with curved faces has the highest potential of realistically predicting foam behavior. This expectation is based on a typical previously-reported response of metal foam shown in Fig. 32.8.

It is clearly visible that models 1 and 2 were not able to capture the densification characteristic of metallic foam. T. Miyoshi, M. Itoh [3] has mentioned an average cell wall thickness of 0.162 at ¼ L and 0.132 at ½ L, where L is the edge length of a cell wall (average cell wall thickness of 0.147).And cell wall thickness used in unit cell of tetrakaidecahedral cells with curved faces is 0.147 mm.

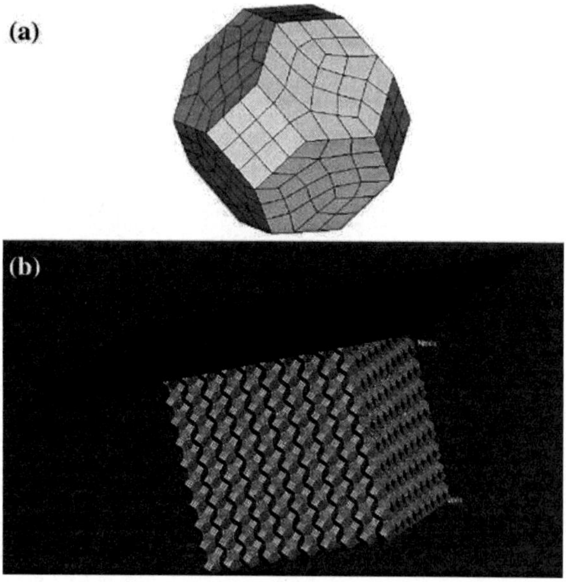

Fig. 32.4 **a** BLT shell-based finite element model of a single cell of the shape of a *tetrakaidecahedron* (also called as Kelvin cell); **b** multi-cell finite element model based on tetrakaidecahedra (density: 0.27 gm/cc)

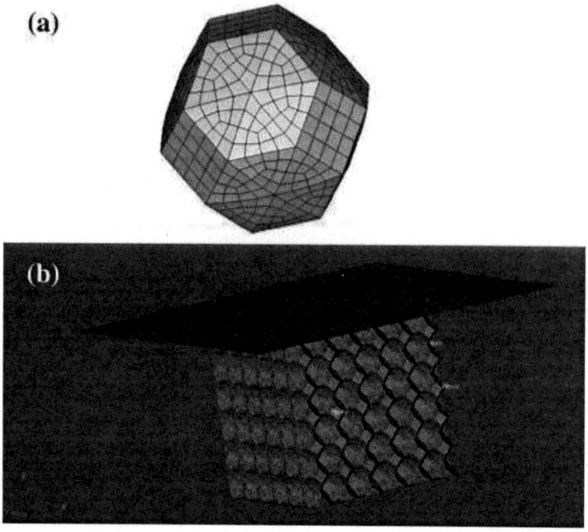

Fig. 32.5 **a** BLT shell-based finite element model of a single cell of the shape of a *tetrakaidecahedron* with curved faces; **b** multi-cell finite element model based on tetrakaidecahedra with curved faces (density: 0.27 gm/cc)

32.4 Conclusions

Different shapes have been explored to analyze the effect of cell shape on stress strain behavior of aluminium foam. It is observed that even though model no 4 and 5 are able to capture the densification characteristic of metallic foam properly, the

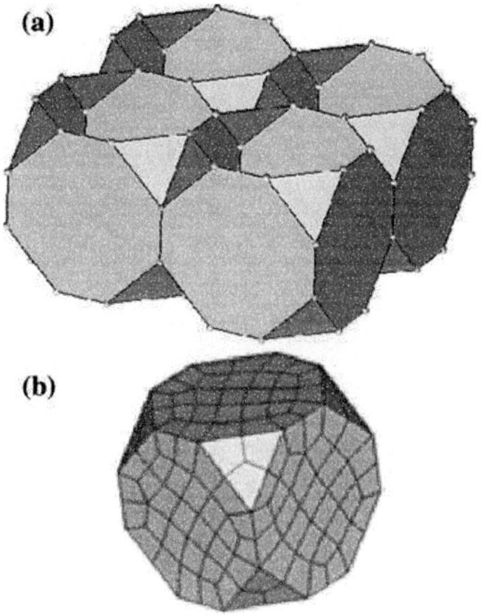

Fig. 32.6 **a** Santosa-Wierzbicki representation of foam and **b** current finite element model of a single cell

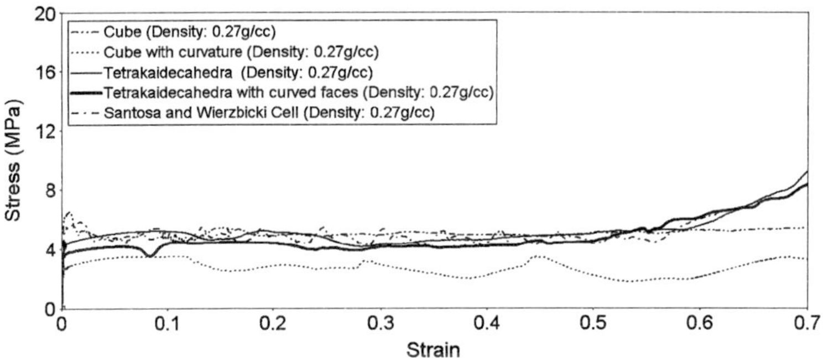

Fig. 32.7 Stress–strain responses predicted with uniform Kelvin cells of different sizes for constant foam density (0.27 gm/cc)

Table 32.1 Comparison of different cell shape with cell wall thickness

No	Cell shape	Cell wall thickness (mm)
1	Cube	0.063
2	Cube with curved face	0.060
3	Tetrakaidecahedron	0.148
4	Tetrakaidecahedron with curved face	0.145
5	Wierzbicki model	0.050

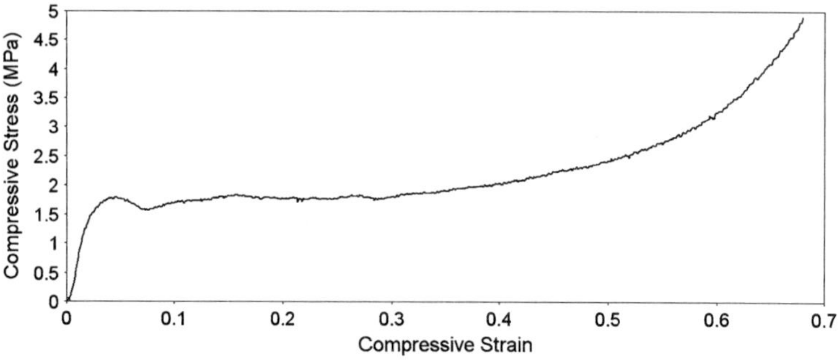

Fig. 32.8 A typical experiment-based stress–strain behavior of a metal (aluminium) foam [4]

former gives a further realistic cell wall thickness and failure behaviour when uniform thickness is used. Hence tetrakaidecahedral cells with curved faces as ideal candidate for numerical representation of aluminium foam.

References

1. Gibson LJ, Ashby MF (1997) Cellular solids, structure and properties. Cambridge University Press, 2nd edn. Cambridge
2. Santosa S, Wierzbicki T (1998) On the modeling of crush behavior of a closed-cell aluminum foam structure. J Mech Phys Solids 46(4):645–669
3. Miyoshi T, Itoh M, Mukai T, Kanahashi H, Kohzu H, Tanabe S, Higashi K (1999) Enhancement of energy absorption in a closed-cell aluminium by the modification of cellular structures. Scr Mater 41:1055–1060
4. Ramamurty U, Paul A (2004) Variability in mechanical properties of a metal foam. Acta Mater 52:869–876

Chapter 33
An Action Effectiveness Measure for Manufacturing Process Performance

Suman Devadula, K. Ramani, Praveen Uchil, Srinivas Kota, Monto Mani and Amaresh Chakrabarti

Abstract The diversity in manufacturing processes has resulted in different kinds of performance measures that suit specific purposes. Performance measures amenable to comparing process alternatives are essential to characterize manufacturing performance. The comparison of process alternatives is increasingly being subjected to broader considerations, like environment and society. This requires a measure amenable to be considered under different perspectives. On analyzing the life cycle processes of a wooden-graphite pencil from a producer's perspective a generic performance measure of 'resource-use effectiveness' is arrived at. Interpretations of the measure under different scenarios and perspectives are presented.

33.1 Introduction

What cannot be measured cannot be controlled. Measuring the performance of manufacturing processes is necessary to control them. The varied nature of manufacturing processes has resulted in numerous specific performance measures. Given the alternatives for any single process it also becomes essential to compare processes. Literature mentions of measures for comparing processes like Energy Efficiency [1] and Specific Energy Consumption (SEC) [2]. However, there is a lack of consensus on the definition of Energy Efficiency [3] post the manipulative adoption of the established thermodynamic concept of energy efficiency. Disagreement over the diversion from the thermodynamic ratio of energy efficiency has been discussed in

S. Devadula · K. Ramani (✉) · P. Uchil · S. Kota · M. Mani · A. Chakrabarti
Centre for Product Design and Manufacturing, Bangalore 560012, India
e-mail: ramani@cpdm.iisc.ernet.in

A. Chakrabarti (ed.), *CIRP Design 2012*, DOI: 10.1007/978-1-4471-4507-3_33,

detail [4] in this context. SEC, in its forms like process SEC, site SEC etc., does not consider the duration of the process in total and hence does not account for any delays, idle times etc. in the process which are also part of process definition. Literature [2] also mentions of process comparison models across machine types, distributed manufacturing situations etc. for manufacturing processes like material removal. However, process performance measures, if generic enough, i.e. in lacking unnecessary distinction [5] but being able to discriminate good and bad performance [6], may be used as indicators to compare processes, not only of subtractive but also of additive and transformative manufacturing processes. Towards this we propose an action effectiveness measure for measuring performance of manufacturing processes. Also, as indicated in literature any performance indicator has the tendency to be run-down due to the effects of positive learning, perverse learning, selection and suppression [5]. To counter this we suggest an approach of perspectives within which only can a measure take any meaning while the perspective itself is subject to change in response to changing contexts [7] of manufacturing. One example of such a change of context is globally distributed manufacturing wherein local resources are leveraged for serving requirements of consumers elsewhere. Further, as changing contexts for manufacturing demand a shift of perspective from profiteering to that of environmental and societal benignity [8], conventional efficiency measures need to be gradually replaced by effectiveness measures through appropriate design of performance measurement systems. A trend in sustainability performance measurement literature is contrary to this need [9–11] and hence, we pitch the proposed measure as an indicator of 'resource-use effectiveness', whatever the resource be, thereby readily accommodating a necessary perspective to start with.

33.2 Measuring Performance of Production Processes

Measurement of organizational processes is generally done through measures as a substitute performance measure. It is essential for organizations to measure the performance of their processes to monitor, control and steer them towards short-term and long-term goals. Production, as an activity of any manufacturing set-up, is similarly required to manage its processes to report its overall performance aligned to the organization's reporting commitments. These goals have to trickle down to the design of performance measurement systems, the major considerations of which are timeliness of information, the range of coverage, the appropriateness of data for decision support and the cost of implementation [12].

Given the diverse nature of production processes, the classic goal to produce the required quantity at the required time by the best and cheapest method remains valid today [6]. However, such a generic goal needs to be viewed in possible perspectives that are being demanded by different stakeholders. This is illustrated in Table 33.1 through the production processes of a 'wooden graphite-pencil'. Literature mentions performance measurement models that have been proposed within which performance measures have largely been classified under time, cost,

Table 33.1 Production process measures of wooden-graphite pencil

Production phase of pencil LC: component	Process	Specification	Parameter	Performance measure
Slat	Slat making	Width and thickness	Area exposed	Conforming sample lengths
	Grooving	Diameter	Area exposed	Conforming sample lengths
Lead	Kneading	Physical composition	Mass	Conforming sample lengths
	Lead extrusion	Diameter and length	Area exposed	Conforming sample lengths
	Lead baking	Tensile strength	Density	Conforming sample lengths
	Lead assembly			Mass handled per unit time and energy
Pencil slat	Slat assembly			Mass handled per unit time and energy
Pencil body	Cutting	Width	Area exposed	Conforming body lengths
	Shaping	Shape	Area exposed	Conforming body lengths
Pencil	End preparation	Shape	Area exposed	Conforming body lengths
	Painting	Hue and surface finish	Mass	Conforming body lengths
	Printing	Legibility	Mass	Conforming body lengths

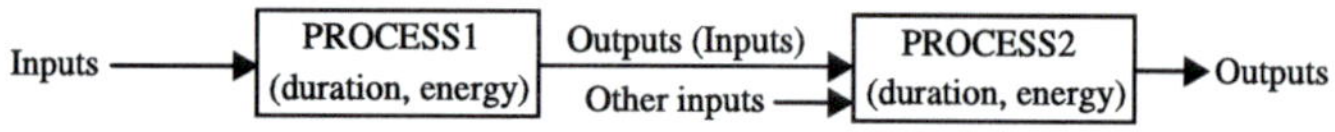

Fig. 33.1 Network of processes

quality, flexibility and productivity. The earlier measures were framed to address the concerns of the investors. Later the concerns shifted to engineering process performance [6]. This has resulted in a proliferation of performance measures. Performance measures are only indicators and there is a tendency to improve measurement rather than improving the underlying performance. This actually decreases the distinction between various performance measures [5] leading to the need for more abstract performance measures.

A generic performance measure that can incorporate multiple perspectives is proposed in this chapter. The proposed measure is defined as 'resource-use effectiveness' indicating how effectively the process uses input material, energy and total process duration. Integrating energy use over the process duration gives 'action' and for this reason one can refer to the measure as action effectiveness of production. Process duration is the elapsed time between the process begin–end and such durations, end-to-end, account for the duration of the whole production system. However, process begin and end are defined in the sense of a unit process i.e. a level of process definition detail beyond which further division is of no interest. The duration for which energy is consumed within any process is equal to or less than the process duration.

33.3 Methodology

Various manufacturing processes of a wooden graphite-pencil are studied from a process performance perspective. Characteristic physical quantities of the production processes necessary for measuring their performance are determined. These are also determined for the other life-cycle processes and a generic performance measure of "resource-use effectiveness" is proposed as an indicator for measuring the performance of any manufacturing process. Further, to portray the measure under different scenarios of process alternatives and perspectives, hypothetical examples are worked out illustrating the use of the proposed measure.

Manufacturing is conceived as a network of processes since 'process' is considered the fundamental unit of manufacturing. The terms manufacturing and production are used interchangeably. A process is defined here as the action towards change of input towards intended output (Fig. 33.1).

With the goal of finding characteristic physical quantities across performance measures in the manufacture of a Wooden graphite-pencil various processes identified from patent databases, literature and educative videos published by pencil manufacturers are observed. The list in Table 33.1 is only indicative and not representative of any specific pencil manufacturer.

Within the production of a Wooden graphite-pencil, Table 33.1 lays out the different production processes. For any given process the process measures, specification, parameter and performance follow. *Specification* is defined here as a label of quality to bet met. *Parameter* provides handles to control quality and a *Performance measure* is taken as an indicator allowing the evaluation of how good a process is according to pre-determined criteria.

For a process, say Kneading of the lead dough, the process specification is physical composition because it is essential for the dough to be homogeneous with respect to the required grade of pencil before lead extrusion. The parameter for Kneading is mass because any non-conformance with respect to the specification being met, will be handled by changing the relative proportions by weight of the dough components i.e. graphite, clay and binder. The performance measure for Kneading is 'conforming sample volumes' because homogeneity is statistically determined over sample volumes taken.

For a process, say Lead Extrusion of the dough after Kneading, the process specification is diameter and length of the extruded lead because these are important for conforming to the pencil length and groove diameter. The performance measure of Lead Extrusion is 'conforming sample lengths' because any rejection based on statistically evaluated samples can be in terms of pencil lengths only. The physical quantities identified necessary are mass, energy and duration combining which a relevant performance measure for Lead Extrusion is the ratio of conforming sample lengths per unit input mass, energy and duration. From the production processes tabulated and described above, it is observed that a performance measure of 'resource-use effectiveness' is central to all of them.

Further, considering the processes included in the Life-cycle, refer Table 33.2, of a Wooden-graphite pencil other than production e.g. for the process of Mining graphite, the specification is concentration, because it is essential to know the relative proportion of mineral that has to be concentrated. The parameter for this is density, as it indicates the mineral proportion so that concentration can be controlled. The performance measure is weight of ore concentrated as required per unit energy and time expended. As illustrated in the Table 33.2, the performance measures of all the life-cycle processes may be generalised in terms of the basic physical quantities of mass, energy and duration of the process. This suggests that the proposed performance measure for production is generic across life-cycle phase processes also. This is proposed as an Action Effectiveness Measure (AEM).

$$\text{AEM} = \frac{\text{Output mass}}{(\text{Total Input mass}) \times (\text{Total Input Energy}) \times (\text{Process duration})}$$

While comparing processes, the measure should be calculated with reference to unit mass output of the process. A hypothetical set of calculations is worked out in Table 33.3 for illustrating the interpretation of AEM.

SEC is defined as the energy required per unit output. Inclusion of process duration and the total material input in the AEM accounts for all the input resources and this helps in distinguishing process performance beyond SEC. As

Table 33.2 Process measures for LC processes involved in the manufacture of wooden pencil

Life-cycle Phase	Material		Process	Specification	Parameter	Performance measure
Material	extraction	Pencil				
	Clay		Mining	Concentration	Density	Mass per unit time and unit energy
	Graphite		Mining	Concentration	Density	Mass per unit time and unit energy
	Wood		Harvesting	Girth	Mass	Mass per unit time and unit energy
			Seasoning	Tensile strength	Density	Mass per unit time
Distribution	Material		Process			
	Packaging		Packing	Zero damage		Conforming packages
			Transporting	Zero damage		Conforming packages
Usage	Sharpener		Sharpening	Diameter measured at lead tip	Area	Mass of pencil material removed for achieving a conforming tip
	Writing surface (paper)		Writing	Length	Mass	% stained length
After usage	Stubs	Binder (clay and graphite are non-degradable)	Disintegration	Naturally non-disintegrable organic form	Area	Disintegration rate
		Body	Disintegration	Naturally non-disintegrable organic form	Area	Disintegration rate
		Paint	Disintegration	Naturally non-disintegrable organic form	Area	Disintegration rate
	Lead Dust	Binder clay and graphite are non-degradable)	Disintegration	Naturally non-disintegrable organic form	Area	Disintegration rate
	Shavings		Disintegration	Naturally non-disintegrable organic form	Area	Disintegration rate

Table 33.3 Comparison of AEM and SEC for process alternatives

	Time (s)	Energy (MJ)	$Matl._{IN}$ (kg)	$Matl._{OUT}$ (kg)	AEM (kg/kg.J.s)	Energy efficiency or SEC^{-1} (kg/J)
Process Alt1	10	100	1,000	800	0.0008	8.00
Process Alt2	15	100	1,000	800	0.0005	8.00
Process Alt3	5	120	900	800	0.0015	6.67

Table 33.4 Perspectives within product life-cycle

Perspectives within LC phases of a product	
Phase	Perspective
Material extraction	Environment, natural resource management (NRM), society, ore producer
Production	Environment, society, safety
Distribution	Environment, producer, market
Usage	Environment, user
After-usage	Environment, producer, NRM

illustrated in Table 33.3 AEM serves to distinguish between the process alternatives 1 and 2 where their SEC values are same. If the AEM values are same for two process alternatives their performances are considered equal.

33.4 AEM and Perspectives

Manufacturing as an organization is required to manage its processes as the requirements arising from changing contexts demand it so. Conventionally what mattered to the organization was functional performance but the current demand is environmental and societal performance also [8]. Environmental and sustainability considerations call for more attention to effectiveness measures with regard to material and energy consumptions and more importantly the wastes. Depleting resources call for meeting the needs of more people within the limited supplies. A trend that emerges from sustainability performance measurement initiatives in terms of environmental performance is that relative indicators, aimed at measuring efficiency rather than effectiveness, are increasingly being proposed for performance measurement [9–11]. Given this requirement for performance measures we propose to augment the interpretation of AEM with regard to perspectives. Perspective can be defined as stake holder's point of view with which the measure is interpreted. Perspectives can be based on worldview, theory, facts, knowledge of interactions etc. In Table 33.1 a producer's perspective is taken for detailing the processes. Some perspectives that can be taken by different stakeholders within the life-cycle phases of any product is explored in Table 33.4 and the process measures arrived at for the Wooden graphite-pencil in Table 33.2 can be specifically arrived at on the basis of any of these perspectives correspondingly. Any performance measure for a process can take meaning only within a perspective that is assumed or explicitly taken.

Considered from the perspective of an environmentalist AEM indicates resource-use effectiveness.

33.5 Discussion and Further Work

The emphasis on resource-use effectiveness in the proposed measure attempts to shift the focus from efficiency to effectiveness. When confronted with multiple process alternatives, the higher the computed AEM value of any process alternative the better is the process compared to others. For example, consider the Buy-to-Fly ratio, defined as the mass of material required to machine a part over the mass of material in the finished part. Input mass use effectiveness i.e. the ratio of output mass to input mass of the process indicates its Buy-to-Fly ratio directly whereas the AEM calculated for the process indicates all incurrence that has gone into achieving this ratio (including energy and time). The possibility of converting all incurrence into monetary value gives resource-use effectiveness per dollar and taken this way the AEM provides further scope for the comparison of processes with similar ratios. For a network of processes, arriving at an aggregate value of the AEM as an equivalent performance measure is yet to be worked out. The challenge in arriving at a computed equivalent performance value lies in accounting for the individual performance values of the processes occurring in series and parallel. However, comparison of processes using this measure is feasible at a black-box level.

Prevailing method of LCA is used for evaluating processes purely from the perspective of their environmental impact. It is observed that LCA uses process as a basic unit of evaluation. The proposed AEM, having been arrived at on considering processes across the LC, also has the potential of accommodating multiple perspectives. In this sense, the local arguments from all affected stakeholders can be included in qualifying the results of processes and this is necessary for broader assessments like sustainability. As the choice of perspectives can align to the demands of changing contexts of manufacturing the notion of good and bad can be accommodated dynamically and this can possibly prevent the running down of performance measures. One way to interpret AEM with perspectives is to introduce weights for the resources that are determined with some predefined basis. In that case, the AEM will take the form,

$$\text{AEM} = \frac{\text{Material}_{\text{OUT}} * \text{w1}}{(\text{Material}_{\text{IN}} * \text{w2}) * (\text{Energy} * \text{w3}) * (\text{Process duration})}$$

where, w1, w2 and w3 are specific non-zero weights defined with respect to the perspective taken by the stakeholder. The weights provide a means for incorporating quality attribute of the resources with reference to a particular perspective.

Energy if considered under specific perspectives, e.g., from a resource-depletion perspective, energy that is harnessed from renewable resources can be given a favourable weight over other sources. The rationale for qualifying the output in conjunction with the weights, as above, needs further study. The introduction of perspectives also lead to multiple values leading to multi-criteria decision making in choosing the appropriate process alternative.

33.6 Conclusion

With stakeholders of diverse backgrounds demanding performance under wider considerations, it is necessary for manufacturers to include perspectives in their performance management systems. Towards this an abstract process performance measure of 'resource-use effectiveness' that includes perspectives is proposed. These perspectives can be taken by various stakeholders in contexts that are local to them. The extendibility of the proposed AEM through various perspectives is illustrated.

Acknowledgments The authors acknowledge the contribution of The Boeing Company for providing financial support under contract PC36018 at SID, IISc.

References

1. EuP Directive (2005) Establishing a framework for the setting of energy requirements for energy using products. s.l.: European Commission
2. Kara S, Li W (2011) Unit process energy consumption models for material removal processes. CIRP Ann Manuf Technol 60:37–40
3. Patterson MG (1996) What is energy efficiency? Concepts, issues and methodological issues. Energy Policy 24:377–390
4. Gavankar S, Geyer R (2010) The rebound effect: state of the debate and implications for energy efficiency research. Bren School of Environmental Science and Management, Santa Barbara
5. Meyer MW (2002) Rethinking performance measurement. Cambridge University Press, Cambridge
6. Hon KKB (2005) Performance and evaluation of manufacturing systems. CIRP Ann 54(2):139–154
7. Schlossberg E (1977) For my father [book auth.] In: Brockman J (ed) About Bateson. Dutton, New York
8. Elkington J (1997) Cannibals with forks: the role of triple bottom line in 21st century business. Capstone Publishing Ltd, Oxford, p 416. ISBN: 9781900961271
9. Azzone G et al (1997) A stakeholders' view of environmental reporting. Long Range Plan 30:699–709

10. Olsthoorn X et al (2001) Using environmental indicators for business—a literature review and the need for standardisation and aggregation of data. J Clean Prod 9:453–463
11. Measuring Environmental Performance of Industry. MEPI [Online] [Cited: 3 Dec 2000]. http://www.environmental-performance.org
12. Neely A (2002) Business performance measurement: theory and practice. Cambridge University Press, Cambridge, p 380. ISBN: 978-0521803427

Part VI
Sustainable Design and Manufacturing

Chapter 34
Product-Service Systems Design Using Stakeholders' Information

G. V. Annamalai Vasantha, R. Hussain, M. Cakkol and R. Roy

Abstract Product-Service System (PSS) provides added value to customers through integrated product and service offerings. Currently, the development of PSS solutions is ad hoc and the roles and involvement of stakeholders at the various stages of PSS development have not yet been fully defined. In this paper, we propose a PSS-Design framework and illustrate the involvement of the customer and suppliers along with manufacturers to develop functional and sustained conceptual PSS solutions. An industrial laser system case study is used to explain the stakeholders' involvement and proposed techniques. The rigorous definition of the involvement of stakeholders fosters co-production and the definition of roles helps to depict how co-creation could occur—this would lead to the development of effective PSS solutions.

34.1 Introduction

To increase competitiveness, manufacturers are looking for strategies that will provide substantial business advantage by being difficulty to imitate by competitors and that will lock customers into longer-term relationships. Product-Service Systems (PSS) is one of the strategies which are perceived to provide such advantage. The merits of PSS have been widely discussed in literature [1] and, importantly, PSS aims to make the best use of capital-intensive products so that more value can be fostered and more revenue generated per unit cost of the product throughout its lifecycle. Also, PSS can increase competitive advantage by

G. V. Annamalai Vasantha · R. Hussain · M. Cakkol · R. Roy (✉)
Cranfield University, Cranfield, Bedfordshire, MK43 0AL, UK
e-mail: r.roy@cranfield.ac.uk

A. Chakrabarti (ed.), *CIRP Design 2012*, DOI: 10.1007/978-1-4471-4507-3_34,

improving knowledge of the customer's business which would give an increased insight as to how they actual use products and services to fulfil their business needs.

Even though PSS can provide the afore-mentioned advantages, the financial losses incurred by some manufacturing firms that have offered PSS solutions have been also pointed out [2]. Accordingly, it has been noted that, currently, PSS could be a high risk strategy considering the cost incurred in the agreed contract period and the minimum involvement of customers and suppliers in the design process. In addition, conceptual PSS design in practice is ad-hoc and lacks a systematic approach [3]. We also recently observed from our industrial studies that, with the exception of a few mature industries, most of the industries lack a systematic framework for the development of PSS solutions. The focus of our research is to develop a systematic framework along with a case study for developing conceptual PSS solutions with the view of emphasizing the integration of stakeholders during PSS development. In this paper, the customer's involvement during the initial stages and the suppliers' involvement in the solution assessment stages along with that of the manufacturer are described in detail. The validation results with three other companies are also discussed.

34.2 Related Research

Various definitions are proposed in literature to define PSS [1]. Stressing that PSS should design systems that will adapt to changing expectations and circumstances and uses, we have defined PSS design as a process to synthesize and create sustained functional behaviour through tangible products and intangible services where sustained functional behaviour should represent how the system achieves its purpose continuously. A review of the objectives of PSS design methodologies proposed in literature revealed that PSS design should focus on integrating business models, products and services together throughout the lifecycle stages to create an innovative value addition for the system [4]. Commonly, PSS business models are classified into product-, use-, and result-oriented [5]. This classification helps to understand how the roles of stakeholders can change and contrasts the various business models against to that of the stand-alone product. Initial process models have been proposed for the co-creation of PSS between customer, manufacturer and suppliers [3, 6] and for the PSS design process, the integrated development of products and services has been widely discussed [7, 8].

It has been noted that the service design process is broadly similar to its equivalent in the product field and adapting existing product design processes to account for the special characteristics of technical services would lead to a greater acceptance of application within the enterprise [3, 8]. Visualizing the whole system in PSS development that will produce intended value addition to stakeholders is also been emphasised [9]. Along with PSS design methodologies, software is also in development to support PSS development. Komoto and Tomiyama [10] argue that in the PSS design processes, designers define the activity to meet specified goals and quality, and define the environment in which the activity takes place. They have developed ISCL (Integrating Service CAD with a life cycle simulator) which has functions to support quantitative and probabilistic PSS design using life cycle simulation. Sakao et al. [11] proposed Service Explorer which incorporates a service model consisting of four sub-models: 'flow model (who)', 'scope model (what)', 'scenario model (why)', and 'view model (how)'. Detailed comparison of PSS design methodologies proposed in literature along with our new framework is described in Table 34.1. Overall, PSS design is still in its initial stages of development and substantial research is required to develop a practical PSS design methodology. Foremost is the rigorous definition of the co-production process between stakeholders through assigning roles of stakeholders at the various stages of PSS development is required. Industry is primarily looking for detailed strategies to evaluate PSS solutions in terms of functionality and business sustainability.

34.3 Research Questions and Methodology

The overall aim of our work is to develop a systematic framework along with a case study for the conceptual PSS design solutions which considers the existing and potential service network capability and incorporates past knowledge from a product-in-use. We argue that designing for the customer processes, capabilities and use context could lead to the development of innovative PSS designs. To stress the importance of the capabilities of the stakeholders, we propose a PSS-Design framework which illustrates the involvement of the customer and suppliers along with the manufacturer to develop functional and sustained PSS solutions [12]. The research questions answered here are:

- How could the customer involvement to be described to facilitate designing PSS for their business process?
- How could the suppliers capability will be assessed at the PSS solutions evaluation stage?

These questions are addressed in the proposed framework by incorporating the techniques of Service Blueprinting and AHP. The description of this framework and the application of these techniques have been explained through an industrial laser system case study. The information provided about this case study is drawn from semi-structured interviews which were conducted with three laser job shop managers and two senior sales people of laser system manufacturers. Figure 34.1 provides the contextual scenario of this case study. In this case, the laser job shop is a customer of the laser system manufacturer. Generally, the laser systems purchased by laser job shops have two years of warranty. Laser cutting has now been commoditized and is driven largely by price and speed of delivery. Laser cutting systems are expensive to procure and operate (Fig. 34.2). The application

Table 34.1 Design stages comparison of PSS-design framework with other related research

Proposed framework	Komoto and Tomiyama [10]	Shimomura et al. [7]	Alonso-Rasgado and Thompson [3]	Maussang et al. [9]
Customer system gap identification and analysis through 'need behind the need', capabilities and life cycle activities, status of products, services and business model	Define goal(s) as specified by receivers	Receiver state parameters (RSP) define the stage change of the receiver	Business ambitions of client	Customer expectations, needs and specifications involved in the whole lifecycle
Identification of new solutions to fill gaps focusing on new or modified activities, objects, parameters, values of parameters and relationships between these elements which leads to additional capability development	Define quality as specified and evaluated by receivers in terms of the parameters of channels and contents	Flow model expresses the sequential chain of agents	Potential business solutions: clearer view of the hardware and/or services that will constitute the final Total Care Product (TCP) provision	External functional representation with the function definition has been realised
Derivation of new business models through feasibility analysis, assigning stakeholder's responsibilities leading to network formulation	Define Environment consists of provider(s), receiver(s), channel(s)and content(s)	Scope model specify a range of a service spanning from an initial provider to a final receiver for effective service design	Core definition of potential of TCP plus TCP options. The solution that will best satisfy the customer's business needs is identified.	Functional deployment: the ways to ensure the result required by the customer
Creating and evaluating synthesized solutions to envelop the whole life cycle avoiding sub-optimized solution to any single activity	Define set of Activities to deliver contents from providers to receivers with channels in environment	A view model expresses the mutual relationships among RSPs and function parameters	Enhanced definition of potential TCP: defined in increasing detail, and consequently a more accurate price can be proposed	Scenarios and Functional Block Diagram (FBD) to design consistent PSS alternatives

(continued)

Table 34.1 (continued)

Proposed framework	Komoto and Tomiyama [10]	Shimomura et al. [7]	Alonso-Rasgado and Thompson [3]	Maussang et al. [9]
Defined representation module of the chosen solution(s) for common interpretation across stakeholders	Quantitative and probabilistic descriptions to determine economic and environmental feasibility from a life cycle perspective	Scenario model represents receivers themselves and their behaviours in receiving the service	Business case validation and evaluation of alternatives	Use of the FBD to extract the specifications of the physical elements to design

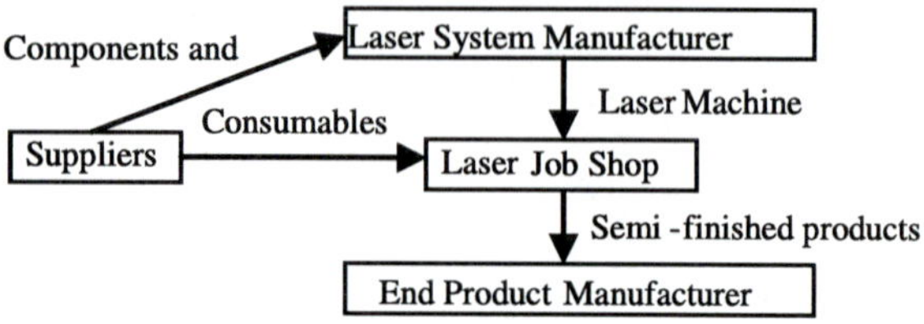

Fig. 34.1 Network of stakeholders and flow of components between them

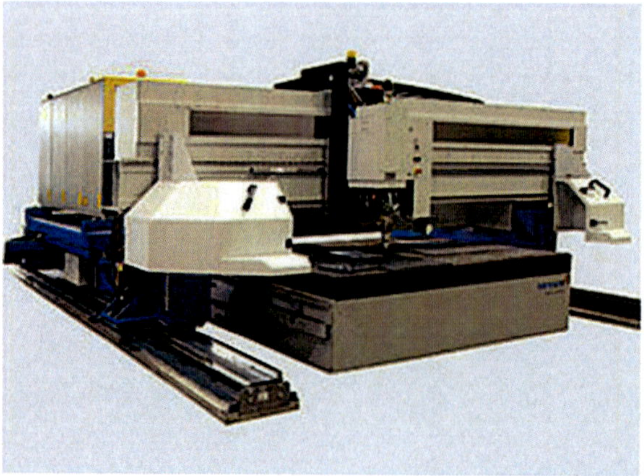

Fig. 34.2 5-axis laser cutting machine

of the proposed framework and the answers to the research questions are now addressed within this industrial context. The framework validation results with three other companies are also discussed.

34.4 PSS-Design Framework

Figure 34.3 illustrates the proposed ten steps PSS-Design framework and highlights the focal areas of this paper. The proposed framework is quickly described by classifying into five different stages namely: Customer system gap identification and analysis (Steps 1–3), Identification of new solutions to fill the gaps (4–5), Derivation of new business models (6–7), Creating and evaluating synthesized solutions to envelop the whole life cycle (8–9) and Common representation module of the chosen solution(s) (10). These stages are highly dependent on each other. Even though the starting point of this framework could be at any stage, this framework ensures that all steps are followed as feedback loops exist between every step.

This framework is focused mainly for the business to business context where products are mature and widely operated by the customers. In this environment the foremost tasks to be carried out is to identify the gaps within the established system which fails to satisfy customer needs. This gap analysis helps to pinpoint

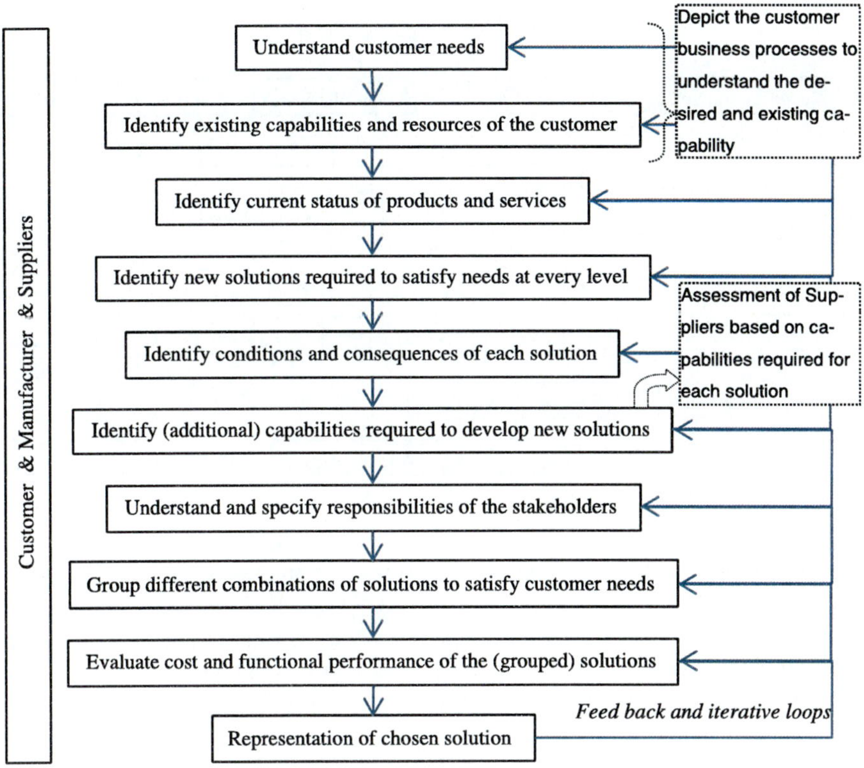

Fig. 34.3 A PSS-design framework and the focus of this paper (*dotted rectangle*)

where value is lost in the system. The Service Blueprint depicted should map the process in which the PSS is embedded—this way, the 'need behind the need' of the PSS is depicted. The second stage generates solutions to fill these gaps. The solutions could solve the symptoms of the gap and/or the underlying root causes of the gap. New solutions could be triggered by focusing on life cycle activities, influencing entities, parameters, changes in the values of parameters and relationships. PSS design should also create business models which are aligned to the capabilities of the involved stakeholders. For example, in the third stage of the framework, the business models are generated by assigning responsibilities taken at various levels of the solutions (these are activities, entities, parameters, and values).

The fourth stage is to improve the generated PSS solutions at the overall system level and avoid sub-optimizing towards any single activity. This life cycle consideration involves synthesizing the identified solutions for every gap and activity and evaluating the functional and cost performance of the grouped solution for the intended period of the contract. The final stage is to represent the developed solution through a common extended IDEF0 representation to avoid misinterpretation between the stakeholders. Table 34.1 compares the design stages of our

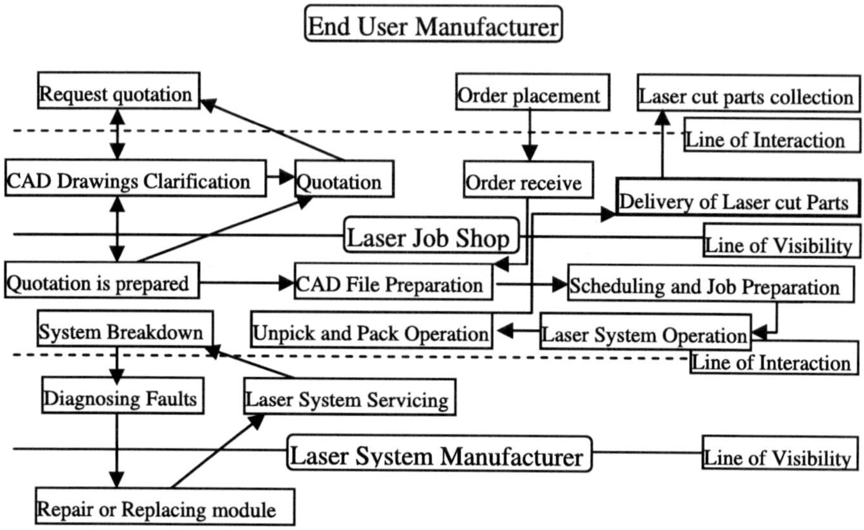

Fig. 34.4 Service blueprint represents current laser job shop processes

framework to other methodologies proposed in literature. All the discussed methodologies bring novel steps to the PSS design methodologies.

The highlights of the proposed framework are as follows:

- Primarily helps to develop PSS solutions which are aligned to all the stakeholders' capabilities by avoiding starting the design from PSS types such as product-, result- and use-oriented.
- It helps to design right product and service mix and combination to satisfy the needs of customers.
- Improves overall life cycle system level by avoiding sub-optimizing any individual activity.
- The sensitivity of important variables in the performance of the proposed solutions is highlighted through the solutions synthesis process.

34.5 Designing for the Customer's Business Process

The foremost aim of PSS design is to improve knowledge of the customer's business through an increased insight as to how and under what circumstances they actually use products and services to fulfil their business needs. To achieve this purpose, the customer's business process in which the PSS is embedded needs to be depicted at the first stage of the design process. The primary outcome of this process is to identify gaps between the actual and desired performance parameters

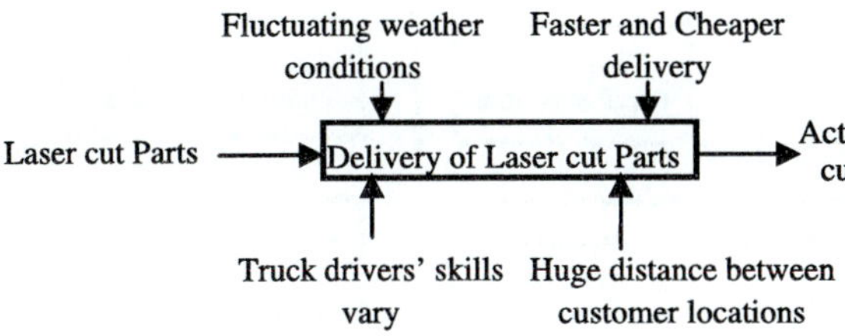

Fig. 34.5 Representation of the delivery process by input, outcome, environment and capabilities

Table 34.2 Illustration of supplier capability requirements for different solutions

Solution No.	Description of the solution	Required gas supplier capability
Solution 1	Providing on-site laser capability for customer	Reliability, technical competence and organisation profile
Solution 2	Setting up a laser cutting facility close to the end product manufacturer's site	Reliability, technical competence and cost
Solution 3	Moving the laser system manufacturer assembly unit within or near to the laser job shop premises	Reliability, responsiveness and technical

of the processes within the system which prevents the fulfilment of customer needs. Identified gaps should depict where the symptoms have manifest as well as where the underlying causes are located. To identify and represent gaps, we have extended the Service blueprint [13] to incorporate following elements: Overall customer needs; provided products and services; agreed business models; the stakeholders' environment, resources, competences and constraints; each process to represent a sub-capability required to achieve the overall capability; and the actual parameterized outcomes from each sub-capability compared against the desired parameterized outcomes. The difference of these will allow the capability gaps to be determined—the largest ones being those to be addressed.

Figure 34.4 illustrates the service blueprinting diagram representing laser job shops processes for laser cutting. The line of interaction separates the customer action area from the provider activity area, representing the direct interactions among customers and providers. The line of visibility differentiates the visibility of the provider's activities to the customer. In the laser system case study, the higher level laser job shops' needs are quoting consistent pricing, to provide laser cut parts quickly with high quality and at low cost. The reliability, speed, uptime and the ease of use of laser cutting machines was extremely high. From the interviews, it was apparent that some of the laser job shop's clients were unhappy with the length of time it took from instructing the job shop to the time that the parts were delivered. For these clients, the costs were also high which made the job shop uncompetitive. This was an issue because, from interviews with three laser job

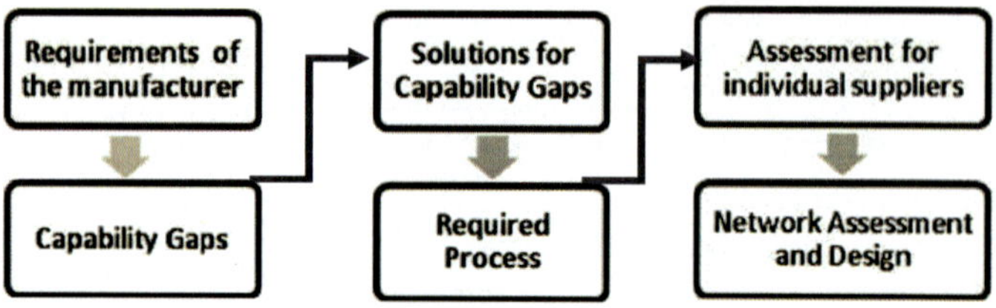

Fig. 34.6 Supplier assessment process followed in laser case study

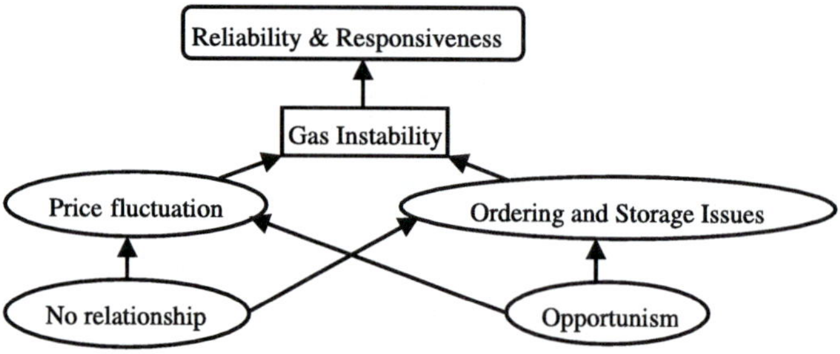

Fig. 34.7 The root cause analysis of the gas supplier problems

shops, it appears that laser-cut parts have been commoditized; generally, the capability to cut laser parts tends to be fairly uniform amongst the laser job shops and so, at the moment, it is just speed and cost that differentiates the job shops.

Mapping each process in Fig. 34.4 with reference to the overall requirements helps to identify gaps that need to be focused upon to generate PSS solutions. Figure 34.5 depicts the capability deficiency within the delivery process by focusing on input, outcome, environment variables, competences, resources and customer needs. It highlights that rather than focusing on speeding up the laser system, there is a need to address the delivery service process. This modified service blueprint notes parameterized values such as span (the length of time that it takes for the capability to be effected) and cost (the cost of the capability). The difference between the desired parameter values and the actual parameter values were then calculated as a difference in span of ‘x’ hours and a difference in cost of ‘y’: this gave the capability gaps.

34.6 Supplier Assessment in PSS Solution Selection

The identification of PSS solutions could be addressed at various levels: activities, entities, parameters and values. Each of these elements could be visualized in terms of addition, substitution, customisation or elimination of the offending sub-capability. Alternatively, an adjustment to the environment (such as better roads, in this case) may lead to an improvement. To retain the larger end product

Table 34.3 Relative importance of each capability with respect to required supplier capability

	Rel.	Resp.	Tech.	Cost	Org.	Rel. Mgt.	Asset Mgt.
Reliability		1.14	1.26	1.08	1.16	1.12	1.62
Responsiveness			1.23	1.11	1.25	1.37	1.64
Technical competence				1.03	1.19	1.23	1.46
Total cost					1.17	1.13	1.36
Organisation profile						1.06	1.26
Relationship management							1.13
Asset management	Inconsistency: 0.00						

manufacturer, the PSS design solutions could be adopted by the laser job shop to resolve the delivery issues are presented in Table 34.2.

In order to better execute a particular PSS solution, an overall supplier capability assessment is needed. We have defined supplier capability as the ability of a particular supplier to make use of a team of resources to perform a task or activity at the required standards according to the aims of the business model designed for the customer. Supply Network Capability on the other hand, is the overall ability of the whole network. We have identified that measurement of a supplier capability should be evaluated through the following capability parameters: *reliability, responsiveness, relationship management, technical competence, organisation profile, total cost and asset management*. Figure 34.6 illustrates steps to execute the supplier network assessment.

Requirements of the manufacturer: Our respondents have pointed out the following summary in Table 34.2 which illustrates the changing requirements of laser job shops—from its gas suppliers—according to different design solutions proposed. These solutions are presented from the identified supplier capability parameters. *Identifying Capability Gaps*: For any PSS solution generated, gas supplier plays a vital role. Since the laser job shop chose to implement solution 1, the relationship between the gas supplier gains more importance compared to traditional operations. Currently, the relationship between gas suppliers and laser job shops are problematic in terms of responsiveness and price fluctuation. In order to address these issues, a root cause analysis is conducted to understand the actual reasons for the problems. As a result, we identified that the relationship between gas suppliers and laser job shops are very transactional and opportunistic behaviour exist in both sides of the relationship. As a consequence of this transactional relationship, price fluctuation problems with ordering and storage issues are experienced rapidly. These then cause gas instability which then reflects back to customers in terms of fluctuating cost, non-reliable services and non-responsive business approach (Fig. 34.7).

Proposing Solutions for Capability Gaps: In order to address this major problem, we argue that there is a need to create a long-term, collaborative relationship with a reliable gas supplier which will then decrease both the uncertainty and risk involved in gas provision. *Identifying required process to select suppliers*: We have used the AHP technique to assess the individual suppliers and also the

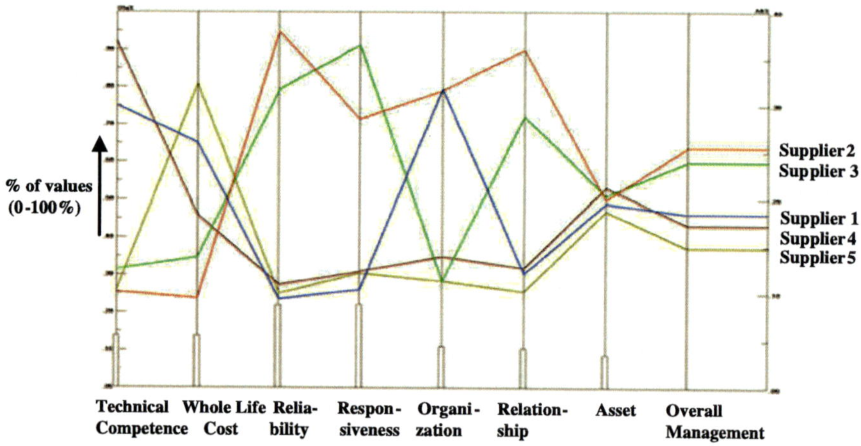

Fig. 34.8 AHP results for Tier 1 suppliers against the capability parameters for solution 1

network for a particular PSS solution because it analytically helps practitioners in the decision making process to better understand the complex characteristics of potential suppliers. Additionally, this technique is heavily used in the industry.

Assessment for individual suppliers: At this point there is a need to evaluate and assess the possible suppliers for the long-term solution proposed. In order to conduct the AHP process:

- First, the weighted averages of parameters for the solution are entered by experts (Table 34.3). In this example, reliability and responsiveness have the highest score: because of the health and safety issues and the need for long-term relationship.
- Secondly, potential suppliers are identified and their values are entered manually in terms of capability parameters.

Expert Choice software is used to perform this calculation systematically by manually entering the capability parameters for each supplier. The use of software allows us to see the performance of each supplier. Figure 34.8 represents Tier 1 suppliers' assessment results against the capability parameters. In this example, supplier 2 and 3 are the best options for tier 1. Columns positioned at the bottom of the figure demonstrate the relative importance of parameters for the solution; in this case reliability and responsiveness are the most important ones. *Assessment of Network*: Tier 1, 2 and 3 suppliers are responsible for gas manufacturing, installation and storage respectively. The evaluation starts with Tier 1. The same AHP process is followed to assess the suppliers in the next two tiers which lead to the design of entire supply network.

To satisfy needs of the laser job shop to deliver laser cut parts faster to the end product manufacturer, Solution 1—Providing on-site laser cutting capability has been chosen. The responsibilities assigned for this solution are: end product manufacturer will provide sufficient space for laser machine and consumables storage, supply minimum guaranteed parts to laser cut and insure the infrastructure within their premises; laser job shop lease the laser machine for intended contract period from the laser machine manufacturer and operate it with their operators; and supplier 2 or 3 will provide guaranteed gas supply at required period with fixed price.

34.7 Validation and Conclusions

The proposed PSS-Design framework has been validated by going through step-by-step application with six industrial experts from three companies and four academicians. The overall feedback is encouraging and potential applications of this framework are highlighted. The notable benefits mentioned are it helps to develop PSS solution from a holistic system approach which aids to change the designer's mind-set from product-centric to PSS; the framework could be used as a general problem solving approach incorporating the co-production process between involved stakeholders; and prospective usefulness, completeness, usability and clarity are highly graded. Limitations noted are trust and transparency needed for this framework are not matured in most of the industrial sectors; need to measure intangibles benefits; to perform more quantitative data analysis to access gaps and getting right information required within each step could be a problem. We believe that the proposed framework could be rigorously developed by applying to industrial case studies to develop functional and sustained PSS solutions.

Acknowledgments This research was funded by the Cranfield Innovative Manufacturing Research Centre IMRC (EPSRC Grant EP/E001874/1) as part of the project "Capability based PSS conceptual design".

References

1. Meier H, Roy R, Seliger G (2010) Industrial product-service systems—IPS^2. CIRP Ann Manufact Technol 59(2):607–627
2. Neely A (2008) Exploring the financial consequences of the servitization of manufacturing. Oper Manage Res 1:103–118
3. Alonso-Rasgado T, Thompson G (2006) A rapid design process for total care product creation. J Eng Des 17(6):509–531
4. Annamalai Vasantha GV, Roy R, Lelah A, Brissaud D (2010) A review of product-service systems design methodologies. J Eng Des DOI:10.1080/09544828.2011.63971210.1080/09544828.2011.639712
5. Cook M (2004) Understanding the potential opportunities provided by service-orientated concepts to improve resource productivity. In: Bhamra T, Hon B (eds) Design and manufacture for sustainable development. Wiley, New York. ISBN 1860584705

6. Morelli N (2006) Developing new product service systems (PSS): methodologies and operational tools. J Cleaner Prod 14:1495–1501
7. Shimomura Y, Hara T, Arai T (2009) A unified representation scheme for effective PSS development. CIRP Ann Manufact Technol 58:379–382
8. Aurich J, Fuchs C, Wagenknecht C (2006) Life cycle oriented design of technical product-service systems. J Cleaner Prod 14(17):1480–1494
9. Maussang N, Zwolinski P, Brissaud D (2009) Product-service system design methodology: from the PSS architecture design to the products specifications. J Eng Des 20(4):349–366
10. Komoto H, Tomiyama T (2009) Design of competitive maintenance service for durable and capital goods using life cycle simulation. Int J Autom Technol 3(1):63–70
11. Sakao T, Shimomura Y, Sundin E, Comstock M (2009) Modeling design objects in CAD system for service/product engineering. Comput Aided Des 41:197–213
12. Annamalai Vasantha GV, Hussain R, Roy R, Tiwari A, Evans S (2011) A framework for designing product-service systems. In: Proceeding of 18th International conference on Engineering Design (ICED11), Vol. 4, pp. 67–76
13. Shostack LG (1984) Design services that deliver. Harvard Bus Rev 84115:133–139

Chapter 35
Importance of User and Usage for Eco-Design

Srinivas Kota, Daniel Brissaud and Peggy Zwolinski

Abstract There are methods and technologies developed for generation and evaluation of product proposals for better environmental efficiency throughout the lifecycle. But in use phase these are not effective due to complexity of user behaviours and usage scenarios. Literature review revealed the need to develop products with the help of user to achieve sustainability. An outline of a conceptual system to include user in design is proposed in this chapter consisting of different dimensions based on the interactions and influences. It is important to consider how and where the user uses the product in reality while designing by including user also in design process. When different elements like people, product and environment come together, their individual characteristics, context in which they operate determine the interactions and influences among them. The questions need to be answered are identified to develop the proposed system.

35.1 Introduction

Sustainability is development of society with efficient use of economy within the carrying capacity of ecology. Sustenance of the environment is essential for sustenance of all living things on this planet; it is, however, under considerable stress

S. Kota (✉) · D. Brissaud · P. Zwolinski
Grenoble-INP / UJF-Grenoble 1 / CNRS, G-SCOP UMR5272,
46, avenue Félix Viallet, 38031 Grenoble Cedex 1, France
e-mail: Srinivas.Kota@g-scop.inpg.fr

D. Brissaud
e-mail: Daniel.Brissaud@g-scop.inpg.fr

P. Zwolinski
e-mail: Peggy.Zwolinski@g-scop.inpg.fr

A. Chakrabarti (ed.), *CIRP Design 2012*, DOI: 10.1007/978-1-4471-4507-3_35,

due to the various activities performed by humans. The air, water, soil are getting increasingly polluted and resources are being depleted. Systematic Design [1] can play an important role in reducing the stress on the environment as the decisions we make in the design stages have a predominant effect on the later stages of the product development and the product life cycle. If the environmental impacts of potential designs can be assessed while designing; design can be changed then and there so as to reduce any later impacts [2, 3].

There is a need to consider the environment as a central criterion in product lifecycle design, so as to reduce stress on the environment. Designers need support for developing environmentally friendly products in an efficient manner. This requires an in-depth understanding of the product life cycle, the environment, the product development process, the needs of designers, and the links between these.

Eco-design is an approach to design where impact on ecology is taken into account while developing products. We need to consider the whole life cycle of the product in design to determine the impact on environment. Life cycle assessment [4] is a method to assess environmental impact throughout the life cycle of a product. Life cycle approach or thinking is also effectively used in explaining the product failures throughout the life cycle [5].

Product life cycle consists of the following phases: Material, Manufacturing, Distribution, Use and After-Use. There are different stakeholders responsible for impact on environment in each phase. Till now, there has been lot of work done and data available on material, manufacturing, distribution and after-use phases with industries in material, production, transportation and recycling sectors. There is very little data about the use phase except the designed energy and material consumption. But in actual practice the data is going to vary depending on who the user and the context in which he uses the product. The information about the actual practise is not fed back for analysis due to which there are lot of gaps while calculating the actual environmental impact during use.

The objective of this chapter is to establish the importance of user and the usage context in eco-design of products. A through literature review is done and the findings are discussed in the following sections to support the importance of an eco-design method based on user and the associated use.

35.2 Literature Review

Eco-design

A number of guidelines were created for assisting designers in the choice of materials [6], fasteners, processes, etc. The application of extended producer responsibility has enabled manufacturers to engage in end of life of their products: disassembly, reuse, recycling and proper disposal [7]. Later, the efforts became directed on the product lifecycle as the basis for thinking, addressing full product lifecycle, from material to after-use. There are many collections of general guidelines like [2] for designing environmentally friendly products.

Considerable effort has been spent in developing 'Design for X' tools for each specific phase of the product lifecycle, like Design for remanufacture [8], design for manufacture and assembly [9], design for disassembly [10], design for reuse [11], and design for recycle [12, 13]. An iterative use of LCA during product development has been reported to be advantageous [14]. These tools have, however, been developed in isolation for manufacturing and after use phases.

Industrial design tools with practical applications [15, 16] and methodological support to eco-design in processes have been developed and applied in different contexts [17]. The link between all these initiatives is to clarify the relationship between product development and environmental concerns by proposing laws and design rules for producing products with less environmental impacts [18].

Energy Efficiency in Use

The introduction of energy using equipment in homes has increased significantly the environmental impact of energy production by increasing demand. Despite the improvement in energy efficiency of white goods, the proliferation of small devices has also resulted in the increase of energy demand over the last two decades. Finding the ways for proper use of durable goods is a key challenge to the sustainable industrial production and consumption.

Government regulations forced manufacturers to produce and sell products with energy efficiency labels, eco labels but these seem not enough [19]. Methods like [20] are targeting the use phase but only for design of energy efficient products under the context of designed use. Use phase is seen to be having a major role in the environmental assessment but not catered adequately because of the behavioural dimension. Detailed analysis of the diversity of patterns of use [21] and diversity in characteristics of users [22] (cultural [23] and geographical [24] diversity) need to be done.

More recent studies highlight the importance of the construction of valid alternative use phase scenarios: the lack of rigor and justification for this step can indeed be significantly detrimental to the validity of LCA results [25]. Studies [26, 27] show that for systems with smaller and medium-term prospects (lifetime of the products from 1 to 40 years), the difficulty is rather the proliferation of products with complex features making the acquisition of reliable data more tedious on the use phase. Most of the authors and practitioners point to the need for methods to define usage scenarios; none has been explained previously in LCA studies. This is an obstacle to the truth of results of some LCA.

Design and User

From the literature [28] we can see that user can be considered in design in different ways:

- *User-centred design* in terms of ergonomic study, interaction design, experience design, universal design and participatory design.
- *Behavioural based design* in terms of design scripts, affordances, persuasive design, ambient and intelligent technologies.
- *Practice oriented design* in terms of observing the user in real scenario to develop products for use.

- *Co-creation* in terms of using user generated content, open source and social innovation.
- *Co-design* in terms of including co-creation with in the design process with user and designer working together.

There are different techniques developed to include users in design in the areas of information and communication technologies (ICT) [29] and working tools: task analysis, focus groups, formal, informal expert reviews, direct test of the prototypes, field investigations and acceptance in use, participatory design [30]. In [31] following strategies discussed: Eco-feedback (the object communicates to encourage the user to change its state), scripts (role-playing between the user and the purpose to educate behavior officials) and features forced (it does not voluntarily make available to the user functions the least responsible). This research assumes that all users have the same reaction to the product and that these changes and ergonomic design are sufficient for the user to have a sustainable behavior.

Some researchers [32] have focused on guiding the user behaviour. However, most of them consider in the design, product general usability only as a way to change the behavior of the user. So many of them offer feedback solutions in the product, communicating better use of products through manuals, screens to encourage users to perform action which results in efficient use of products [33, 34, 35]. In practice, as we can see from the literature [36, 37] users have their own way in using products and it results in varied impacts. When the product is re-designed to cater to this, the only option found in the literature is the simple deletion of functions "unsustainable" [38, 39].

Human behaviours frequently confound design intentions to reduce product-related environmental impacts [40, 39]. A "persona" is a generic one defined by a limited number of ethnographic characteristics [41]. Finally, the user can be approximated by the more personal aspects: representations and emotions.

First an attempt should be made to model the properties of the user like values [42], characteristics to help the product designer in his decisions and then an integration of the user himself as an actor in the design process. The complexity is twofold: to model all aspects of a user and create an effective design process to consider these. Today we lack a unified model of the user incorporating all the facets in the design process.

35.3 Discussion

From the literature on product life cycle thinking, an imbalance of scientific and technical knowledge between the life cycle phases arises. The early phases of product life, material, production and distribution have been widely discussed in the impact studies, and the systems are defined relatively accurately. Knowledge and data to model these phases are indeed part of the skill of the producer. The later phases of the life of a product have been addressed more recently by environmental experts, as knowledge and data from sources scattered and sometimes conflicting environmental interests: local government, business recovery, recycling. However, the information to model all phases of the life cycle are available with a limited number of famous actors and professionals. It is unfortunately not the case for the use phase of a product that appeals to everyone. As a result, today, the central phase of life cycle, use is seen as having the poor relation in the environmental analysis: The data used are often extremely simplified and defined arbitrarily. This is due to the difficulty of taking into account: the behavioral dimension of the user as a social being and the diversity of patterns of use and users.

There is an increasing willingness of users to have products to suit their lifestyle and these are going to have a major role in determining the environmental impact. It is important to instrument designers with tools that model the user characteristics and use phase scenarios of the product. It is important to develop an eco-design method which takes care of the actual environmental benefits achieved by including user behaviour and context of use [43]. This requires a more comprehensive understanding of 'users' as social beings, and the role of consumption in everyday life. Normally requirements are considered while developing solutions, but the solutions depend on many factors including context, the actual actions and the impact is going to change based on variety of factors which needs to be considered while developing solutions.

A product that has integrated its environmental requirements into design to improve environmental performance throughout its life cycle is necessary for sustainability. It is therefore essential to instrument designers with tools that model the use phase of a product. The network of actors in use is much more diffuse and difficult to identify, we do not yet have data on the use of methods to supplement the current eco-design. When we talk about use and users, we introduce a cultural and territorial thinking. Adapting the supply of products to the attitudes and expectations of users and consumers appears to be a powerful way to increase the diffusion of sustainable technologies and enhancing the resulting environmental benefits. This requires the development of a user and usage-centered eco-design method incorporating the social, economic and environmental dimensions.

The eco-design method needs to support the design of products tailored to the user's environmental expectations and abilities, rather than providing him a product to which he has to adapt. It should explore the potential of changing the design of product according to the user rather than the behavior of the user based on product. With this research, we want to develop an eco-design method by investigating and understanding the environmental benefits actually achieved in use, without which the development only partially fulfills the objective.

Online store, amazon has come up with a new idea for reducing packaging "hassle free packaging". In this idea unnecessary packaging is removed at the delivery centre and disposed accordingly, which reduced both the burden on the user and environmental impact. Indian Railways has come up with a solution "virtual reservation message" to reduce the paper used for tickets by offering the consumers to save the ticket in their mobile phones and they need to show this

Table 35.1 Relation between product, user and use

	Product	User	Use	Result
1	Bad	Buy	Bad	Bad
2			Good	Bad
3	Good	Buy	Bad	Bad
4			Good	Good

with an identity card to the ticket checking officer. This initiative will reduce the use of paper, printing ink, and energy used to make them. Both these initiative will be a success if the user opts for those and user characteristics are going to determine the success of these new innovations and needs to be considered while designing the product/service. Table 35.1 shows how product, user and usage are connected with respect to environmental benefit. If the product or use is bad then the net result is bad and both needs to be good and user needs to buy that product to get net result as good. Because the pre-use phases are all with-in the manufacturer's control, the impact can be regulated but use cannot be controlled fully as it depends on users motifs and they are divergent. After-use can be controlled but products need to be properly discarded from use and it depends on user again.

To develop a method for integrating user characteristics and context of use in product design to lower environmental impacts throughout its life cycle, we need to introduce requirements in the early stages of product design. We need to link upstream eco-design methods for material, production and distribution, to downstream eco-design methods for end-of-life by providing an eco-design integration method for the use phase. It should have both the dimensions of use: the user and the context in which the user operates. It requires to model user with all the characteristics, use context with all its characteristics. Various representations of users with highlighted behavioral characteristics which influence the environmental performance of products in use phase need to be identified to help in producing, consuming and disposing products in a sustainable way.

35.4 Proposed System

Our goal is to help designers to determine user preferences and incorporate those in design to reduce environmental impact during usage and assess the impact in actual usage based on the context of use. Different domains need to be investigated thoroughly to see how these influence each other: designer, user, product life cycle, product development, and environment. Figure 35.1 shows the conceptual representation of the proposed system. There are five dimensions: People dimension (designer, user), Design dimension (synthesis, analysis), Product dimension (product/service/system), Life cycle dimension (use), Platform dimension (technology, method)

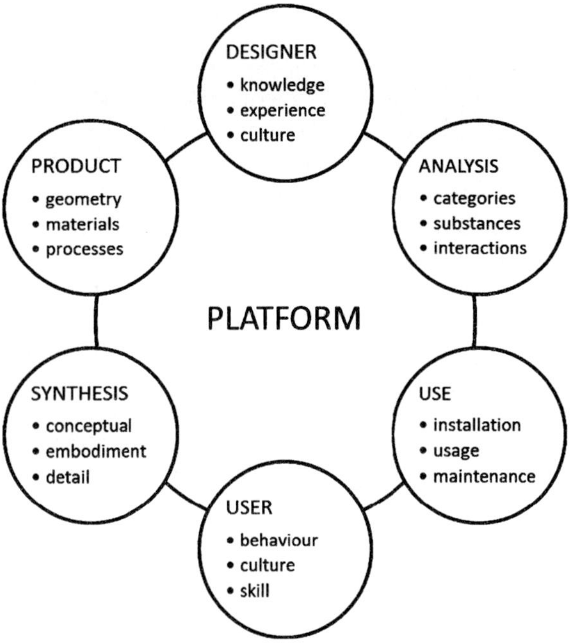

Fig. 35.1 Conceptual representation of the proposed system

People dimension: This dimension needs to get necessary inputs from different stakeholders on their requirements.

Design dimension: This requires information in different stages of design for both generation and evaluation of solutions.

Product dimension: This is the representation of product or service or system in terms of their constituents.

Life Cycle dimension: As use phase is not dealt adequately other than energy efficiency our concentration is on usage and user behaviour.

Platform dimension: We need to see requirement of generic or specific method and corresponding technology required to realise the support.

To realise the proposed system we need to conduct research on solving the following questions:

- How to characterise individual user behaviour? What kind of behaviour models exist?
- How to characterise the usage scenarios (context)? What kind of usage scenarios (context) exist?
- How to develop/construct shared understanding between designers, environmentalists, users and other stakeholders?
- How to integrate *user* and *designer* for *synthesis* and *analysis* of *product* and its *usage*?
- What should be the characteristics of the combined Platform?
- What kind of technology is in need for the combined Platform?

35.5 Conclusions

In conclusion, we say that we need an eco-design method which considers the user in his multi-faceted decision-making relevant to reduce the environmental impact of products. There are models developed in other domains like user interaction (UI) and information and communication (IC) which should be studied thoroughly for their suitability and adaptability to the eco-design domain. We propose to integrate the user in design to develop a suitable product by developing several product platforms or architectures with similar features but adapted to the profiles of end users and the context in which they need to be used. We support the idea that designers should develop the product as clean as possible with the help of user feedback and then simply train the user to use in best way naturally.

Acknowledgments This work is a part of ECO-USE project funded by Agence Nationale de la Recherche (ANR, The French National Research Agency) with project number: 1081C0212/ANR-10-ECOT-006-01.

References

1. Pahl G, Beitz W (1996) Engineering design—A systematic approach. Springer, Berlin
2. Brezet H, Hamel Van C (1997) Eco design—A promising approach to sustainable production and consumption, UNEP Manual, Delft University of Technology, 1997
3. Mueller K, Lampérth M (2001) Parameterised inventories for approximate life-cycle assessment. In: International Conference on Engineering Design, ICED'01, Vol 1. Glasgow, pp 669–676, 21–23 Aug 2001
4. Consoli F (Ed) (1993) Guidelines for life-cycle assessment: a 'code of practice'. Society of environmental toxicology and chemistry (SETAC), 1993, Brussels
5. Luca Del F (2011) Product failure: a life cycle approach, ICED'11, Denmark, 15–18 Aug 2011
6. Shibaike N (2001) Incorporating environmentally conscious materials selection in CAD system. In: Second international symposium on environmentally conscious design and inverse manufacturing, Ecodesign'01, Tokyo, pp 1098–1101, 11–15 Dec 2001
7. Rosemann B, Meerkamm H, Trautner S, Feldmann K (1999) Design for recycling, recycling data management and optimal end-of-life planning based on recycling-graphs. In: International conference on engineering design ICED'99, Vol 3. Munich, pp 1471–1476, Aug 1999
8. Okumura S, Morikuni T, Okino N (2001) Life design for remanufacturing durable products. In: Second international symposium on environmentally conscious design and inverse manufacturing, Ecodesign'01, Tokyo, pp 275–280, 11–15 Dec 2001
9. Boothroyd G, Dewhurst P, Knight WA (1994) Product design for manufacture and assembly. Marcel Deckel, New York
10. Cser L, István Z (1996) Assembly and disassembly—An unified approach? Proceedings of the 28th CIRP international seminar on manufacturing systems, Johannesburg, 1996, pp 61–65
11. Hoffmann M, Kopacek B, Kopacek P, Knoth R (2001) Design for re-use and disassembly. In: Proceedings of second international symposium on environmentally conscious design and inverse manufacturing, Ecodesign'01, Tokyo, pp 378–381, 11–15 Dec 2001
12. Seliger G, Mueller K, Perlevitz H (1999) Product and process planning for recycling. In: International symposium on environmentally conscious design and inverse manufacturing, Ecodesign'99, Tokyo, pp 989–994, 1-3 Feb 1999

13. Oyasato N, Kobayashi H, Haruki K (2001) Development of recyclability evaluation tool. Second international symposium on environmentally conscious design and inverse manufacturing, Ecodesign'01, Tokyo, 2001, pp 303–306, 11-15 Dec 2001
14. Ritzen S, Norell M (2001) Environmental consciousness in integrated product development. In: Second international symposium on environmentally conscious design and inverse manufacturing, EcoDesign'01, Tokyo, pp 346–350, Dec 2001
15. Lofthouse V (2006) Eco design tools for designers: defining the requirements. J Cleaner Prod 14:1386–1395
16. Luttropp C, Lagerstedt J (2006) Eco design and The Ten Golden Rules: generic advice for merging Into Environmental aspects product development. J Cleaner Prod 14:1396–1408
17. Vezzoli C, Sciama D (2006) Life cycle design: from general methods to product type specific guidelines and checklists: a method adopted to develop a set of guidelines/checklist handbook for the eco-efficient design of NECTA vending machines. J Cleaner Prod 14:1319–1325
18. Karlsson R, Luttropp C (2006) Eco Design: what's happening? An overview of the subject area of eco design and of the papers in this special issue. J Cleaner Prod 14:1291–1298
19. Horne RE (2009) Limits to labels: the role of eco-labels in the assessment of product sustainability an routes to sustainable consumption. Int J Consum Stud 33(2):175–182
20. Navarro TG, Rizo SC, Gisbert PF, Bono GLV, Romero MAS (2001) Design for energy efficiency. In: International conference on engineering design, ICED'01, Glasgow, pp 613–620, 21–23 Aug 2001
21. Redstrom J (2008) RE: Definitions of use. Des Stud 29(4):410–423
22. Redstrom J (2006) Towards user design? On the shift from object to user as the subject of design. Des Stud 27(2):123–139
23. Liang B, He Y (2011) The effect of culture on consumer choice: the need for conformity vs. the need for uniqueness. doi:10.1111/j1470-6431.2011.01013.x
24. Abeliotis K, Koniari C, Sardianou E (2010) The profile of the green consumer in Greece. Int J Consum Stud 34(2):153–160
25. Reap J, Roman F, Duncan S, Bras B (2008) A survey of unresolved problems in life cycle assessment. Int J Life Cycle Assess 13:374–388
26. Unger N, Schneider F, Salhofer S (2008) A review of eco design and environmental assessment tool and their appropriateness for electrical and electronic equipment. Prog Ind Ecol Int J 5(1–2):13–29
27. Scharnhorst W, Kohler A, Rebitzer G, Hischier R, Jolliet O (2004) Progress in modern life cycle assessment: practice and research. Int J Life Cycle Assess 9:143–151
28. Kakee S, Conny B, Jaco Q (2011) Designing change by living change. Des Stud. doi:10.1016/j.destud.2011.08.002
29. Jomaa I, Poirson E, Da Cunha C, Petiot J–F (2011) A methodology for designing a recommender system based on customer preferences. ICED'11, Denmark, 15–18 Aug 2011
30. Vredenburg K, Mao J, Smith PW, Carey T (2002) A survey of user-centered design practice in Minneapolis. Minnesota, Proceedings of CHI' 2002, ACM press, pp 471–478
31. Zachrisson J, Boks C (2010) When to apply different design for sustainable behaviour strategies? Knowledge collaboration and learning for sustainable innovation ERSCP-EMSU conference, Delft, 25–29 Oct 2010
32. Lockton D, Harrison D, Staton NA (2009) The design with intent method: a design tool for influencing user behavior. Appl Ergon 41(3):382–392
33. Tang T, Bhamra T (2008) Changing energy consumption behavior through sustainable product design, DESIGN 2008, Croatia, 19–22 May 2008
34. Jelsma J, Knot M (2002) Designing environmentally efficient services; a 'script' approach. J Sustainable Prod Des 2:119–130
35. Järvi P, Paloviita A (2007) Product-related information for sustainable use of laundry detergents in Finnish Households. J Cleaner Prod 15:681–689
36. Beal CD, Stewart RA, Fielding K (2011) A novel mixed method smart metering approach to reconciling differences between perceived an actual residential end use water consumption. J Cleaner Prod. doi:10.1016/j.jclepro.2011.09.007

37. Geppert J, Stamminger R (2010) Do consumers act in a sustainable way using their refrigerator? The influence of consumer real life behaviour on the energy consumption of cooling appliances. Int J Consum Stud 34(2):219–227
38. Rodriguez E, Boks C (2005) How design of products affects user behaviour and vice versa: the environmental implications, environmentally conscious design and inverse manufacturing eco design 2005, Tokyo, pp 54–61, Dec 2005
39. Wever R, van Kuijk J, Boks C (2008) User-centered design for sustainable behaviour. Int J Sustain Eng 1(1):9–20
40. Lilley D (2009) Design for sustainable behavior: strategies and perceptions. Des Stud 30(6):704–720
41. Pruitt J, Adlin T (2006) The persona lifecycle: keeping people in mind THROUGHOUT product design, 1st edn. Morgan Kaufmann, Los Altos
42. Pepper M, Jackson T, Uzzell D (2009) An examination of the values that motivate socially conscious and frugal consumer behaviours. Int J Consum Stud 33(2):126–136
43. He L, Chen W (2011) Usage context-based choice modeling for hybrid electric vehicles. ICED'11, Denmark, 15–18 Aug 2011

Chapter 36
Approaches for Sustainability Assessment in the Conceptual Design Phase

Kai Lindow, Robert Woll, Masato Inoue, Haruo Ishikawa and Rainer Stark

Abstract This paper presents two different approaches to support the development of sustainable products. One approach is management-oriented. A sustainability dashboard has been developed to visualize the impact of different design alternatives on sustainability dimensions as well as the technical performance. The second approach directly assists the designer to narrow the design solution space when taking sustainability demands and technical performances into consideration. Both approaches are demonstrated on the example of a technical product. The paper concludes with a critical evaluation and a comparison of the two approaches.

36.1 Introduction

In an issue of The Times of India from last year Narayani Ganesh wrote in his column on sustainable development:

> Natural systems take millions of years to evolve to become sustainable, as Darwin observed while visiting the Galapagos Islands more than a century ago.

K. Lindow (✉)
Industrial Information Technology, Technische Universität Berlin,
Pascalstr. 8-9, 10587 Berlin, Germany
e-mail: kai.lindow@tu-berlin.de

R. Woll · R. Stark
Virtual Product Creation, Fraunhofer IPK, Pascalstr. 8-9,
10587 Berlin, Germany

M. Inoue · H. Ishikawa
Department of Mechanical Engineering and Intelligent Systems, The University
of Electro-Communications, 1-5-1 Chofugaoka, Chofu-shi, Tokyo, 182-8585, Japan

A. Chakrabarti (ed.), *CIRP Design 2012*, DOI: 10.1007/978-1-4471-4507-3_36,

Due to the fact that every product has environmental and social impacts the basic aim is to minimize the negative impacts and, at the same time, improve positive impacts in an effort to create a more sustainable world.

From an engineering point of view it does not remain millions of years to make technical systems sustainable. Particularly product development contributes to the principle of sustainable global development. Development processes specify all product properties and characteristics, so that succeeding product life phases are broadly predetermined [1]. At the present time, there are hardly scientifically proven and generally accepted methods or tools that serve to analyse the sustainability of different product alternatives. In general, engineers are not aware of the interdependence of individual product life cycles and they do not have the opportunity to estimate the impact of their design alternatives on the entire product life [2]. The engineer solely relies on personal knowledge of both product technology and development methods. He/she has to balance and to evaluate the effects, alternative development decisions might have on the product‘s sustainability. Due to increasing product complexity and diversity this mission is rather impossible and the engineers have needs for special sustainability related methods and tools. At the same time, companies have recognized that sustainable product development is essential for long-term success [3].

In order to better understand the method and tool support for sustainable product development, the following categories can be drawn (cp. [4, 5]):

- Checklists and guidelines: They are of qualitative nature and used for verification in order to proof if a product meets the pre-determined properties (e.g. materials checklists and Eco-design guidelines).
- Analytical tools: They are quantitative tools for evaluating and assessing a product's environmental and social performance (e.g. Life Cycle Assessment).
- Software and expert systems: They are computational tools used to support the application by providing user guidance and managing the huge amount of data and information

36.1.1 Challenges

The challenges of implementing sustainable thinking within a company can be met on different corporate levels (Fig. 36.1). The different levels can be understood as a framework which is based on the idea that sustainable product development is an internal process within a company. The company is in turn embedded into the society.

Starting from the principle of sustainable development sustainability demands have to be incorporated into the company's philosophy and responsibilities at a normative level. The main drivers are legislation and a growing awareness among the customers regarding a sustainable lifestyle. Against this background the

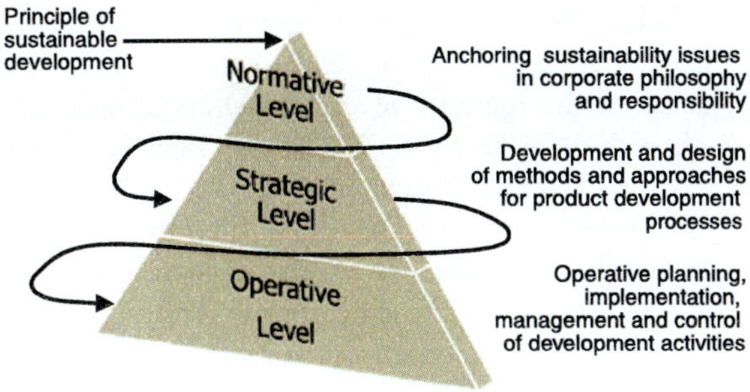

Fig. 36.1 From the principle of sustainable development to engineering development activities

manufacturer's responsibility of its products grows throughout the product life cycle [6].

Companies consider themselves as responsible to transfer sustainability demands from a normative to the strategic level. At this level methods and approaches for sustainable product design and processes are developed. The main challenge on this level is to consider simultaneously and equally the objectives of sustainability at all three dimensions (social, environmental and economic dimension) without compromising other essential engineering criteria such as product's performance and functionality [7]. At the same time, traditional performance indicators such as time, costs and quality cannot be neglected (in order to ensure a continued company's existence).

Eventually, at the operative level, the planning, execution and control of engineering activities is carried out. Since essential attributes and characteristics of products are already determined in product design and development process, it is necessary to integrate sustainable aspects into the product development process. Nowadays, designers rely on their product knowledge and methods in order to evaluate the impact of design alternatives on the product life cycle.

Moreover, the major challenge within companies is the selection of appropriate tools and methods in order to fulfil the company's strategic aims [5]. Companies usually define their sustainability responsibilities on the normative level: this, however, does not imply that strategic aims, guidelines, methods and tools also contribute to operational product development process level. Although product design is the most promising process to achieve corporate sustainability aims, design engineers only have little awareness about sustainability issues [2].

36.1.2 Focus of Research

In order to meet the needs for sustainability evaluation during the conceptual design phase, this paper presents two different approaches which should be understood as a complementary toolset within a company. One is management-oriented and can be settled between the strategic an operative level.Accordingly, a decision-making method for the early phases of design and a sustainability dashboard are presented. The other approach provides a method for decision-making on the engineering level (respective the operative level) by addressing economical, environmental and social goals including the trade-off between them. Both approaches are demonstrated by using the example of an alternator.

The alternator of a car was selected as an exemplarily object for a sufficient complex technical product. On the basis of three different scenarios, which vary in terms of the end-of-life situation, different sustainability goals are defined. The technical (e.g. roller bearing—friction bearing), social (e.g. re-manufacturing in Sierra Leone—India), monetary and environmental conditions are modified. The design of an alternator on the basis of re-manufacturing processes describes the use case for this research.

36.2 Preference Set-Based Design Method for Sustainable Product Creation

36.2.1 Classification and Categorization

The previous series of our research have proposed a preference set-based design method (PSD) which enables the flexible design and robust decision-making under various sources of uncertainties and, additionally, under sustainability demands. The PSD generates a ranged set of design solutions satisfying different sets of performance demands [8].

The preference set-based design method for sustainable product creation can be settled at the operative level of a company. Therefore, the main user is a product designer or engineer. He/she elaborates different design alternatives, uses the method and makes a design decision. According to the categories of sustainable product development, the method can be divided in analytical tools.

36.2.2 Objectives

The multi-objective satisfactory design including not only physical performances but also sustainable aspects from a global environmental protection perspective is necessary. When it comes to a sustainable life cycle, decision-making at the early phases of design is important. The design phase contains multiple sources of uncertainties in describing design. Therefore, handling the uncertainties in the design phase has a great importance. The preference set-based design (PSD) method enables the flexible and robust design under various sources of uncertainties while capturing designer's preference [9, 10].

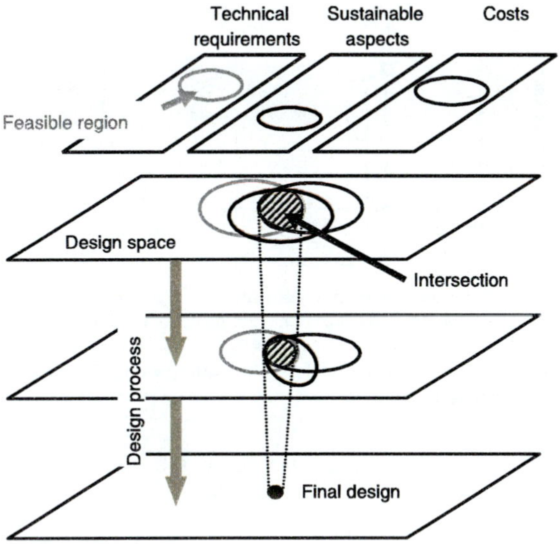

Fig. 36.2 Procedure of the set-based design method

36.2.3 Approach

The PSD method consists of the set representation, set propagation and set narrowing which are represented as the different layers in Fig. 36.2. In this case the solution space for technical requirements, sustainability aspects and costs are aligned and a final design solution was identified. A detailed description of the PSD method can be found in [8, 9]. The designer's design preference and the robustness of design solution are greatly important. A high design preference means that there are large feasible design subspaces within the designer's required performance spaces. On the other hand, design robustness includes the accuracy, convergence and stability of design. A high accuracy of design means that minimizing variations of a performance causes variations of design variables. A high convergence of design means that designers can find the preferable design solution easily. However, a high stability of design means that a low probability of design modification occurs. This approach eliminates infeasible design subspaces by evaluating the design preference and robustness.

36.2.4 Results

In this part the applicability of the PSD method for sustainable product development will be presented and discussed using the example of an alternator design. Due to simplification, only two different EoL scenarios were taken into account.

In a first step, different officially-accepted sustainability indicators were investigated and it was identified which indicators are related to the product

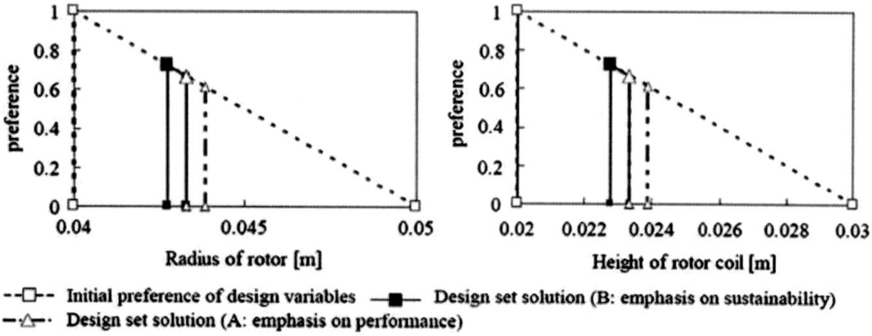

Fig. 36.3 Comparison of design set solutions between performance-orientation and sustainability-orientation [9]

development process. Thereafter, the proposed method was applied to a multi-objective design problem including technical performances and sustainable issues. The proposed set-based design method was applied to a multi-objective design problem including technical performances, e.g. power output of the alternator, and sustainable issues, e.g. CO_2 emissions.

Difference performance values between performance-oriented solutions and sustainability-oriented solutions can be achieved. The design values of the performance-orientation can achieve high power but heavy performance and high CO_2 emissions. Meanwhile sustainability-orientation can achieve light and low CO_2 emissions but less power output.

If a decision between performance-oriented or sustainability-oriented solution is made in middle management, the designer can obtain the specific technical parameters from the design set solutions according to the chosen scenario (Fig. 36.3).

36.2.5 Remarks

Different officially accepted sustainability indicators have been investigated. It became obvious that official sustainability indicators have only a weak link to the specifics of product development processes. The approach described clearly shows that the complex relations between decisions in product creation and their effects on the sustainability of products and processes in the product creation process have to be examined in a balanced way. Nevertheless, further research about product-related sustainability indicators must be carried out. Thereby a major challenge is to quantify the social dimension of sustainability. Additionally, the PSD method must be extended by further economic and environmental, as well as social aspects.

36.3 Sustainability Dashboard

36.3.1 Classification and Categorization

The Sustainability Dashboard can be settled between the strategic and the operative level of a company. At the operative level, designers or engineers have to prepare and evaluate different design alternatives whereas at the strategic level the middle management has the possibility to compare different design alternatives and make a strategic decision. According to the categories of sustainable product development, it is a composition of analytical tools and software support.

36.3.2 Objectives

The main research question is about how middle management can be supported in decision-making process when it comes to the evaluation of different design alternatives. Since design for sustainability is already a complex issue for a designer, making decisions at middle management is getting even more multifaceted because corporate aims and system development have to be considered as well. For this reason, it is necessary to communicate design impacts on the sustainability dimensions appropriately. Based on the House of Quality a so-called House of Sustainability was developed. This approach performs through different steps (which will be explained in the next chapter) where at the end a Sustainability Dashboard provides a visualized overview of different design alternatives for different scenarios. That way, middle management would be able to make reasonable decisions.

36.3.3 Approach

The literature review revealed that the Quality Function Deployment (QFD) in particular the House of Quality (HoQ) is the most promising method to align product properties (parameters) and process parameters against the background of sustainability demands. The basic idea of this approach is to replace the Voice of Customer by the Voice of Sustainability. Even though a method and process description of the approach was established, this paper focuses on the tools which were developed. In fact, the method is very close to the tools which are described and can be conceived by the sequence of the tools used.

On the basis of the QFD approach a tool chain for this research activity was developed. First of all, a House of Sustainability has to be established. The operation purpose is to get a ranking of the importance of product parameters for sustainability issues. Therefore, different HoQ have to be established according to

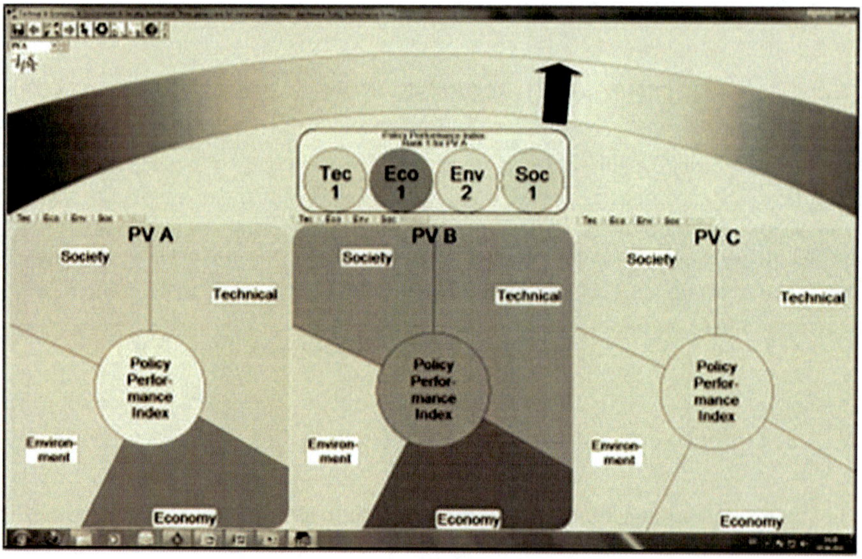

Fig. 36.4 Sustainability dashboard

each sustainability dimension and scenario. For instance, the interference between waste products and the height of rotor coil and has to be evaluated. Depending on the location, the waste products have to be evaluated as well (e.g. automated processes in Germany vs. Sierra manual processes Leone).

The import, merging and prioritising of the inserted information is done automatically by the tool. In a second step, different product alternatives can be configured. In this case there are three different design alternatives A, B and C. The user has the possibility to set values for different design parameters; e.g. in the case of the alternator the height of stator can be varied from 250 to 350 mm. The tool offers the possibility to choose a small (250 mm, scenario A), medium (300 mm, scenario B) and large variant (350 mm, scenario C). When all preferences are set, the data is automatically exported to the Sustainability Dashboard in a last step. The Sustainability Dashboard visualises the provided product and sustainability information (Fig. 36.4). This gives the user an appropriate overview in order to make suitable decisions. The Dashboard will be explained in the next chapter.

36.3.4 Results

The approach is implemented into an IT supported assessment and visualization method. The method is validated on the example of an alternator. In a first step, a set of sustainability indicators was established in order to assess the sustainability of the alternator. Furthermore, various End-of-Life (EoL) scenarios were developed to investigate their effect on the product shape. For example, materials were replaced by other materials (e.g. steel to plastics) and principle solutions were changed (e.g. roller bearings to friction bearing).

This was followed by a current state of the art Life Cycle Sustainability Assessment (LCSA, consisting of Life Cycle Assessment, Life Cycle Costing and Social Life Cycle Assessment) for the different scenarios.

The results are depending on the scenarios and set of indicators vary in intensity from one dimension of sustainability. The results were finally provided with a dashboard. Thus, the different design alternatives are compared with each other quickly. That way, conclusions about the sustainability impact of each design alternative can be drawn respectively to the EoL situation. The method helps to capture the complex relationships between design alternatives and their impacts on the entire product life cycle, select sustainable product alternatives and thus create sustainable value effectively and efficiently.

36.3.5 Remarks

The following remarks should be seen as a critical evaluation of the Sustainability Dashboard research. Basically, there are three aspects:

- The limitation to quantifiable product parameters excludes the explicit consideration of product characteristics which are of high relevance for sustainability assessment (e.g. materials can only defined by their (physical) properties).
- Reasonable weighting of process parameters demands results of a LCA of a similar product. This limits the application of the method to the development of different product alternatives (and not only the comparison of pre-determined design alternatives).
- The weighting of relations between product and process parameter is very complex and difficult to retrace due to indirect relations. This demands a systematic way of conducting the assessment (e.g. expert panels or questionnaires).

36.4 Conclusion

This paper describes two different approaches to support the development of sustainable products. At the strategic company level a sustainability dashboard has been developed in order to visualize the impact of different design alternatives on sustainability dimensions as well as the technical performance. The second approach directly assists thc dcsigncr to narrow the design solution space when taking sustainability demands and technical performances into consideration (operative company level). Both approaches are demonstrated on the example of a car alternator. That way, the research questions about the assessment of a product's sustainability within the conceptual design phase can be answered on the according corporate level.

It can be concluded, that improved supporting tools and methods have to be developed in order to better align strategic objectives from a normative corporate's level with sustainability goals at an engineering level. The aim must be to develop an integrated methodological approach that can be used throughout the various levels of the company. The integrated approach has to draw on a common information base. Furthermore, future research activities have to focus on the development of assistance systems which should be integrated into the working environment of the engineer in order to achieve products that are economically more successful, environmentally more viable and socially more responsible.

Eventually, the development of only sophisticated tools in academia is not enough against the background of industrial needs. A systematic approach on how to implement and run supporting tools within the design process has to be established. The major industrial needs do not have to be neglected. These basic needs are (according to [2, 11]):

- Easy to implement within the design process,
- Easy to lean and understand by the employees,
- Delivering results as accurate as possible,
- Reducing the amount of required information and
- Reducing the resources for evaluation and assessment.

Acknowledgments We thank Marzia Traverso and Erwin Schau from the Department of Sustainable Engineering, Technische Universität Berlin, for the lively exchange of knowledge. This research was supported by the German Research Foundation, DFG PAK 475/1. Furthermore, research about this topic is supported by the Japan Society for Promotion of Science, Grant-in-Aid for Young Scientists (B) No. 21760109 and Grant-in-Aid for Scientific Research (C), 20560124, in 2009.

References

Devanathan S, Ramanujan D, Bernstein WZ, Zhao F, Ramani K (2010) Integration of sustainability into early design through the function impact matrix. J Mech Des, p 132

Endris K, Marco T, Sergio T (2011) Integration of sustainability in NPD process: Italian experiences. In: Proceedings of the IFIP WG 5.1 8th international conference on product lifecycle management (PLM2011), 11–13 July 2011, Eindhoven, Netherlands

Hauschild M, Jeswiet J, Alting L (2005) From life cycle assessment to sustainable production: status and perspectives. CIRP Ann—Manuf Technol, pp 1–21

Baumann H, Boons F, Bragd A (2002) Mapping the green product development field: engineering, policy and business perspectives. J Cleaner Prod 10. Elsevier Science Ltd, Netherlands, pp 409–425

Pigosso DCA, Rozenfeld H, Seliger G (2010) Ecodesign maturity model: criteria for methods and tools classification. In: 8th GCSM—Global conference on sustainable manufacturing, Abu Dhabi, UAE, pp 22–24

Niemann J, Tichkiewitch S, Westkämper E (eds) (2009) Design of sustainable product life cycles. Springer, Berlin

Weenen JC (1995) Towards sustainable product development. J Cleaner Prod. Elsevier Science Ltd, Netherlands, pp 95–100

Inoue M, Ishikawa H (2009) Set-based design method reflecting the different designers intentions. In: 19th CIRP Design conference competitive design, Cranfield University, 30, 31 Mar 2009, pp 404–411

Inoue M, Lindow K, Stark R, Ishikawa H (2010) Application of sustainable aspects to the set-based design method. In: Proceedings of the 11th international design conference, design, Dubrovnik, Croatia, 17–20 May

Inoue M, Lindow K, Stark R, Ishikawa H (2010) Preference set-based design method for sustainable product creation. In: Proceedings of the 17th international conference on concurrent engineering, CE2010, Cracow, Poland, 6–10 September

Hallstedt S, Ny H, Robèrt K-H, Broman G (2009) An approach to assessing sustainability integration in strategic decision systems for product development. J Cleaner Prod 18. Elsevier Science Ltd, Netherlands, pp 703–712

Chapter 37
Integrating Low Carbon and Energy Efficiency Constraints in Sustainable Product Design

S. S. Krishnan, P. Shyam Sunder, Venkatesh Vunnam and N. Balasubramanian

Abstract Designing sustainable products is an important element of the low carbon economy that is emerging around the globe. Low carbon materials utilization and energy efficiency in manufacturing and in the consumption phase of a product's life cycle are key components of sustainable product design. However, integrating these factors into the design of consumer products needs further development. The technical, economic and policy aspects of sustainable manufacturing are vital drivers for integrated low carbon and energy efficient products. In this paper, we present some of our work on the technical and policy aspects of sustainable product design. We present a view of the energy efficiency of select appliances in the market and also show an example of integrating low carbon and energy efficiency in various stages of a product life cycle. This work is being extended further to be able to develop an integrated platform for the design of sustainable products.

S. S. Krishnan (✉) · P. Shyam Sunder (✉) · V. Vunnam (✉) · N. Balasubramanian (✉)
Center for Study of Science, Technology and Policy, Bangalore 560001, India
e-mail: ssk@cstep.in

P. Shyam Sunder
e-mail: shyam@cstep.in

V. Vunnam
e-mail: venkat@cstep.in

N. Balasubramanian
e-mail: nbalu23@cstep.in

A. Chakrabarti (ed.), *CIRP Design 2012*, DOI: 10.1007/978-1-4471-4507-3_37,

37.1 Introduction

Life cycle stages of consumer products today directly or indirectly impact the environment in the form of energy and materials usage throughout their lifecycle. Consumer durables such as refrigerators, air conditioners, ceiling fans, televisions and other various products are observed to be fast moving energy consuming products utilized in residential and commercial buildings.

37.2 Sustainable Design and Manufacture

Manufacturing is sometimes thought of as a simple open system into which various resources flow in for conversion and products and wastes flow out. However, one could take a much more extensive view of this problem [1]. If we take the systems view of manufacturing, and track the consequences of manufacturing and design decisions throughout the entire product development cycle, this would take us through (1) raw materials production, (2) manufacturing, (3) the use phase, and finally to (4) the end-of-life phase. This is a far broader view of manufacturing than the one that simply looks at the consumption, wastes and pollutants occurring at the factory. It has become clear that integrating manufacturing into a sustainable society requires the broader systems view [1].

A Process model on sustainable manufacturing has been developed by researchers at the Center for the Study of Science and Technology (CSTEP) [2]. In this model, shown in Fig. 37.1, the major process activities are represented. Each of these activities has an impact on the environment where an impact can be defined as a material or energy flow in either direction. Some activities such as raw material mining, energy production, manufacturing, use phase, recycling and others have a direct impact on the environment. For example, a car has a direct impact during its use phase. Some activities such as the design process and the maintenance and end-of-life analysis have an indirect impact in that these activities have the potential to substantively alter the direct impact of other activities. This study shows that for a completely sustainable manufacturing model, all the processes must interact with the environment through the sustainable infrastructure layer. Some activities listed under sustainable analysis are energy efficiency, material an energy flow, waste flow, total environmental impact and their associated technical, economic and other analyses. The Sustainable Manufacturing (SM) Process Model also details the sequence of processes that occur in the life cycle of manufacturing. The raw material mining and energy production feed the manufacturing plant with required inputs. Inputs also come in from design processes which drive the manufacturing process. The product is also subject to routine and periodic maintenance analysis checks which may feedback with retrofit activities that modify or upgrade the plant.

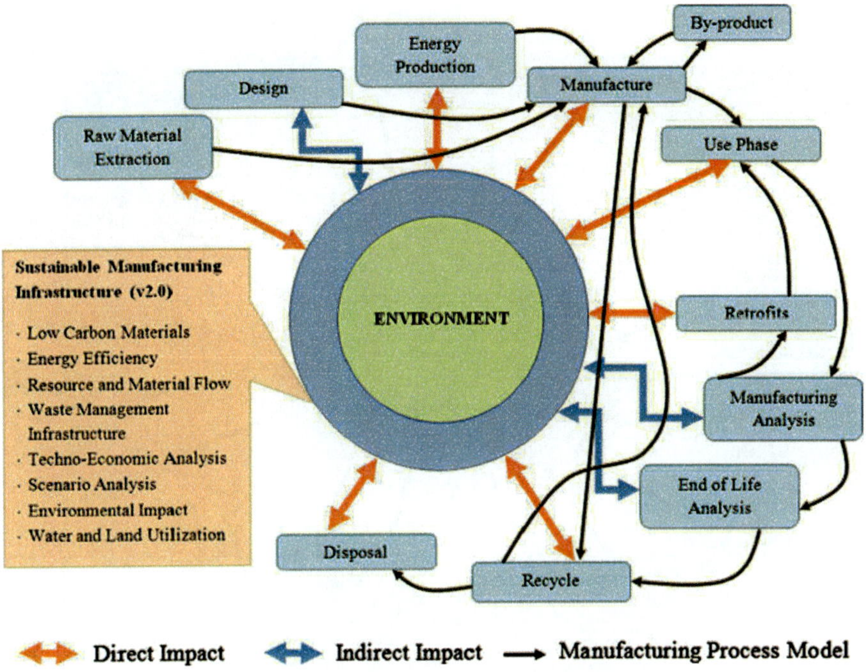

Fig. 37.1 Sustainable manufacturing process model

The CSTEP researchers [2] also provided a component view of the sustainable manufacturing infrastructure which is shown in Fig 37.2. This view represents all of the stakeholders who comprise the SM infrastructure. Each of the institutional stakeholders forms an aggregation relationship, in Unified Modeling language (UML) terminology, with the SM infrastructure. This report states that sustainability cannot be described as a separate activity that can be taught, trained, learned or practiced independent of the target domain. Sustainability has to be integrated into the various activities that comprise the current economic processes of human endeavor. In SM, sustainability analysis has to be incorporated into the different components shown in the SM infrastructure model. Each activity of the SM process model has to perform its entire repertoire of sub—activities while treating sustainability considerations as an additional factor.

This may be treated in various formulations by different components as an optimization function, a hard constraint, a soft constraint, a policy option, a policy mechanism guideline, a compliance target parameter or in other ways such as a modification of societal preferences, value systems and demands. However, the fact remains that in a systems view of the SM process model, sustainability needs to be urgently integrated into the current set of activities.

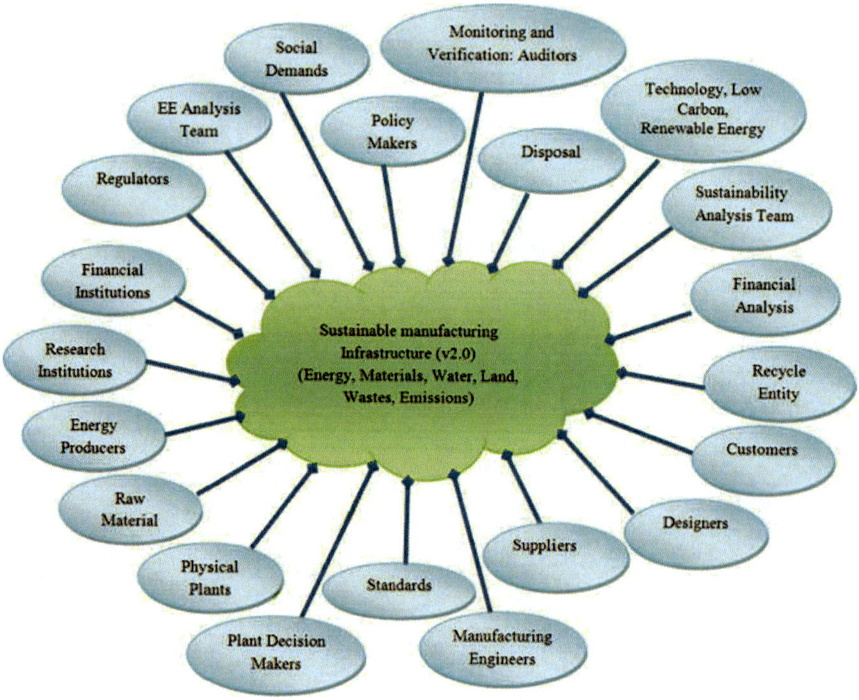

Fig. 37.2 Sustainable manufacturing infrastructure: component view

37.3 Global Initiatives for Sustainable Product Design

India initiated the National Mission for Enhanced Energy Efficiency (NMEEE), one of the 8 missions under the national action plan on climate change enunciating principles of achieving national growth objectives through qualitative directions that enhance ecological sustainability and mitigation of GHG emissions. Bureau of Energy Efficiency (BEE), is the principle organization under the Ministry of Power implementing the star labeling system for appliances. BEE has covered major building appliances and also agricultural pumps in the labeling systems [3]. From 2006, BEE is responsible for the star labeling system for residential and commercial domestic appliances in India. Table 37.1 shows the numbers of models which have been rated till Dec 2010. Energy star program by the U.S. Environmental Protection Agency and the U.S. Department of Energy (DoE) first initiated the energy efficiency among consumer durables helping consumers save money and protect the environment through energy efficient products and practices.

In 2009, Energy savings from consumers was able to avoid GHG emissions equivalent to those from 30 million cars—all the while saving nearly $17 billion on their utility bills according to DoE. Super Efficient Appliances Deployment (SEAD) initiatives are current developments from the US, DoE., where SEAD

Table 37.1 Total number of models from each appliances categories rated till Dec 2010 by BEE [3]

Appliances category	Model rated (2010)	% Growth rate (2010)[a]
Frost free refrigerator	633	5
Air conditioner	1,508	5
Tubular lighting	64	10
Direct cool refrigerators	599	5
Water heaters	325	9
Television	60	30

[a] IndiaStats

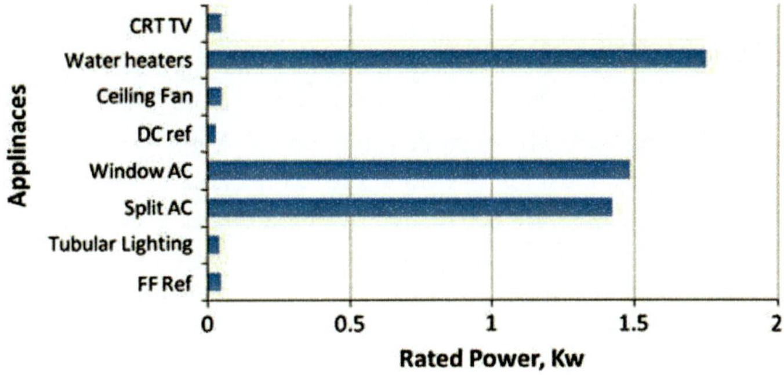

Fig. 37.3 Average rated power of appliances

partners will work together to "pull" super-efficient appliances and equipment into the market by cooperating on measures like manufacturer incentives and R&D investments. At the same time, partners will "push" the most inefficient equipment out of the market by working together to bolster national or regional policies like minimum efficiency standards. For example, SEAD will identify opportunities for strengthening appliance and equipment efficiency standards through international cooperation.

37.3.1 Product Classification and Energy Rating

Each durable has a factor called load or power needed to function, which is expressed in wattage (w) or Kilo watts (kW). A typical tubular lighting needs 30–40 watts, and water heater needs 1.5 kW or simply 1,500 W. From the yearly models rating information data, Heating Ventilation and Air Conditioner (HVAC) are the major energy consuming durables with load of 1,200–1,800 W on average. Other appliances are observed to be less than 200 W on average as shown in Fig. 37.3.

Table 37.2 Average hours of use of appliances

Appliance	Hours in use
FF refrigerator	8,760
Tubular lighting	2,920
Split AC	3,360
Widow AC	3,360
Direct cool refrigerator	8,760
Ceiling fan	2,400
Water heaters	1,095
Television	2,920

37.3.2 Product Usage Patterns

Duration of utilization varies from product to product. A refrigerator a most widely used appliance is continuously in operation round the clock in order to preserve the food. Typical refrigerators either frost free or direct cool are assumed to be in use for 24 h per day which is 8760 h per year [3, 4]. Some are seasonal like ceiling fan which are mostly used in summer or during hot conditions. Table 37.2 summarizes the assumed operating time in hours.

37.3.3 Annual Energy Consumption and CO_2 Emissions

Table 37.3 shows the comparative energy consumption scenarios of selected appliances between best and the average rated appliances today. The estimations is based on the current average energy rating and assumed consumption hours as explained in the earlier section.

37.4 Life Cycle Assessment

A Life cycle assessment (LCA) tool can be used to evaluate the environmental aspects and potential impact associated with a product and services throughout its life span. The generic approach to evaluate the LCA of a product is by cradle to grave. The basic aim of LCA is to:

1. Identify the energy and materials used and waste released to the environment.
2. Assess the impact of energy and materials used and wastes released
3. Identify the opportunities for reducing environmental impacts.

The whole framework follows ISO principles and requirements for conducting and reporting life cycle assessment studies [5]. For all appliances evaluated, it is

Table 37.3 Average energy consumption and CO_2 from average and best technology appliances in 2010

Appliance	Technology	Per unit kWh	Total volume energy GWh	Primary energy (PJ)	CO2 emission (Kg, per unit)
Frost free refrigerator	Average	663.0	5,151.9	58.5	543.7
	Best	397.4	3,088.5	35.1	325.9
Direct cool refrigerator	Average	390.0	12,123.1	137.8	319.8
	Best	256.6	7,975.7	90.7	210.4

significant the use phase dominated the life-cycle impact, with a proportion of more than 90 %. Production represented less than 8 % of the overall environmental burden [6].

Policy interventions and technological improvements have led to substantial reductions in energy consumption of appliances since 1981. The boundary for this study entails raw material extraction, manufacturing, and use phase for a functional appliance unit.

37.4.1 Integrating Low Carbon Materials and Energy Efficiency

We have developed a method to assess the life cycle impact of a product during the design stage by factoring in the low carbon material substitution options and energy efficiency constraints during manufacturing and the service life of common products. In this work, we have assessed the emissions and energy impact for a refrigerator which is one of most commonly used appliances. For a top door or front door refrigerator a typical materials used in the manufacturing is estimated. Figure 37.4 shows an integrated LCA model of a refrigerator. The proportion and kinds of materials used in this model is an approximate estimate and referring to the analysis of Association of Home appliance Manufacturing [1].

37.4.2 Results

The model was demonstrated using the GaBI 4 software. However, the integration of low carbon material substitution and energy efficiency in manufacturing and service life energy consumption are concepts which can be integrated independent of the software analysis tool. We have shown the various materials used as baseline Case 1 and in Case 2 we have substituted a certain amount of steel with a low carbon material. We have also replaced the existing manufacturing process

Refrigerator	Case 1	Case 2	Case 1	Case 2
Materials	Mass (kg)		Emissions (Kg- CO2)	
Aluminum	2.321	2.321	22.715	22.715
Polystyrene expandable granulate (EPS)	6.875	6.875	17.711	17.711
Acrylonitrile-butadiene-styrene granulate (ABS)	5.577	5.577	17.020	17.020
Copper	2.97	2.97	11.555	11.555
Glass	3.256	3.256	7.254	7.254
Cast Iron	5.016	5.016	6.144	6.144
Brass	0.781	0.781	3.539	3.539
PVC	1.111	1.111	2.481	2.481
Styrene-butadiene rubber mix (SBR)	0.187	0.187	0.580	0.580
Steel	52.305	39.22875	56.680	42.510
Low carbon Materials		13.07625		10.627

Fig. 37.4 Integrated LCA model of refrigerator using GaBi 4

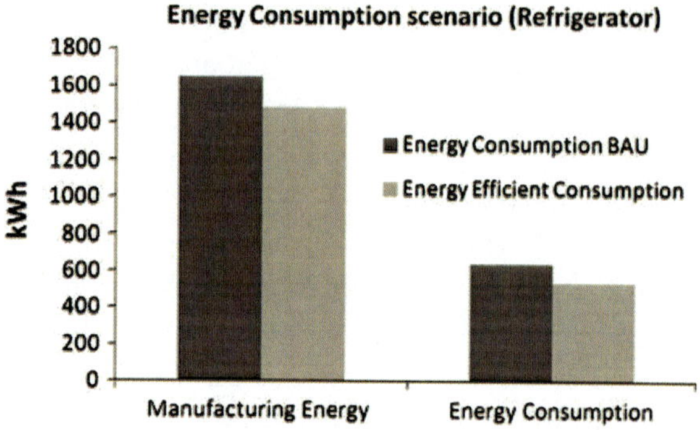

Fig. 37.5 Impact on emissions through low carbon and energy efficient design

with a more efficient process and have improved the design of the product by reducing its energy consumption during usage as well. All three factors need to be incorporated at the design stage in order to efficiently produce competitive products for a low carbon economy.

The results show that the carbon emissions from the metals such as Steel and aluminum emit high quantity of CO_2, 56.68 and 22.71 kg CO_2 respectively per unit manufacturing and plastics materials like ABS and EPS together estimated to emit 34 kg CO_2 as shown in Figs. 37.5 and 37.6 below. Case 1 pertains to the base case and Case 2 is the result of low carbon material substitution and energy efficient manufacturing and service life consumption standards.

Additionally, the particles in the air from typical refrigerator units were contributed by aluminum, glass and other metals. Collectively from metals, about 100–110 g of particle is displaced and on an overall from all the selected materials approximately 170 g of particles is displaced in the air from a single unit manufactured.

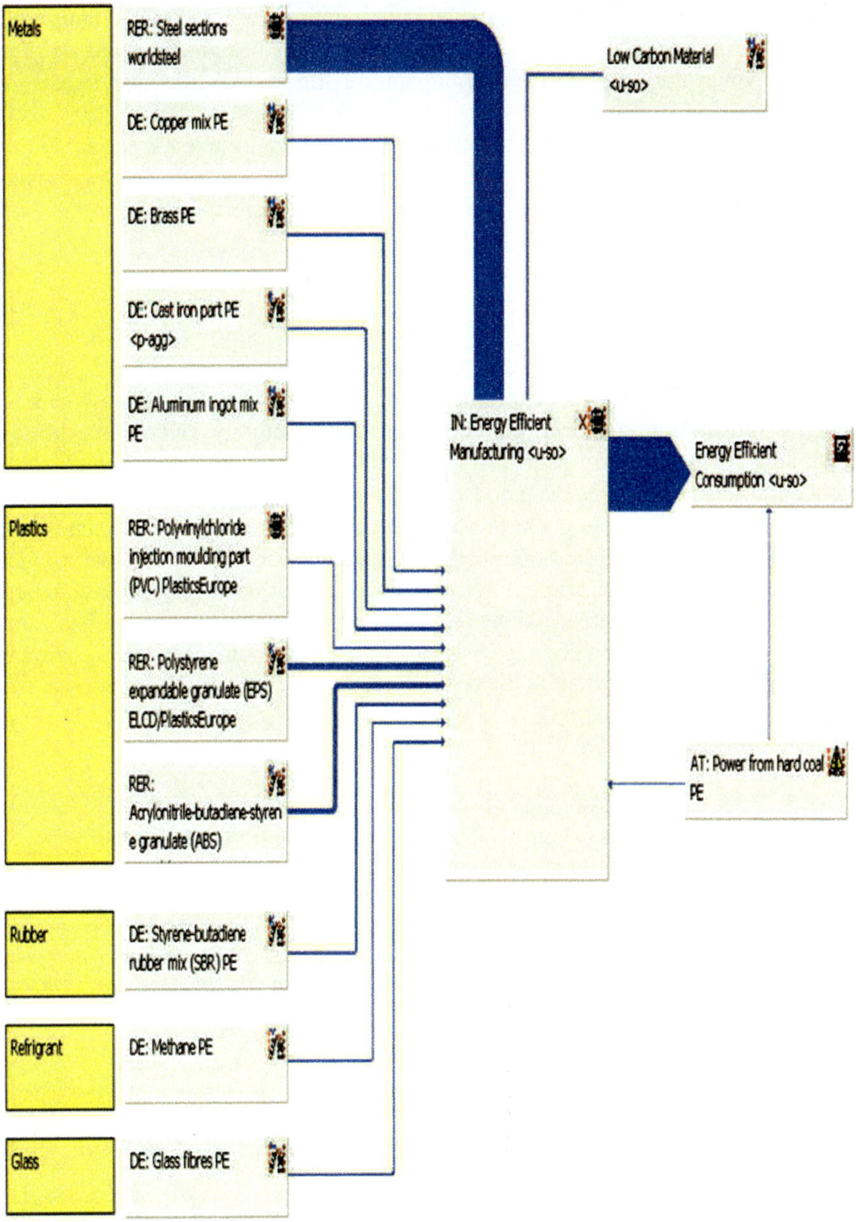

Fig. 37.6 Integrated LCA model of refrigerator using GaBi 4

The snapshot of the constructed LCA model is as shown in the Fig. 37.6. The model shows the materials consumed in manufacturing the refrigerators and the energy consumption per unit at manufacturing level and user end operating level for one year. The objective of the model is to identify the reduction of emissions that result from substituting low carbon materials at the design stage and also the energy savings during manufacture and during the service life due to design decisions taken early on.

Such an integrated approach can help in improving environmental sustainability of large volumes of appliances and the transition to a sustainable low carbon economy while retaining the competitiveness of manufacturing businesses.

37.5 Conclusion

In this work we have presented some of our result on integrating low carbon material substitution and energy efficiency in manufacturing and service life of common appliances. It is seen that both these factors can be integrated into products at the design stage itself and this can results in large energy and emissions savings over the life cycle given the large volumes of the common appliances.

These factors need to be integrated into a framework for sustainable product design for a low carbon economy. These design methods need to become standard practices for increasing the competitiveness of products given the societal and policy moves for sustainable practices and stricter environmental norms. We are continuing the above work with the development of specific design options in the Indian manufacturing context.

References

1. Gutowski TG, Murphy CF et al (2001) WTEC Panel report on environmentally benign manufacturing. Int Technol Res Inst
2. Krishnan SS, Balasubramanian N, Subrahmanian E, Arun Kumar V, Ramakrishna G, Murali Ramakrishnan A, Krishnamurthy A (2009) Machine level energy efficiency analysis in discrete manufacturing for a sustainable energy infrastructure. In: Second annual international conference on infrastructure systems and services, Chennai, India, December
3. Bureau of Energy Efficiency (BEE). Standard and labelling, BEE- Bureau of Energy Efficiency, [Online] (2010). http://www.beeindia.nic.in/content.php?page=schemes/schemes.php?id=2
4. Boegle Alexander, Singh Daljit, Sant Girish (2008) Energy saving potential in Indian households from improved appliance efficiency. Prayas Energy Group, Pune
5. Horie YA (2004) Life cycle optimization of household life cycle optimization of household. s.l.: Center for sustainable systems, University of Michigan, Michigan
6. Otto R, Ruminy A, Mrotzek H (2006) Assessment of the environmental impact of household appliances. s.l.: Appliance magazine

Chapter 38
Eco-Friendly Wood Polymer Composites for Sustainable Design Applications

G. S. Venkatesh, A. Deb, Ajay Karmarkar and B. Gurumoorthy

Abstract Environmental inputs can improve the level of innovation by interconnecting them with traditional inputs regarding the properties of materials and processes as a strategic eco-design procedure. Advanced engineered polymer composites are needed to meet the diverse needs of users for high-performance automotive, construction and commodity products that simultaneously maximize the sustainability of forest resources. In the current work, wood polymer composites (WPC) are studied to promote long-term resource sustainability and to decrease environmental impacts relative to those of existing products. A series of polypropylene wood–fiber composite materials having 20, 30, 40 and 50 wt. % of wood–fibers were prepared using twin-screw extruder and injection molding machine. Tensile and flexural properties of the composites were determined. Polypropylene (PP) as a matrix used in this study is a thermoplastic material, which is recyclable. Suitability of the prepared composites as a sustainable product is discussed.

G. S. Venkatesh (✉) · A. Deb
Center for Product Design and Manufacturing, Indian Institute of Science, Bangalore, India
e-mail: venkatesh@cpdm.iisc.ernet.in

A. Deb
e-mail: adeb@cpdm.iisc.ernet.in

A. Karmarkar
Institute of Wood Science and Technology, Malleswaram, Bangalore, India
e-mail: ajay@icfre.org

B. Gurumoorthy
Deptartment of Mechanical Engineering, Indian Institute of Science, Bangalore, India
e-mail: bgm@mecheng.iisc.ernet.in

A. Chakrabarti (ed.), *CIRP Design 2012*, DOI: 10.1007/978-1-4471-4507-3_38,

38.1 Introduction

Wood plastic composite (WPC) is a very promising and sustainable green material to achieve durability without using toxic chemicals. Plant fibres as reinforcements for polymers are currently the fastest growing type of polymer additives. The use of plant fibre reinforced composites has been increasingly studied in recent times. With increasing concern regarding environmental issues, scientists all over the world are seriously looking at natural fibres (plant fibres) as alternatives to replace synthetic fibres [1–4]. Construction, aerospace, packaging industry and automakers are aiming to make every part either recyclable or biodegradable. Bio-based composites will enhance mechanical strength and acoustic performance, reduce material weight and fuel consumption, lower production cost, improve passenger safety and shatterproof performance under extreme temperature changes, and improve biodegradability for the auto interior parts [5]. It is necessary to explore new ways to create greener and environmentally friendlier chemicals and materials for a variety of applications. Natural fibres are lighter than conventional fibre materials and therefore can contribute to cost reduction; their production is also more cost effective. Since the 1990s, natural fibre composites are emerging as realistic alternatives to glass-reinforced composites in many applications. Natural fibre composites such as hemp fibre-epoxy, flax fibre-polypropylene (PP), and china reed fibre-PP are particularly attractive in automotive applications because of lower cost and lower density [6]. Joshi et al. [6] conducted a life cycle assessment of natural fibre reinforced (NFR) composites and glass fibre reinforcement (GFR) composites and proposed that NFR composites are likely to be environmentally superior to GFR composites in most applications.

Wood is a renewable resource and "environmentally friendly" compared with other materials [7, 8]. Wood is a much less energy intensive material than most of its substitutes and that the carbon and global warming consequences of substituting wood for other materials, such as steel, aluminum, cement or bricks, could be significant. Wood products are easily recycled and it contributes less greenhouse gas emission than non-renewable steel and concrete. Thermoplastic resins offer the possibility of recyclability [9]. Composites with thermoplastic matrices have attracted considerable attention recently, due to their favorable properties, environmental advantages and processing characteristics [10]. Polypropylene (PP) is a resin of the polyolefin family and is one of the most widely used plastics in the world today, with a density of only 0.9 gm/cm^3, it is also the lightest of the widely used thermoplastics.

The interest in natural fiber-reinforced polymer composite materials is rapidly growing both in terms of their industrial applications and in terms of fundamental research. They are renewable, cheap, completely or partially recyclable, and biodegradable. Overall, the variety of bio-based automotive parts currently in production is astonishing; DaimlerChrysler is the biggest proponent with up to 50 components in its European vehicles being produced from bio-based materials. Uses of natural-fiber reinforcement have proven viable in a number of automotive

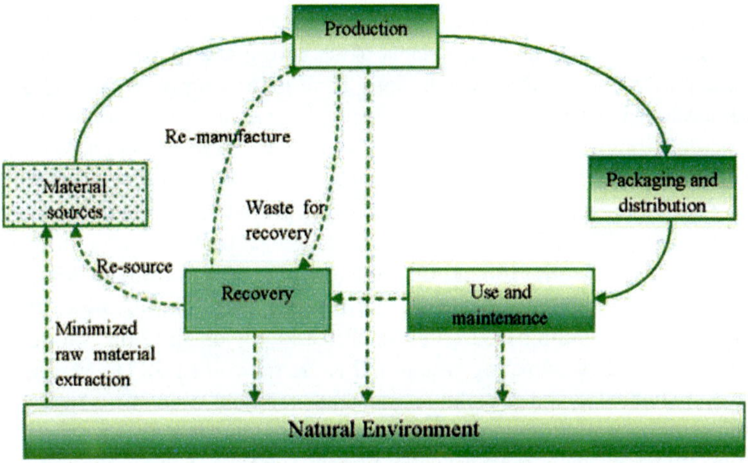

Fig. 38.1 The evolution of sustainable manufacturing concepts and practices

parts. Flax, sisal, and hemp are processed into door cladding, seat back linings, and floor panels. Coconut fiber is used to make seat bottoms, back cushions, and head restraints, while cotton is used to provide soundproofing, and wood fiber is used in seat back cushions [11].

In the current study, pine wood flour is used as reinforcement and polypropylene is used as a matrix in the preparation of wood polymer composites. WPC composites were obtained using twin-screw extruder and injection molding. Tensile and flexural tests were conducted on WPC composites to know the strength and stiffness's as high strength and high stiffness materials are required in design applications.

In Sect. 38.2, types of the materials and the manufacturing process were described with emphasis on sustainable manufacturing. Mechanical properties of the material along with cradle-to-cradle concept and applicability of material to the design is explained in Sect. 38.3. Life cycle of the WPC composite is briefed in Sect. 38.4. Finally, in Sect. 38.5, it is concluded highlighting the WPC as a sustainable product highlighting its weaknesses and the improvement required in the quality of the product.

38.2 Sustainable Manufacturing

Sustainable manufacturing is the developing of technologies to transform materials with reduced emission of greenhouse gases, reduced use of non-renewable or toxic materials and reduced generation of waste. Figure 38.1 shows the evolution of sustainable manufacturing concepts and practices.

Requirement for sustainable manufacturing include; Use of less material and energy, use of toxic to non-toxic and non-renewable to renewable input materials, reduce unwanted outputs with the use of cleaner production, industrial symbiosis and convert outputs to inputs by means of recycling and all its variants.

In the current study Pine wood flour is used as reinforcement. One of the requirements of sustainable manufacturing is the use of minimum raw material to be extracted from the environment (Fig. 38.1). Since waste wood could be used as reinforcement and less felling of trees are required, there is a minimum burden on the use of forest resources and the nature. Polypropylene is used as a matrix. In comparison with PVC, PU and FET, PP use fewer problematic additives, have reduced leaching potential in landfills, reduced potential for dioxin formation and reduced technical problems and costs during recycling [12]. Achilias et al. [13] investigated the recycling of LDPE, HDPE and PP on different products. The recovery of plastic in recycling was found to be more than 98 %. Matrix used in this study is nontoxic and recyclable and reinforcement used is renewable, hence both material used contributes to the sustainability.

In this study, we have synthesized a compatabilizer for compatibilizing natural fiber with polypropylene by incorporating isocyanate functionality onto polypropylene. m-Isopropenyl-α, α-dimethylbenzyl-isocyanate (m-TMI) was grafted on isotactic polypropylene by melt phase reaction using reactive extrusion.

The first step in wood–plastic composite fabrication, called compounding, blends organic plant fibers (wood) with an inorganic thermoplastic (PP). The percent of wood fiber used in this processing step is very important as it directly affects the tensile strength and Young's modulus of the product produced. In the current study 0, 20, 30, 40, 50 % by weight of wood is used as reinforcement. Pure polypropylene is termed as PP and WPC composites are termed as PPWx (x takes the values from 20 to 50 with a step of 10).

The major limitation encountered when trying to improve energy efficiency during the compounding process is the poor compatibility of wood fibers with the thermoplastic matrix PP. This poor compatibility stems from trying to combine hydrophilic wood fibers with a hydrophobic polymer (PP). The hydrophilic nature of these fibers causes the wood to swell during mixing and shrink during solidification, creating large aggregates and voids to form in the respective matrices. Traditionally, wood–plastic composites were fabricated using compression molding. The disadvantage of using this process in the manufacture of WPC's is that any part created takes a general form. This means that additional manufacturing processes, which tend to be wasteful and expensive, are required to produce a finished product. The original extruder used in wood–plastic composite forming was the single-screw extruder. During the single-screw extrusion process, friction is developed on the surface of the screw and barrel, which forces the material to flow from the hopper to the die. Despite this friction force being essential to moving the material down the barrel of the extruder, the creation of this force requires a large amount of energy and can have ill effects on the part's properties.

In order to improve upon the material and energy efficiencies of the extrusion process, the twin-screw extruder is used (Fig. 38.2).

Polypropylene homopolymer (3 kg), and m-TMI-grafted-PP (20, 30, 40, 50 wt % on fiber) were dry blended in a double cone blender. The polypropylene

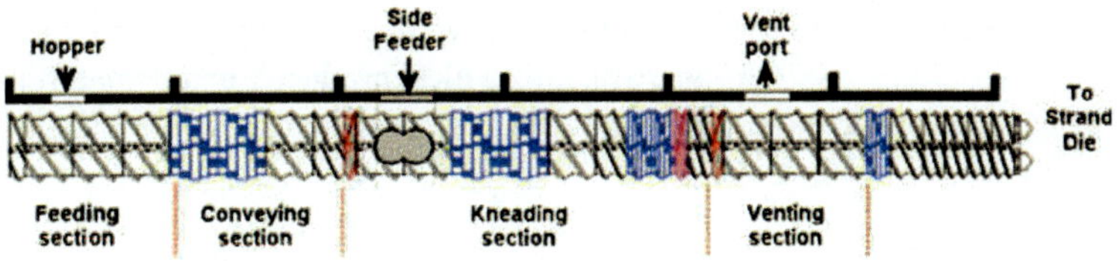

Fig. 38.2 Twin screw extruder

blended with compatabilizer was fed through thc main inlet hopper of the extruder. The granules were introduced at a pre-defined rate using a volumetric pellet feeder. The wood fibers were force-fed into the molten polypropylene using a twin screw side feeder. The wood fibers were fed to the side feeder at a predefined rate using another volumetric feeder. The extruder main screw RPM was set to 150 for all the experiments. The system was started with pure PP feed and the wood fiber feed rate was gradually increased to the set value. In the experiments, feed rate of the polymer and wood fiber were decided depending upon the weight of the wood fiber reinforcement. After approximately 10 min, when steady state conditions were reached, the pure PP feed was changed to mixture of PP, compatibilizer and processing additives. Vacuum venting was used to remove the residual moisture and volatiles produced during WPC production. The product was recovered by guiding the molten extrudate into a standard cold water stranding bath. The cooled strands were fed to a palletizer. The palletizer RPM was adjusted to match the throughput of the extruder so as to produce 3 mm long pallets. The pellets were dried to a constant weight and stored in sealed plastic bags.

Finally, the compounded pallets of nanoclay-PP composites were molded into standard ASTM type specimens using a microprocessor controlled, closed loop injection molding machine.

Due to hydrophilic nature of wood fibers and hydrophobic nature of polymers (PP) there will be a poor compatibility between the two and hence the poor dispersion. To overcome this, a compatabilizer m-TMI-grafted-PP was used which results in better mechanical properties as there would be a uniform dispersion of wood particles in a PP matrix.

To improve the energy efficiency of the forming process, intermeshing twin-screw extruder (Fig. 38.2) was used. The main advantage of this twin-screw extruder, over the single-screw extruder, is the increased materials and energy efficiency gained during forming. In an intermeshing twin-screw extruder with the co-rotating screws, the extruded material is more evenly mixed and spends less time in the extrusion barrel than its single-screw counterpart. This decrease in dwell time is caused by an increase in the material's speed down the barrel, which is allowed due to the elimination of backpressure. Backpressure is present in single-screw extruders because of the friction forces created by the screw. The utilization of two screws in this intermeshing extruder eliminates backpressure by displacing the material using the screws themselves and not friction. This increase in material speed allows for a smaller energy input during forming and lower material temperature rise during extrusion, which produces better material properties in the final product and increased energy efficiency.

38.3 Mechanical Properties

Existing wood–plastic composite (WPC) products are typically used for applications such as moldings, casings, and deck boards—where they are not critical structural elements. For this reason, establishment of nominal design values for structural applications of these products has not been a high priority. There is a need to determine allowable design values for wood plastic composite materials for designing elements in the structural applications. In order to design with WPC material, strength properties must be quantified and the material behavior must be understood. Standard tests for WPC material were proposed and used to determine material properties. The specific strength properties considered are modulus of elasticity (MOE), tensile strength, flexural modulus and strength. The resulting strength values could be used in the design of components in specific applications.

Tensile tests were conducted in accordance with ASTM D638 [14], with specimen type I. Crosshead speed was 50 mm/min. Flexural (three point bending) strength was measured as per ASTM D790 [15] using a support span of 102 mm and a cross head speed of 2.8 mm/min. Specimens for the test had the dimensions $127 \times 12.7 \times 3.2\ mm^3$. Five replicates were run for each composition. Tensile and flexural tests were carried out using Schimadzu make Universal testing machine equipped with an extensometer at room temperature. Tensile modulus and tensile strength increased continuously (Table 38.1) as the wood content increased from zero to 50 %. Increase in tensile modulus and tensile strength was found to be 236 and 49 % respectively for PPW50. Similarly, flexural modulus and flexural strength also increased as the filler content was increased. Increase in flexural modulus and flexural strength was found to be 272 and 97 % respectively for PPW50. Fu et al. [16] determined the tensile properties of short fiberglass reinforced polypropylene composites. They determined the tensile strength of glass fiber reinforced composites as a function of fiber volume function. Fiber volume fraction was varied from 5 to 25 % by weight and the tensile strength varied in the range of 46–50 MPa. Tensile strength of the WPC composites was found to be slightly less than glass reinforced PP composites. Since strength of the WPC composites are only slightly less than the glass reinforced PP composites, WPC composites could be used in better design applications.

Durability of wood and wood products is scarcely limited under dry conditions that are, at moisture content below 15 % where they are rarely attacked by wood destroying fungi. Wood products like WPCs absorb less moisture compared to solid wood. Therefore, they exhibit higher dimensional stability as well as fungal resistance when exposed to moisture [17]. The most important factors determining the durability of WPCs that are exposed to the environment for a long time are the

Table 38.1 Tensile and flexural properties of PP and WPC composites

Sample	Tensile modulus (MPa)	Tensile strength (MPa)	Flexural modulus (MPa)	Flexural strength (MPa)
PP	1755	35.2	1176	34.4
PPW20	2732	38.2	2096	47.3
PPW30	3425	43.4	2826	56.4
PPW40	4234	48.4	3467	61.4
PPW50	5891	52.7	4375	67.6

surrounding humidity or moisture as well as heat (temperature and sun light). Weathering of WPCs over years leads to a change in color, thereby affecting their aesthetic appeal. More importantly, weathering also has an impact on the mechanical properties of WPCs, which can be impaired by swelling and shrinking stresses of the wood fibers that are exposed to moisture.

Fullrich and Muller [18] investigated the mechanical characteristics of WPCs that were produced employing thermally modified wood fibers as reinforcement material, but without use of a coupling agent. After 5 years of storage, mechanical characteristics of WPC produced from thermally treated wood fibers improved, evident from increased tensile strength, Young's modulus, MOE as well as reduced strain and MOR. The positive effect of the long-term storage was attributed to improved physico-chemical interactions between the hydrophilic polar surface of the wood fibers and the polymeric thermoplastic materials, which presumably because of oxidative degradation reactions became polar during storage.

In the present study, treatment of wood fiber in the manufacture of WPC's is not considered. However, treatment of wood fiber might be necessary to maintain long-term mechanical properties of WPC's.

38.4 Life Cycle of WPC Composites

Life cycle of WPC composites is shown in Fig. 38.3. Life cycle of the WPC starts with cultivation of wood, as wood flour/particles are needed in the preparation of composites. The natural fiber component uses 45 % less energy, and results in lower air emissions [6]. Other material used in the production of WPC composites is polypropylene and is produced from monomers. PP could be mechanically recycled several times using conventional equipment. The existing technically viable recycling techniques for thermoplastic composites comprise mechanical, thermal and solvent routes [9]. WPC composites are manufactured sustainably without wasting any material and without any rework by twin-screw extruder and injection molding. After the useful service of the product WPC composites can be re-extruded or the constituents to be separately collected and recycled. Youngquist

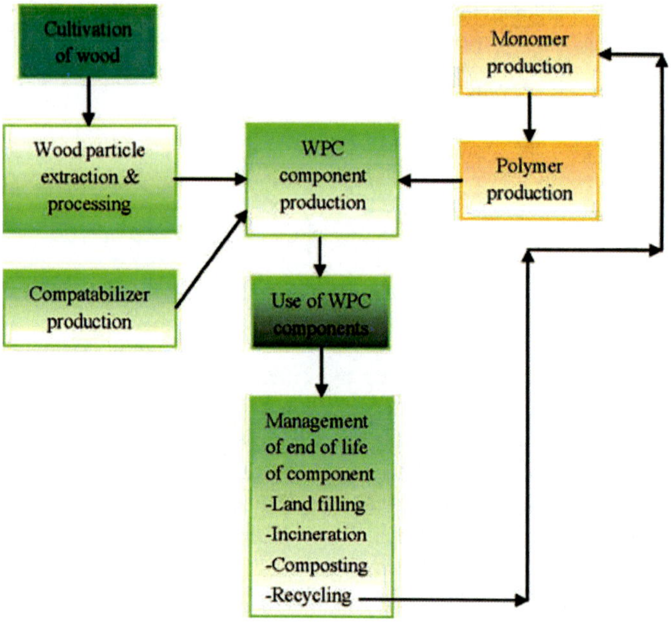

Fig. 38.3 Life cycle of WPC composites

et al. [19] showed that newsprint WPC systems could themselves be recycled (re-extruded and injection molded) numerous times with little or no apparent loss in mechanical properties. PP like other thermoplastics is capable of being recycled, for a multitude of reasons some of it will end up in landfill. PP is inert under landfill conditions; it does not contribute to leaching in landfill sites, does not generate methane gas and is stable within the landfill [20].

The Cradle to Cradle is a concept, developed by McDonough and Braungart [21], considers the completely industrial system and gives a new vision on sustainable design. "Cradle to Cradle" postulates the intelligent design and can eliminate the concept of waste. C2C goes beyond basic principles of sustainable development. Products are developed to be considered as a resource at the end of their life cycle instead of just waste. With C2C, the material cycle is closed and infinite recycling is possible without quality loss, waste or accumulation. Recycled WPC systems could themselves be recycled (re-extruded and injection molded) numerous times with little or no apparent loss in mechanical properties. Stark [22] found that reinforcing polypropylene with wood fiber derived from recycled hardwood and softwood pallets provided improved bending and tensile strength over the use of wood flour. Therefore, it can be concluded that WPC follows C2C concept and it is a potential sustainable development material.

38.5 Conclusions

Rapidly rising indiscriminate industrial processes can damage the ecology and environmental balance of the planet. There is a growing feeling in the international community that future technology development will need concepts such as manufacturing sustainability, minimum use of energy and renewable raw materials. WPC composites are the promising sustainable materials as it can be manufactured sustainably and is sustainable as regard to its use, recyclability and disposal. Mechanical testing was carried out on WPC composites and there is an increase in properties as the filler content increased. Increase in tensile modulus and tensile strength was found to be 236 and 49 % respectively and increase in flexural modulus and flexural strength was found to be 272 and 97 % respectively for 50 % reinforcement of wood as compared to pure PP. As WPC composites possess good mechanical properties, it could be used in many design applications. Additional research improve properties and processing and thereby increase the number of potential applications. To do this we must learn to improve melt-blending processes to achieve better fiber dispersion with minimal fiber breakage, improve the bonding between the wood fiber and plastic matrix especially, as this is related to more variable recycled materials. Improving systems to moisture, to biodegradation, and to fire are important properties that also need to be addressed.

References

1. Hill CAS, Abdul Khalil HPS, Hale MD (1998) A study of the potential of acetylation to improve the properties of plant fibre. Ind Crops Prod 8:53
2. Varma IK, Krishnan SRA, Krishnamoorthy S (1988) Effect of chemical treatment on mechanical properties and moisture regain of jute fibres. Textile Res Inst 58(9):537
3. Satyanarayana KG, Sukumaran K, Kulkarni SGK, Rohatgi PK (1986) Fabrication and properties of natural fiber reinforced polyester composites. Composites 17:329
4. Bisanda ETN, Ansell MP (1991) The effect of silane treatment on the mechanical and physical properties of sisal epoxy composites. Compos Sci Technol 41:165
5. Ashori Alireza (2008) Wood–plastic composites as promising green-composites for automotive industries! Bioresour Technol 99:4661–4667
6. Joshi SV, Drzal LT, Mohanty AK, Arora S (2004) Are natural fibre composites environmentally superior to glass fibre-reinforced composites? Composites: Part A 35:371–376
7. Lippke B, Wilson J, Garcia JP, Bowyer J, Meil J (2004) CORRIM: life-cycle environmental building materials. Forest Prod J 54(6):8–19
8. Puettmann ME, Wilson JB (2005) Life-cycle analysis of wood products: cradle-to-gate LCI of residential wood building materials. Wood Fibre Sci 37:18–29
9. Otheguya ME et al (2009) Recycling of end-of-life thermoplastic composite boats. Plast, Rubber Compos 38:406–411
10. Ijaz M, Robinson M, Gibson AG (2007) Cooling and crystallisation behaviour during vacuum-consolidation of commingled thermoplastic composites, Composites: Part A 38:828–842
11. Holbery J, Houston D (2006) Natural-fiber-reinforced polymer composites in automotive applications. JOM 58(11):80–86
12. A Review of the Availability of Plastic Substitutes for Soft PVC in Toys (1999) Report Commissioned by Greenpeace International

13. Achials DS, Antonakoue E, Roupakias C, Megalokonomos P, Lappas A (2008) Recycling techniques of polyolefins from plastic wastes. Global NEST J 10:114–122
14. American Society for Testing and Materials (1995a) Standard test method for tensile properties of plastics. D638-94b. Philadelphia
15. American Society for Testing and Materials (1995b) Standard test method for flexural properties of unreinforced and reinforced plastics and electrical insulating materials. D790-92 Philadelphia
16. Fu SY, Lauke B, Mader E, Yue CY, Hu X (2000) Tensile properties of short-glass-fiber- and short-carbon-fibre-reinforced polypropylene composites. Composites: Part A 31:1117
17. Clemons C (2002) Wood–plastic composites in the United States—the interfacing of two industries. Forest Prod J 52(6):10–18.25
18. Follrich J, Muller U (2010) Effects of long-term storage on the mechanical characteristics of wood plastic composites produced from thermally modified wood fibres. J Thermoplast Compos Mater 23:845–852
19. Youngquist JA, Myers GE, Harten TM (1992) Lignocellulosic–plastic composites from recycled materials. ACS Symp Ser 476:42–56
20. Environmental Information Document for Polypropylene Relevant to Australia. Available via http://www.lyondellbasell.com/NR/rdonlyres/C2ED0A47-6430-45FA-87A4-D40181088 14D/0/AusPPEnvirostatementSep09.pdf
21. McDonough W, Braungart M (2002) Cradle to cradle: remaking the way we make things. North Point Press, New York
22. Stark NM (1999) Wood fiber derived from scrap pallets used in polypropylene composites. Forest Prod J 49(6):39–46

Chapter 39
Understanding Needs in Eco-Design Learning for Novice Designers

Flore Vallet, Dominique Millet and Benoît Eynard

Abstract The paper explores how engineering students approach eco-design as novices to the topic. Only few observational studies focus on novice eco-design practice, whilst it is a classical topic in traditional design research. This gap is addressed by an observational study of six multidisciplinary groups of European engineering students. The eco-design task consists in the environmental performance improvement of a vacuum cleaner. Two viewpoints are introduced and compared: a traditional life-cycle and a stakeholder perspective. Main results show that (1) novices tend to start the task by searching for solutions whatever the method; (2) environmental framing of the task is easier with a stakeholder viewpoint than with a life cycle one.

39.1 Introduction

Education in sustainability is acknowledged to be a key challenge to move towards a sustainable society [1]. In the research field, sustainability in higher education is addressed through different perspectives such as: regional applications, teaching and learning insights, tools for sustainability education, implementing and reporting

---

F. Vallet (✉) · B. Eynard
Université de Technologie de Compiègne, 60203 Compiègne, France
e-mail: flore.vallet@utc.fr

B. Eynard
e-mail: benoit.eynard@utc.fr

D. Millet
SUPMECA Toulon, 83000 Toulon, France
e-mail: dominique.millet@supmeca.fr

A. Chakrabarti (ed.), *CIRP Design 2012*, DOI: 10.1007/978-1-4471-4507-3_39,

sustainability in academia [2]. Restricting ourselves to environmental impacts of products, we focus on eco-design education and more specifically on how eco-design is implemented by novices in eco-design with the support of dedicated tools. In the tradition of design research, a first move to get insight into the eco-design practice among novices was reported in [3]. Collado and Ostad-Ahmad-Ghorabi [4] emphasized that 'soft' environmental information are useful to engineering design students in order to foster the generation of eco-innovative concepts. However, both studies do not address how core eco-design activities, namely environmental assessment and solution finding, are conducted by novices. Thus the research question we intend to address in this paper is the following: "How is the eco-design process performed by students who are novices in eco-design?"

Key features of eco-design education and tools are detailed in Sect. 39.2. The research method is exposed in Sect. 39.3. An experiment in redesign of a medium complexity consumer product with novices in eco-design is detailed (Sect. 39.4). Results concerning the main difficulties encountered by novice designers are presented. These are discussed and put into perspective of future work in Sect. 39.5.

39.2 Views on Eco-Design Education: Process and Tools

To clarify the main steps of an eco-design process, Brezet and Van Hemel state that, with the integration of environment in design, new activities have appeared in the traditional product development process [[5], p. 49]. These new activities deal with: the search for environmental information, initial environmental evaluation and environmental strategy. The last two seem to differentiate eco-design from traditional design according to [4]. Thus developing environmental skills for product designers supposes the dissemination across higher education of both approaches, namely improvement and environmental assessment. Some practical educational case studies involving this dual approach to eco-design products can be found in [6; 7]. After a presentation of several eco-redesign projects, Lloveras [6] emphasizes the benefit for engineering design students to implement quantitative impact eco-points to compare initial and final concepts in an objective manner. In [8], students are provided a first assessment of a kettle. Through an iterative generation of concepts, predetermined environmental guidelines are refined to better match the product. Lastly [9] illustrate an eco-design program for engineering, business and industrial design students. In this context, eco-design projects are developed using qualitative assessment and practical design guidelines.

But this analysis should not be disconnected from the tools used within curricula, because of the important role played by those in supporting eco-design education. Independently of their aims (improvement and/or assessment), the presentation of tools for eco-design education varies from simple tools to eco-design guides or dedicated web platforms. Systematic step-by-step approaches seem to be well-adapted to foster eco-design learning for product design students. This need is covered in eco-design guides or manuals, for instance: 'PROMISE' guide [5], Okala course guide [10]. 'Environmental improvement' guide [11], D4S guide [12]. As pointed out by [13], eco-design guides refer to multiple eco-design tools. They also notice a clear domination of environmental assessment against creativity, decision making and cost accounting tools. The MET (Material Energy Toxicity) matrix and the LiDS Wheel (or Eco-Design Strategy Wheel) of [5] appear to be the most mentioned specific eco-design tools. Besides, eco-design tools can also refer to the 'stakeholders' of the product value chain. Stakeholders, who are the individuals or organizations "who can help or hinder the pursuit of [your] objectives" [14] appear to be closely associated with the phases of the product life cycle. Exchanges between stakeholders (i.e. information or material flows) often cause to environmental impacts, for instance in negotiations along the supply chain [11]. That is why stakeholders are taken into consideration in eco-design guides: stakeholder network [11]; stakeholder interaction matrix [12].

39.3 Research Method

A participant observation activity without coding scheme was organized in an academic context. It is based on a classic experimental research plan with equivalent groups, identical pre and post-test [15]. Besides, this is a typical case of "small-scale design experiment" as discussed in Cash et al. [16]. It is meant to question the limits of the life-cycle vision presented in the ISO 14062 framework in an educational context [17].Two viewpoints are introduced in this experiment, defining the independent variables. The first one is related the traditional life-cycle of the product, whereas the second is centered on stakeholders, whose role appear to be predominant as shown in Sect. 39.2. The starting point consists in a set of structured data compliant with the level of expertise of students. This includes: EuP executive summary on vacuum cleaners [18]; feedback from Rowenta on the Shock Absorber model [19]; an eco-design process broken into four classical stages (Framing; Initial environmental assessment, Environmental Strategy definition and Solution finding) [5].

39.3.1 Participants and Eco-Design Brief

The study involved 24 European students who participated in the 2011 Winter School of the ECIU (European Consortium of Innovative Universities) program. Half the participants were undergraduate mechanical engineering students from the French hosting university UTC. The rest of the group was formed by seven management students from Strathclydde-UK, two students from Tallyn-Estonia and two students from Twente-Netherlands. There is a fair balance with regard to gender representation, with 10 female for 14 male students. The great majority of the participants are not familiar with eco-design (19/24, i.e. 79 %), whilst 4 students claim to have previous notions of eco-design. One single student has been trained to eco-design and life cycle assessment through a dedicated course and project. For the sake of the program, this participant was not left out, being aware that the responses of her group could be biased. It was also decided to mix local and foreign students in each team. The eco-design brief deals with the redesign of a classical consumer product of medium complexity including around a hundred parts: a vacuum cleaner. Participants are asked to analyse the problem from an environmental viewpoint in order to redesign a vacuum cleaner. The task is introduced as a realistic design study for a European white good company.

39.3.2 Hypotheses, Data Collection and Analysis

In order to investigate our research question, two hypotheses H1 and H2 are formulated. A success factor is defined for both hypotheses. Based upon documents and video recordings, the number of environmentally relevant propositions is examined and evaluated by two eco-design experts.

H1 *The viewpoint has an influence on the way novice eco-designers perform the eco-design process.* A life-cycle representation is expected to be more difficult to handle for novices than a stakeholder representation

H2 *There is a difference in the way novice eco-designers address the eco-design steps of the process.* In line with the literature review, environmental assessment is expected to be more difficult for novices in eco-design than solution finding

The whole group was broken into 6 mixed and supposedly homogeneous teams of 4 persons to 'balance the conflicting opinions' [16]. Two teams were asked to follow a traditional life cycle perspective based on a MET matrix, whilst two other teams followed a stakeholder perspective. We introduced a modified MET matrix where life-cycle stages are replaced by stakeholders. A control condition was added for the last two teams, where students were free to adopt whatever approach they wished. A two-hours training was organized on the day before the actual experiment. All participants were introduced to basic notions of eco-design (e.g. environmental impacts, life cycle, Functional Unit) and to main steps of the eco-design process by means of several examples. Students from the 'Control' and 'Stakeholder' groups were treated with a specific presentation of stakeholder concepts. A quick pre-test (Q1) was made to figure out the origin initial environmental knowledge of each participant. On the following day, the experiment (1h30) started with a warming up activity, followed by the redesign task. It was facilitated by a researcher who was allowed to respond to students' questions. Students filled up a flash questionnaire (Q2) to give an immediate feedback on the process. A last questionnaire (Q3) was sent by e-mail 3 weeks later. This is summarized in Fig. 39.1. Four types of data were collected: observational notes of the experimenter, video recordings; filled-up questionnaires and responding sheets.

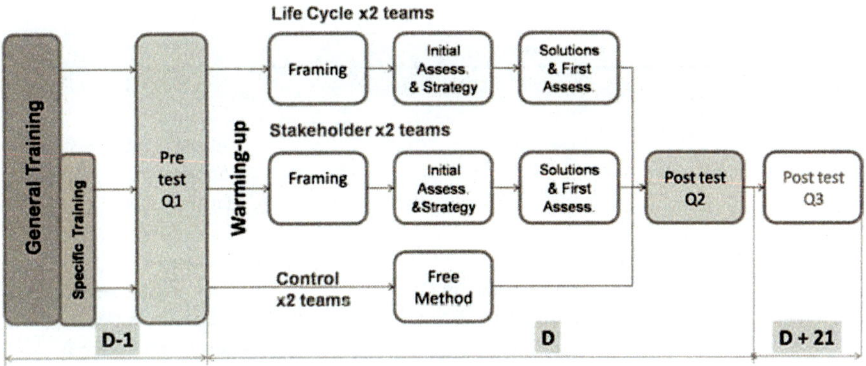

Fig. 39.1 Agenda of the experiment

39.4 Empirical Results

In this section, we provide an insight on the reasoning and documenting of the different teams for the three main stages of the task: Framing; Initial environmental assessment and Solution finding. Results extracted from the questionnaires are not exposed in this paper for brevity.

39.4.1 Environmental framing

Control groups do not appear to frame the problem, which implies that the boundaries of the study are not clearly defined. Life Cycle groups (abbreviated L1 and L2) are asked to map the life cycle of the vacuum cleaner, propose a Functional Unit and suggest the most impacting life cycle stages. Both groups identify fossil oil and metals as part of material flows. L2 proposes the following Functional Unit: "480 h hovering" and ranks by order of importance: Use, Disposal and Material Extraction. Stakeholder groups are asked to map the different stakeholders of the project and to site them. Answers given by the two groups are classified into internal (to the company) and external stakeholders by the authors (Table 39.1). Three stakeholders are quoted by both teams (in bold): consumer, marketers and logisticians. It can be seen that the second group gives more accurate answers and covers a wider range of stakeholders. The quotation of energy provider is especially relevant for this study. Concerning this step, the number of environmentally relevant propositions evaluated by eco-design experts are: C1: 0; C2:0; L1: 3; L2: 3; S1:7; S2:15.

Table 39.1 Framing of stakeholders

Groups	Quoted stakeholders	
Stakeholder 1	Internal stakeholders	External stakeholders
	Investors	Greenpeace
	Marketers	Government
		Europe
		Transporters
		Consumer
Stakeholder 2	Management (WE[a])	**Consumer**
	Design Planning (WE)	University Research (WE)
	Production (EE[b])	Competitors (WE)
	Marketing	Legislator (Europe)
	Customer service	Suppliers (EE)
	Parts retailer (WE)	**Logistics** (WE)
		Retailer (Europe)
		Recycler (WE)
		Energy provider (WE)

[a] Western Europe
[b] Eastern Europe

39.4.2 Initial Environmental Assessment

There is no evidence of an initial assessment of environmental impacts of the vacuum cleaner in the Control groups. Both teams start searching for solutions as soon as they receive the brief. Both Life Cycle groups indicate qualitative answers in the MET matrix. The items indicated in "Materials" and "Energy" rows are quite similar, and are correctly appreciated. Since the production processes are not known by students, their answers remain generic, e.g. "Process energy". More confusion is visible in the "*Toxicity*" category, where CO_2 emissions, electricity or paper bags are quoted. For Stakeholder groups, the first step was to select the 3 or 4 main stakeholders for the study. Then the adapted MET matrix was introduced through the following question to the students: "What are the expectations and requirements of the stakeholder St_i regarding environmental impacts or values?" An example of the results by the first Stakeholder group is given in Fig. 39.2. The double entry on Impacts and Values seems confusing for the students, and the presentation of the matrix should be improved to that respect. As previously noticed, toxicity is the most difficult concept to understand, especially for Stakeholder 1 team. It also has to be noted that the Stakeholder 1 group appears to mix statements and concept propositions (e.g. "Recommended by" sticker), pointing out the interlinked character of problem and solutions. For the Initial Assessment phase, the number of environmentally relevant propositions evaluated by eco-design experts are: C1: 0; C2: 2; L1: 17; L2: 16; S1:17; S2:13.

Impacts/ Values	ST1: European Union	ST2: Consumer	ST3: Transporters	ST4: Investors/Marketors
Material	European renewable ressources	Light Ecological Best Quality/Price	Volume, lighter	Sustainable materials (to appear sustainable for customers but short life span)
Energy		Energy consumption (bill) Efficient (using 1h instead of 2h)	Env. friendly boats with huge kites	
Toxicity		No toxic material No burn plastic No bags	Sea life	
Other	Rules, Laws, Standards of faire trade (social impacts) Fair contracts for workers	Long cable Attractive packaging Easy to store Not to expansive Cheapest Low noise level Design Used on all surfaces	Social impacts: transporters with EU contracts (under EU laws)	Connect product image to appeal to the market demographic Brand image collaboration with Dyson brand to increase product sales « Recommanded by » sticker

Fig. 39.2 Modified MET matrix by stake holder1 group

39.4.3 Solution Finding

The last step of the task was to propose several ideas of improvement (Table 39.2) and to map them against two dimensions: innovation and environmental performance. The most tackled issues are: bag consumption, energy consumption and noise associated with a customer viewpoint. Control 2 group implements a large morphological matrix with 9 criteria (in rows) and 3–5 ideas (in columns). It has to be added that solutions from Stakeholder 1 group are mostly driven by shape. The number of environmentally relevant concepts evaluated by two eco-design experts are included in Table 39.2.

39.5 Conclusions and Further Work

Through this experiment, the difficulties and needs experienced by novices in eco-design have been illustrated in the case of redesign of a vacuum cleaner. The level of achievement of the task is summarized in a semi-quantitative way for each group in row, and each step in column (Table 39.3). 'Minimal answer', 'Partial answer' or 'Full answer' are classified by the number of relevant answers within each category. For instance, the highest number of relevant solutions being 10, between 6 and 10 corresponds to a 'Full answer', between 2 and 5 is a 'Partial answer' and one or no answer is a 'Minimal answer'.

The main findings of this study are the following. Firstly, H2 ('The viewpoint has an influence on the way novice eco-designers perform the eco-design process')

Table 39.2 Environmental solutions addressed by teams

Groups	Quoted/Sketched solutions	Nb relevant concepts
Control 1	High efficiency motor; smart filter (ionizing particles); round brush (for power saving); renewable manufacturing energy; recycle packaging; dust container instead of bags; integrated circuits to prevent current leakage; energy recovery	8
Control 2[a]	Move: hovering; Materials: natural fibers, composites; Energy: battery; Particles: water…	3
Stakeholder 1	Baby-shaped, plant-shaped (with batteries in a pot) vacuum cleaner; "Dyson" concept with water filter	2
Stakeholder 2	Hybrid vacuum cleaner; solar energy; silent vacuum cleaner; bacteria filter; kids vacuum cleaner; retractable tube; water/liquid vacuum cleaner; "Xbox"	6
Life cycle 1	Plaited natural fiber pipe; no bags; wood panel structure (also vibration absorber); energy efficient motor	4
Life cycle 2	Standard component; compostable housing; solar energy; electromagnetic field; no bags; eco-friendly materials; KERS[b] on the nozzle; reward for reusable materials; "dust claimer" house; musical program to reduce noise; artificial spit	10

[a] Only selected concepts among several are mentioned here
[b] Kinetic Energy Recovery System

Table 39.3 Achievement of eco-design stages by teams

□ minimal answer; ◩ partial answer; ■ full answer

Groups	Framing	Strategy	Initial Asses sment	Solution fin ding
Control 1	□	□	□	■
Control 2	□	□	◩	◩
Life Cycle 1	□	□	■	◩
Life Cycle 2	□	◩	■	■
Stakeholder 1	◩	□	■	◩
Stakeholder 2	■	□	◩	■

appears to be partially validated. If the stakeholder perspective is more efficient to frame the problem, there is no significant difference observed between groups in environmental assessment and solution finding phases. Concerning framing, the representation of the life-cycle of a vacuum cleaner appears to be extremely tedious for students, as they express difficulties in figuring out the level of detail they are supposed to describe. By contrast the stakeholder approach seems easy to access for both students in Product Design and Business Management. However,

Stakeholder and Life Cycle groups experience a common puzzling topic: the notion of *toxicity*.

Secondly, our experiment seems to validate H2, stating that 'There is a difference in the way novice eco-designers address the eco-design steps of the process'. All groups tend to start with the eco-design task by searching for solutions. This could be explained by the fact that looking for environmental improvement requires far less expertise than environmental assessment [20]. This also points out the need for novices to enter eco-design problems though idea generation, which could be considered in the development of eco-design curricula.

Finally, some complementary observations can be made. The step-by-step approach introduced in the training session is not referred to any of the control groups. As designers traditionally do, students chose to implement and/or customize known methods in order to limit the cognitive load due to the learning of a new practice. Surprisingly, and contrary to the conclusions of Collado and Ostad-Ahmad-Ghorabi [4], the medium-level of information provided could not be used appropriately by novices. They were not able to decide what to look at and what information to select for the rest of the task.

Some limits have to be reported. Firstly Le Pochat et al. [20] underline the difficulty in drawing conclusions about learning and knowledge creation within a limited amount of testing time. Secondly, according to Cash et al. [16], the experiment could be improved with a placebo to refine the understanding of influencing variables on eco-design learning.

The objective of the future research work is to document guidelines on eco-design competences by means of comparing novice and expert practices.

Acknowledgments The authors express their gratitude to Philips Gerson for his precious help, and to the participants of the ECIU Winterschool on Eco-Design hosted by the Université de Technologie de Compiègne in February 2011. The reviewers are also thanked for their helpful comments.

References

1. Vezzoli C (2003) A new generation of designers: perspectives for education and training in the field of sustainable design. Experiences and projects at the Politechnico di Milano University. J Cleaner Prod 11(1):1–9
2. Lozana Garcia FJ, Kevany K, Huisingh D (2006) Sustainability in higher education: what is happening? J Cleaner Prod 14(9–11):757–760
3. Vallet F, Millet D, Eynard B (2009) Investigating the use of eco-design guides: presentation of two case studies. ICED'09, International conference on engineering design, Stanford, USA, 24–27 Aug 2009
4. Collado D, Ostad-Ahmad-Ghorabi H (2010) Influence of environmental information on creativity. Des Stud 31:479–498
5. Brezet JC, Van Hemel C (1997) Eco-design—A promising approach to sustainable production and consumption. United Nations Publications (UNEP), NY
6. Lloveras J (2007) Ecodesign: a subject for engineering design students at UPC. In: Takata S, Umeda Y (eds) Advances in life cycle engineering for sustainable manufacturing businesses-proceedings

of the 14th CIRP conference on life cycle engineering, Waseda University, Tokyo, Japan, 11–13 June 2007. Springer, Berlin, pp 101–106
7. Telenko C, Seepersad CC, Webber WE (2008) A compilation of design for the environment principles and guidelines. In: Proceedings of the ASME international design engineering technical conferences—design for manufacturing and the life cycle conference, August 3–6, New York City, USA
8. Telenko C, Seepersad C (2010) A methodology for identifying environmentally conscious guidelines for product design. J Mech Des 132(9):1–9
9. Baeriswyl MC, Eppinger SD (2011) Teaching design for environment in product design classes. In: Proceedings of international conference on engineering design, ICED'11, 15–18 Aug. Copenhagen, Denmark
10. White P, St Pierre L, Belletire S (2007) Okala course guide. IDSA, VA
11. Mc Aloone T, Bey N (2009) Environmental improvement through product development. [online] http://www.ipu.dk/English/IPU-Product-Development/Publications/Guide-til-miljoeforbedringer-gennem-produktudvikling.aspx. Accessed 23 Feb 2010
12. Crul M, Diehl J (2009) Design for sustainability, a step-by-step approach. United Nations Publications (UNEP), NY
13. Diehl JC, Brezet JC (2005) International ecodesign education: personalized knowledge transfer. In: Proceedings of the international conference on engineering design, ICED'05, 15–18 August. Melbourne, Australia
14. Fussler C, James P (1996) Driving eco-innovation—A breakthrough discipline for innovation and sustainability. Pitman Publishing, London
15. Blessing L, Chakrabarti A (2009) DRM, A design research methodology. Springer, London
16. Cash P, Elias E, Deconinck E, Culley S (2011) Methodological insights from a rigorous small scale design experiment. Design Studies. doi:101016/jdestud.2011.07.008
17. ISO/TR 14062: 2002 (2002) Environmental management, integrating environmental aspects into product design and development. Technical Report, ISO, Geneva
18. AEA Energy and Environment (2009) Work on preparatory studies for eco-design requirements of EuPs (II), Lot 17 Vacuum cleaners. TREN/D3/390-2006, Issue 1, Feb 2009
19. CCI St Etienne, IDP Montreal (2009) L'éco-conception: quels retours économiques pour l'entreprise?, Mar 2009
20. Le Pochat S, Bertoluci G, Froelich D (2007) Integrating eco-design by conducting changes in SMEs. J Cleaner Prod 15(7):671–680

Chapter 40
Multiple Criterion Decision Making Application for Sustainable Material Selection

S. Vinodh and R. Jeya Girubha

Abstract Material selection plays an important role in enabling sustainability, since it facilitates the design and manufacturing process involved in the production of a component. The design and manufacturing process in turn reduces the ecological impacts and improving the social dealings Material selection for an automotive component involves multiple criteria. So in order to select the best material, Multiple Criterion Decision Making (MCDM) method is the best option. To avoid the tribulations because of linguistic variables, fuzzy based MCDM is introduced in this context. Compromise ranking has been widely used as an effective MCDM method. Therefore in this context, a fuzzy based compromise ranking method is used. Compromise solution has been arrived for a threshold value and a single solution is also arrived by setting a different threshold value.

40.1 Introduction

Sustainability attributes in a manufacturing organisation can be mainly fetched by incorporating changes in materials, design and manufacturing process. Since the material selection is the first step and it paves the way for design modification and manufacturing process selection. So in this context in order to achieve sustainability for a component, material selection in a sustainable environment is taken as a first step and discussed. Material selection is a part of sustainable development, refers to the establishment of materials which save resources, pursue cleaner production

S. Vinodh (✉) · R. Jeya Girubha
Department of Production Engineering, National Institute of Technology,
Tiruchirappalli, Tamilnadu 620015, India
e-mail: vinodh_sekar82@yahoo.com

A. Chakrabarti (ed.), *CIRP Design 2012*, DOI: 10.1007/978-1-4471-4507-3_40,

environment and it should be cost efficient. The methodology deals with the selection of materials for an automotive component from a set of materials and the current material which is used by the organisation is Styrene Maleic Anhydride (SMA) plastic. The case study involves the selection of material from Polypropylene (P1), Acrylonitrile Butadiene Styrene (ABS) (P2), SMA (P3), and Poly Vinyl Chloride (PVC) (P4) using VlseKriterijumska Optimizacija I Kompromisno Resenje (VIKOR).

40.2 Literature Review

Generally the sustainable concepts can be classified as economic, environment and social. In particular to manufacturing firm, the concepts can be elaborated as material perspective, product design and manufacturing process. P. Chatterjee et al. [1] explained VIKOR procedure for flywheel and sailing boat mast design material selection. The author obtained full ranking of materials by considering many criteria related to the application of the respective products. Finally the author demonstrated and validated the effectiveness and flexibility of VIKOR. Shanian and Savagado [2] used ELECTRE (ELimination Et Choix Traduisant la REalite) for the material selection process. Rathod and Kanzaria [3] used two methods namely Technique of ranking Preferences by Similarity (TOPSIS) and fuzzy TOPSIS for the phase change material selection. Rao and Patel [4] used a compromise ranking method and AHP for the material selection problem. Bahraminasab and Jahan [5] used a comprehensive method for femoral component of total knee replacement and obtained the ranking of the materials. Since the input details regarding the criterion are vague in nature, it demands to select the material in a fuzzy environment. Therefore it is preferred use one of the most suitable material selection MCDM methodology, compromise ranking method in a fuzzy environment. Opricovic and Tzeng [6, 7] developed VIKOR method; it determines the compromise solution from a set of alternatives. Compromise solution is a feasible solution which is closest to the ideal solution. Compromising ranking method (VIKOR) is used, as it is an effective MCDM tool for material selection in many applications and there is a large reduction of mathematics and an effective correlation of results with the past researches.

40.3 Case Study

The case study has been conducted in automotive parts manufacturing organisation located in India and the study has been conducted for an automotive component (Dashboard). The inputs are gathered from three decision makers of the organisation, who are responsible for the implementation of sustainable concepts. The inputs are acquired in the form of fuzzy numbers in order to avoid vagueness of the

Table 40.1 Excerpt of the weights of the criteria

	DM 1	DM 2	DM 3
C6	MG	G	G
C7	G	G	G
C8	G	MG	MG

input details. The associated fuzzy numbers are trapezoidal numbers. Trapezoidal fuzzy numbers form the most common division of fuzzy numbers and the commonly used triangular fuzzy number is a special case of trapezoidal fuzzy number. Trapezoidal fuzzy number can encompass more uncertainty than the trapezoidal fuzzy number. A trapezoidal fuzzy number can be defined as $\{(n_1,n_2,n_3,n_4)|n_1,n_2,n_3,n_4 \in R;\ n_1 \leq n_2 \leq n_3 \leq n_4\}$ which respectively, denotes the smallest possible, most promising, and largest possible values [8] and the membership function is defined using Eq. 40.1. The linguistic and the corresponding fuzzy sets are as follows Very Poor (VP)—(0.0, 0.0, 0.1,0.2), Poor (P)—(0.1, 0.2, 0.2, 0.3), Medium Poor (MP)—(0.2, 0.3, 0.4, 0.5), Fair (F)—(0.4, 0.5, 0.5, 0.6), Medium Good (MG)—(0.5, 0.6, 0.7, 0.8), Good (G)—(0.7, 0.8, 0.8, 0.9), Very Good (VG)—(0.8, 0.9, 1.0, 1.0) and for the ratings of the criteria, the linguistic and the corresponding fuzzy sets are Very Low (VL)—(0.0, 0.0, 0.1,0.2), Low (L)—(0.1, 0.2, 0.2, 0.3), Fairly Low (FL)—(0.2, 0.3, 0.4, 0.5), Medium (M)—(0.4, 0.5, 0.5, 0.6), Fairly High (MG)—(0.5, 0.6, 0.7, 0.8), High (G)—(0.7, 0.8, 0.8, 0.9), Very High (VG)—(0.8, 0.9, 1.0, 1.0) [8]

$$\mu_A(x) = \begin{cases} \dfrac{x - n_1}{n_2 - n_1}, & x \in [n_1, n_2] \\ 1, & x \in [n_2, n_3] \\ \dfrac{n_4 - x}{n_4 - n_3}, & x \in [n_3, n_4] \\ 0 & \textit{Otherwise} \end{cases} \tag{40.1}$$

40.4 Compromise Solution Methodology

By considering the functionalities of the component, it is evident that the following eight criteria have to be taken into account for the material selection. Cost [9] (C1), Temperature limit [10] (C2), Capable to recycle [11] (C3), S/W (Strength to weight) ratio [9] (C4), Conductivity [10] (C5), Harmful emissions [9] (C6), Strength [9] (C7), Elongating Properties [11] (C8)

The data has been collected accordingly in two forms, first one is the decision maker's view about the criteria (Weights of the criteria) and the second one deals about the decision makers view about the material with respect to each criterion. The excerpt of the weights of the criterion is shown in Table 40.1 and an excerpt of the ratings of criterion against each material by decision maker 1 is shown in Table 40.2.

Table 40.2 Excerpt of the ratings of the criteria against each material by Decision Maker 2

	C6	C7	C8
P1	FH	FH	FH
P2	VH	H	VH
P3	FH	H	FH
P4	H	VH	H

Table 40.3 Crisp values for weight and material ratings

	C1	C2	C3	C4	C5	C6	C7	C8
P1	0.69	0.72	0.69	0.70	0.69	0.71	0.69	0.69
P2	0.87	0.87	0.85	0.85	0.85	0.87	0.85	0.84
P3	0.65	0.61	0.67	0.66	0.67	0.65	0.63	0.64
P4	0.80	0.87	0.85	0.80	0.85	0.85	0.87	0.74

The aggregated values for the weights of the criterion and for the material ratings against the criterion are formed by using the Eq. 40.2 [1] for the trapezoidal numbers.

$$X_{ij} = \{X_{ij1}; X_{ij2}; x_{ij3}; X_{ij4}\} \tag{40.2}$$

Where,

$$\begin{aligned} X_{ij1} &= \min\{X_{ijk1}\} \\ X_{ij2} &= \frac{1}{k}\Sigma X_{ijk2} \\ X_{ij3} &= \frac{1}{k}\Sigma X_{ijk3} \\ X_{ij4} &= \max\{X_{ijk4}\} \end{aligned}$$

After aggregating the values, normalisation is done to stabilise the different units of the criterion. The normalised values are defuzzified using the Eq. 40.3.

$$\frac{a+b+c+d}{4} \tag{40.3}$$

Here a, b, c, d represents the four values of the trapezoidal fuzzy number. After normalising the values, crisp values are obtained as shown in Table 40.3.

Fuzzy VIKOR is employed as the compromise ranking solution method in this context. Inputs are defuzzified and it is computed by Fuzzy VIKOR method. Fuzzy VIKOR is an efficient tool in handling the material selection problem. The following step shows the methodology of VIKOR.

From Table 40.3, the maximum and the minimum values of each criterion are selected as the best and worst values for each criteria and it is shown in Table 40.4. In this methodology, VIKOR index is used to finalise the result. The calculation of utility, regret and VIKOR index is as follows using Eqs. (40.4–40.6) [1] Where f_i^* denotes the

Table 40.4 Best and worst values

	C1	C2	C3	C4	C5	C6	C7	C8
Best value	0.87	0.87	0.85	0.85	0.87	0.92	0.87	0.85
Worst value	0.60	0.60	0.65	0.67	0.70	0.65	0.65	0.65

best crisp value, f_i^- denotes the worst value. S^* and R^* the maximum of utility and regret indices. The utility index and regret index for a material denotes the relative advantage and disadvantage of the material over the other material. VIKOR index is used to rank the materials by computing the relative advantage and disadvantage of the material.

$$S_i = \sum\nolimits_{j=1}^{n} \frac{w_j^o(f_i^* - f_{ij})}{(f_i^* - f_i^-)} \tag{40.4}$$

$$R_i = \max_i \left(\frac{w_j^o(f_i^* - f_{ij})}{(f_i^* - f_i^-)} \right) \tag{40.5}$$

$$Q_i = \frac{v(s_i - s^*)}{s^- - s^*} + \frac{(1 - v)(R_i - R^*)}{R^- - R^*} \tag{40.6}$$

v is introduced as a maximum group utility and $1 - v$ is the weight of the individual regret. It denotes the utility function and regret function contribution to the VIKOR index. Here v is taken as .5 as per the suggestions provided by the decision makers of the organisation. Arranging s_i, R_i, Q_i in increasing order to determine the rank and the alternative which has least VIKOR index value is considered to be the best one. Tables 40.5 and 40.6 shows the calculated indices and ranking of material.

From Table 40.6, it is clear that the VIKOR index is less for the material type 2. Therefore, P2 is selected as the best alternate for the automotive component. Since it is a compromise ranking method, it gives out a set of compromise solutions. The alternate will be considered as the best one if it satisfies two conditions [5].

C1. Acceptable advantage: $Q(A^{(2)}) - Q(A^{(1)}) \geq 1/(m-1)$, where $A^{(2)}$ is the second position in the alternatives ranked by Q.

C2. Acceptable stability in decision making: Alternative $A^{(1)}$ must also be the best ranked by S or/and R. When one of the conditions is not satisfied, a set of compromise solutions is selected.

The compromise solutions are composed of:

1. Alternatives $A^{(1)}$ and $A^{(2)}$ if only condition **C2** is not satisfied (or)
2. Alternatives $A^{(1)}$, $A^{(2)}$, ….., $A^{(m)}$ if condition **C1** is not satisfied. $A^{(M)}$ is calculated using the relation $Q(A^{(M)}) - Q(A^{(1)}) \geq 1/(m-1)$ for maximum M.

In this case, both of these conditions have been satisfied, so the material acquiring least VIKOR index was selected as the best material. In this context, it is concluded that ABS was selected as the best alternate material for the automotive component.

Table 40.5 Calculated indices

	S	R	Q
P1	5.0	0.9	0.9
P2	0.4	0.3	0.0
P3	5.8	0.9	1.0
P4	0.6	0.2	0.0

Table 40.6 Ranks of the materials

	S	R	Q
1	P2	P4	P4
2	P4	P2	P2
3	P1	P3	P3
4	P3	P1	P1

40.5 Results and Conclusions

The material selection for an automotive component in a fuzzy environment using a compromise ranking solution shows the best material. Since the material selection plays a major role in the automotive organisation, this context has gained more attention. The same material selection problem can be solved by some other MCDM methods and can increase the number of decision makers in order to come up with a better result. After the selection of material, design modifications can be brought into the product and manufacturing process can also be changed in order to manufacture the product in a sustainable environment.

Acknowledgments The authors express their special thanks to Department of Science and Technology (DST), New Delhi, India for the financial support extended towards the implementation of project titled "Development of a model for ensuring sustainable product design in automotive organizations" (Ref No.SR/S3/MERC-0102/2009). This research study forms a part of this principal research project.

References

1. Chatterjee P, Athawale VM, Chakraborty S (2009) Selection of materials using compromise ranking and outranking methods. Mater Des 30:4043–4053
2. Shanian A, Savadogo O (2006) A material selection model based on the concept of multiple attribute decision making. Mater Des 27:329–337
3. Rathod MK, Kanzaria HV (2011) A methodological concept for phase change material selection based on multiple criteria decision analysis with and without fuzzy environment 32:3578–3585
4. Rao RV, Patel BK (2010) A subjective and objective integrated multiple attribute decision making method for material selection. Mater Des 31:4738–4747
5. Bahraminasab M, Jahan A (2011) Material selection for femoral component of total knee replacement using comprehensive VIKOR. Mater Des 32:4471–4477

6. Opricovic S, Tzeng GH (2004) Compromise solution by MCDM methods: A comparative analysis of VIKOR and TOPSIS. Eur J Oper Res 156:445–455
7. Opricovic S, Tzeng GH (2007) Extended VIKOR method in comparison with outranking methods. Eur J Oper Res 178:514–529
8. Shemshadi A, Shirazi H, Toreihi M, Tarokh MJ (2011) A fuzzy VIKOR method for supplier selection based on entropy measure for objective weighting. Expert Syst Appl 38:12160–12167
9. Cui X, Wang S, Hu J (2008) A method for optimal design of automotive body assembly using multi-material construction. Mater Des 29:381–387
10. Fohlen V, Johansson A (2004) Development of squeak and rattle design guidelines for the instrument panel area. Department of technology, Mathematics and Computer Science. M019
11. Yates A, Lyon P (1992) Effect of Long Term Exposure on Mechanical Properties of WE43. Magnesium Elektron Internal Report: MR10/Data/236

Chapter 41
A Strategic Approach for Sustainable Product Service System Development

Henrik Ny, Sophie Hallstedt and Åsa Ericson

Abstract Product-Service Systems (PSS) have been justified by a desire to find sustainable solutions that go beyond contemporary approaches. The characteristics of PSS offerings are to link goods and services in development and to provide systemic performance-based solutions to the customers. This paper investigates how established strategic product development tools for socio-ecological sustainability could be adapted for PSS development. An approach is suggested for how to apply these tools in early PSS development phases.

41.1 Introduction

An increasing majority of scientists from various fields agree that our society is currently on a long-term unsustainable course and that it is urgent to deal with this situation [1–4]. This kind of global sustainability challenge calls for a more strategic and holistic approach that brings together several knowledge domains.

Product-Service Systems (PSS) is identified as a window of opportunities to integrate distinct knowledge domains to benefit environmental issues [5]. PSS has a main focus to progress sustainable solutions of products and services and it is commonly described as [6, p. 239]:

H. Ny (✉) · S. Hallstedt · Å. Ericson
Blekinge Institute of Technology, SE-37179, Karlskrona, Sweden
e-mail: henrik.ny@bth.se

A. Chakrabarti (ed.), *CIRP Design 2012*, DOI: 10.1007/978-1-4471-4507-3_41,

> ...a system of products, services, supporting networks and infrastructure that is designed to be; competitive, satisfy customer needs and have a lower environmental impact than traditional business models.

Manzini and Vezzoli [7] argue that PSS development intends to use a service business model to affect the patterns of consumption, and, by this, attack the problems of sustainability at a totally different level. Williams [8] discussed an example of such a changed business model where car manufacturers go from selling cars to providing car leasing. At a first glance, such an approach lowers the number of cars that are produced, i.e. strategically supports sustainability by decreasing production, but does not necessarily affect those cars that are actually being produced, i.e. fewer cars but not by default 'greener' cars. During product development, decisions take place to determine what resources and processes are required to manufacture a product. Product development, especially the early phases, is therefore often pointed out as a potential leverage point to deal with the social and ecological sustainability problems that products may cause [9, 10]. Such problems include systematic poisoning (e.g. through systematic increases in concentrations of man-made substances) and physical degradation of nature, as well as reduction of people's capacity to meet their basic needs. Apparently, PSS has to facilitate proactive decisions in early development phases to meet sustainable concerns. Though, it could be questioned if the PSS vision alone could do this, and if so, by building on existing approaches for sustainability.

Generally, existing approaches for sustainability focus on certain aspects of societal sustainability challenges. For example, Environmental Management Systems (EMSs) [11], Cleaner Production [12] and various ecological indicators like Factor 10 [13]. Eco-design or Design for Environment deals with the integration of sustainability into product development, including tools like matrices, spider webs, checklists, guidelines, and comparison tools [14–16]. These have been developed for different purposes, such as for assessment of environmental impacts, identification of environmentally critical aspects, comparison of environmental design strategies, comparison of product solutions and prescription of improvement strategies [17–19]. Life-Cycle Assessment (LCA) [20] aims to evaluate impacts of materials and products from the "cradle" (resource extraction), through transport, production, and use, to the "grave" (waste management after end use). As the ISO-standard indicates, LCA is one of the rigorous and frequently used sustainability tools but the rigor also means that it is time consuming and requires expensive investments in man-hours [21]. A life-cycle map of a PSS offering would be even more complex than a traditional LCA product system. The PSS offering could be modeled as several layers of interacting activities from the direct 'frontline' (e.g. buying, using and disposing of a car) to the progressively deeper 'back-office' functions of the PSS (Fig. 41.1).

Accordingly, there is a need for approaches that apply a strategic perspective for managing sustainability aspects in early phases. This paper aims to describe strategic tools for sustainable product development and investigate what they may add to current tools for sustainability in order to contribute to early PSS development.

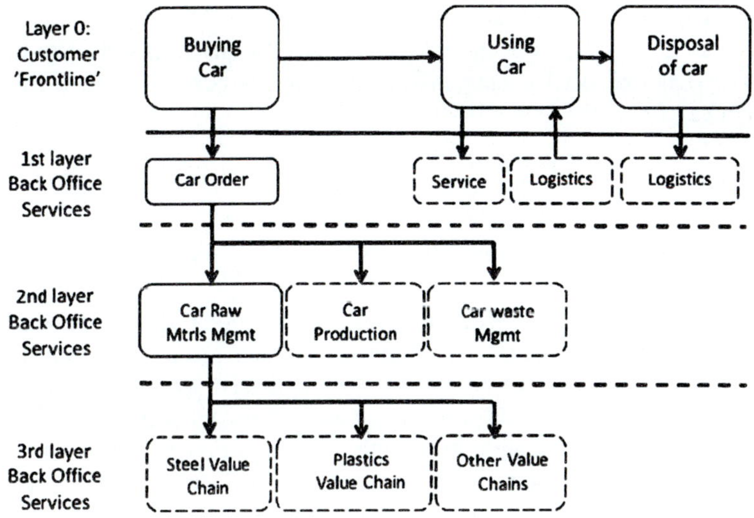

Fig. 41.1 A partially completed life cycle map of a PSS offering and its related material and energy flows through multi-layered services and physical activities (*dashed line* means that not all activities are included)

41.2 Strategic Tools for Sustainable Product Development

A generic Framework for Strategic Sustainable Development (FSSD) has previously been developed to frame strategic planning towards a rigorously principled definition of sustainability [16, 22, 23]. The practical application of this Framework is facilitated by a manual called the ABCD process [22]. The first step (A) of the ABCD includes that the planning team analyzes the planning topic within its system enough to agree on robust principled definitions of objectives and how those relate to the FSSD's basic principles of global socio-ecological sustainability [22, 23]. Then, in step B, they use back casting from the principled objectives to identify significant current problems and assets along the life-cycle of the planning topic in that context. In step C they brainstorm and list desirable future solutions, visions and actions that are possible within the principled constraints of success. Finally, in step D, they prioritize among the actions to arrive at a strategic plan, and the possible need of tools for management and monitoring is identified.

In a product development context this framework strives towards a situation when products do not contribute to systematic degradation of socio-ecological systems. This has led to the design of strategic tools like:

- the Method for Sustainable Product Development (MSPD) [24, 25], intended for product developers throughout the whole development process;
- expert-facilitated Templates for Sustainable Product Development (TSPDs) [26], aimed to complement the MSPD in earlier development phases with an overview of market desires, concepts and stakeholders;

- the Strategic Life-Cycle Management (SLCM) Matrix [22, 27] that uses generic sustainability principles of the FSSD as a lens for quick identification of 'hot-spot' issues that need deeper investigation in product and service life-cycle activities.

41.2.1 Method for Sustainable Product Development: MSPD

The Method for Sustainable Product Development (MSPD) should encourage and aid development of products that support society's transformation towards sustainability. More specifically, it should facilitate: (1) identification of potential problems of present or planned products caused by substances and activities during the product life cycle that are critical with regard to principles for sustainability; (2) guidance in finding solutions to the potential problems by modifications of present or planned products, and (3) stimulation of new products and business ideas based on sustainability aspects [25].

The MSPD includes one manual and three tools:

- An MSPD Manual, with the aim of providing the user with the objectives and the theory of the MSPD, and instructions on how to use its different tools.
- A model of a product development process (PDP), which includes phase-specific questions for various traditional aspects within the phases.
- Sustainability product assessment (SPA) modules, which include strategic guiding questions (Table 41.1) to identify potentially critical substances and activities during the life cycle of the existing or planned product and questions to generate proposals for improvements.
- A prioritisation matrix, which includes questions to facilitate evaluation and choice among proposals. Sustainability aspects are integrated with traditional economic and technical aspects to improve the applicability of the method from a business perspective.

When a project group has read the manual, they can work with the product development questions for a particular phase. Then the SPA modules should be used. The proposals that are generated from SPA modules will be evaluated in the prioritisation matrix. The most suitable proposal(s) will be chosen before continuing working in the next product development phase.

41.2.2 Template for Sustainable Product Development: TSPD

The expert-led TSPD is complementary to the MSPD and it should (1) help product development teams arrive faster and more easily at an overview of the major sustainability challenges and opportunities of a product category in earlier

Table 41.1 Examples of questions for two of the SPA modules in the MSPD

SPA module	Step of the ABCD	Sustainability principles (SPs) 1, 2, 3 or 4	Example questions
Product function	B	1, 2, 3, 4	Is there a dissipative use of the product and does the product consist of; metals; chemicals; resources from ecosystems; resources put in global human need perspective?
	C	1, 2, 3, 4	Are there any product types with no dissipative use of the materials, that can be incorporated into societal cycling of materials (low material losses) or even into tight technical loops with no or very small losses to the environment and that fulfil the customer needs?
Product design	B	1	Are fossil fuels currently needed for product usage?
	C	1	How can the product be designed to use renewable energy sources (directly or via electricity)?

Fig. 41.2 Three templates for sustainable product development (TSPDs)

	Needs	Concepts	Stakeholders
Now (B-step)			
Sustainable Future (C-step)			

development phases and (2) facilitate creative communication between top management, stakeholders, and product developers [26].

The TSPD consists of three question templates, each containing the current situation (B of the above-mentioned ABCD process) and possible future solutions and visions (C of the ABCD) (Fig. 41.2).Template I covers market desires/needs (focusing on market desires and their relation to basic human needs and on identifying the desired product function), Template II covers concepts (focusing on life-cycle sustainability consequences of meeting the market desires with a certain product concept) and Template III covers extended enterprise (focusing on societal stakeholder consequences from the product concept and how they can be influenced).

The issues covered by the three templates are considered to be particularly relevant for the creation of a generic overview. Step D of the ABCD (prioritization of solutions from C) is excluded, as the templates' purpose is not to be prescriptive but to trigger creativity and act as input for later detailed priorities. There are TSPD procedures both for evaluating an existing product and to provide planning support for a new product concept. Both procedures are supposed to be performed by an integrated product development team in a company or company network and require facilitation by an internal or external expert in strategic sustainable development and the template approach.

	SP1	SP2	SP3	SP4
Raw Mtrls				
Production				
Use				
End of Life				

Fig. 41.3 The strategic life-cycle management (SLCM) matrix that scrutinizes the four FSSD sustainability principles (SPs) against product life-cycle phases

41.2.3 Strategic Life-Cycle Management Matrix: SLCM

The Strategic Life Cycle Management (SLCM) Matrix (Fig. 41.3) is an approach that, in line with the B-step of the ABCD, uses the Sustainability Principles as a lens to evaluate product life-cycles, including raw materials, production, packaging and distribution, use and end of life [22].

In practice, several concrete process tools are emerging from the SLCM theory [27]. The process varies but they all revolve around a sustainability performance matrix that scrutinizes violations of each of the four sustainability principles for each activity in the life-cycle.The assessment team, with or without the assistance of an expert facilitator, uses their best knowledge to study current product or service lifecycle stages for potential violations of the sustainability principles. It is important to include people from different departments within the company that together have a good overview. This can be done in a relatively short time without being too detailed. Early overview workshops should be complemented with more widely distributed questionnaires and directed investigations.

41.3 Sketching a New Strategic Approach for PSS Development

This new approach is expected to be particularly useful for PSS developers looking for cost effective ways to get a quick overview of sustainability consequences of various alternative choices in the process, software, hardware and business model design. With this, and the general characteristics of PSS, in mind we suggest that any tool for sustainable PSS development should, in early technical product development: facilitate a service perspective, identify links of goods and services, provide strategic steps to meet long term socio-ecological sustainability.

The MSPD covers the design process, i.e. the early stages of product development and could, thus, be a key to service integration before the characteristics of the goods is settled. The MSPD, TSPD and SLCM form a methodological foundation for sustainable product development, applicable for any business and all product types. Given that we are focused on giving quick advice to how PSS developers could integrate sustainability early in their work processes we will in the rest of this text focus on the simplified TSPD and SLCA tools. We are still

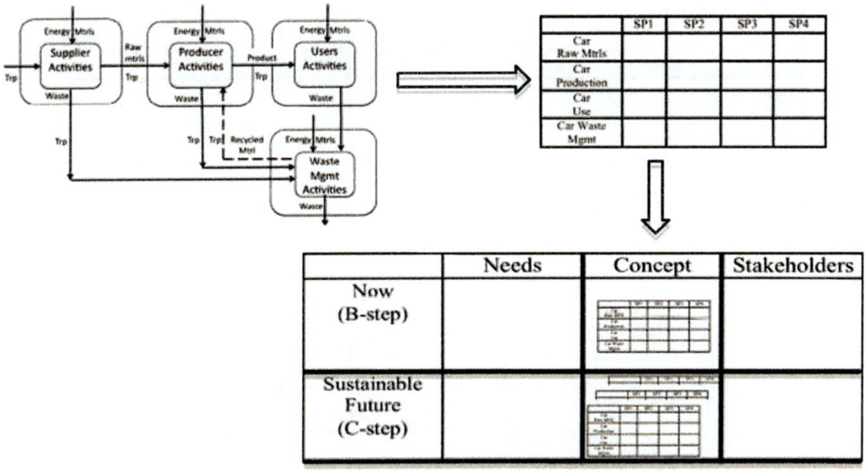

Fig. 41.4 How the traditional product life-cycle perspective phases are covered by the SLCM Matrix in the B- and C-step of the TSPD product concept template

convinced, though, that the MSPD could be very suitable for later stages when PSS developers want to further strengthen their toolboxes for sustainability support.

To provide for synergies between the TSPD and the SLCM Matrix they can be used in combination (Fig. 41.4). The quick and broad sustainability overview of the TSPD makes its use suitable for early stages in the product development (primarily the need and concept phases). More sustainable product designs that the TSPD has suggested can then, using investment calculus and other support tools, be prioritized along different parameters, such as return on investment. The ability of the SLCM Matrix to more deeply evaluate the sustainability life-cycle performance of product concepts, makes it particularly suitable to support the concept template of the TSPD. When evaluating the sustainability performance of both the existing and potential new concepts, for example, the TSPD would describe both the sustainability considerations of the market niche (template I of the TSPD) and the stakeholder relationships (template III of the TSPD) while the physical concept life cycle consequences (template II of the TSPD) from suppliers, producers, users and waste managers would be covered in more detail in the SLCM Matrix.

The TSPD could also be applied for the above described 'multi-layered' PSS (Fig. 41.1). Then the concept template could be supported by a modified multi-layered SLCM Matrix (Fig. 41.5) that systematically scrutinize all the PSS processes, services and hardware life-cycles against sustainability principles (the SPs). This new Matrix can thereby, just as the original one (Fig. 41.3), be used to map out sustainability implications of both a current PSS (B-step of the ABCD process) and of several alternative new future PSS setups (C-step of the ABCD). The TSPD could be used to both evaluate an existing PSS offering (e.g. car leasing) and to provide planning support for a new one (Table 41.2). To ensure

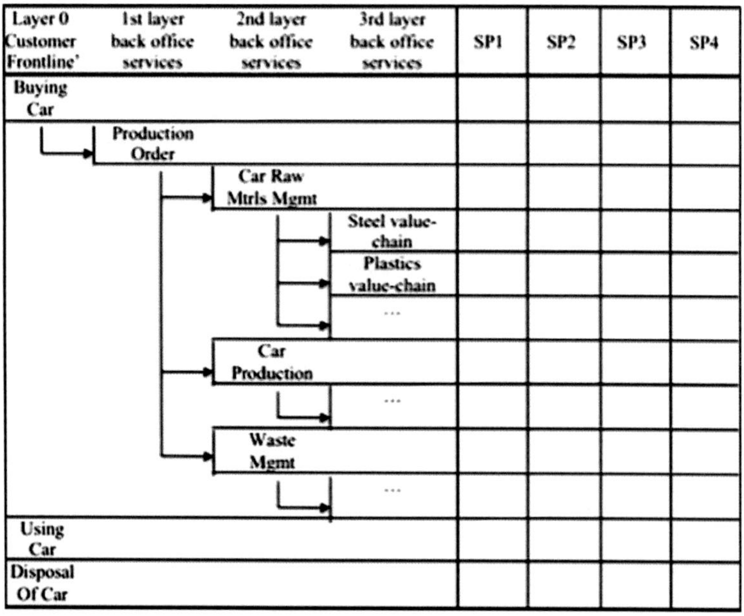

Fig. 41.5 How the new car PSS from Fig. 41.1 can be scrutinized against sustainability principles (SPs) in a new multi-layered SLCM Matrix

Table 41.2 Step-by-step sequences for the approach for strategic sustainable PSS development

For evaluating existing PSS concepts
1. The sustainability expert leads a preparatory dialogue with the development team
2. Triggered by generic TSPD questions (Table 41.1), the sustainability expert use the sustainability principles as a checklist and systematically identify strengths and weaknesses in the current PSS offering activities (step B of the above described ABCD process). This is followed by identification of solutions and visions for future sustainable (step C)
3. If needed, the SLCM Matrices are used to more systematically study each process step and life-cycle stage of current and alternative future sustainable car leasing PSS offerings
4. The development team responds to the TSPD statements and SLCM listings by providing aspects from competence fields, such as product development and management. Simple misunderstandings are sorted outland possible differences of opinion are clarified
5. The sustainability expert gives feedback on the responses to the client
6. Items 3 and 4 are repeated until consensus is reached for this overview assessment
7. A final presentation is given and a consensus report, including the templates, is produced
For planning support for new PSS concepts
1. The sustainability expert leads a preparatory dialogue with the development team
2. Together they identify the desired function and an existing reference product or service that could fulfil it
3. They plan for a new PSS by using the identified reference product/service when walking through steps 2 through 7 of the above procedure for evaluating an existing PSS

maximum effect of the method, the development team should contain a set of competences related to the particular case. The team can also include members not only from the initiating PSS company but also from its supply chain and wider web of stakeholders. Both procedures require facilitation by an internal or external expert ("the sustainability expert") in strategic sustainable development and the template approach.

41.4 Concluding Discussion

We have presented three generic strategic tools for sustainable product development (the MSPD, TSPD and SLCM Matrix) that are able to aid the integration of sustainability aspects in companies' strategic decision-making and product development. We have also shown the rationale for how a combination of the TSPD and SLCM Matrix tools could provide a foundation for competitive and sustainable PSS development. The template approach (TSPD) combining needs (e.g. business model), concept (life-cycle consequences of business model) and extended enterprise (stakeholders) is likely to facilitate a system perspective. It also quickly combines a PSS approach with its sustainability consequences in product and service development and its relations with external stakeholders. The SLCM Matrix is also likely to contribute significantly to a quick and cost effective sustainability assessment of PSS concepts. All in all, our study indicates that, rightly adopted, the new approach can strengthen the strategic sustainability focus of PSS development in industrial practice. This is supported by a case study where the long life light tubes producer Aura light International [28] successfully used the SLCM Matrix to map the ecological and social sustainability aspects when a PSS approach was explored for Long-life Products. More case studies are now needed for further development and successful dissemination of this new approach.

References

1. Millennium Ecosystem Assessment (MA) (2005) Ecosystems and human well-being: Our human planet: summary for decision-makers (Millennium Ecosystem Assessment). Island Press, Chicago
2. Stern N (2006) Stern review on the economics of climate change. HM Treasury, London
3. Intergovernmental Panel on Climate Change (IPCC) (2007) Fourth assessment report. Climate Change 2007: Synthesis Report
4. Rockström et al (2009) A safe operating space for humanity. Nature 461:472–475
5. Mont O, Tukker A (2006) Editorial. Product-Service Systems: reviewing achievements and refining the research agenda. J Cleaner Prod 14:1451–1454
6. Mont OK (2002) Clarifying the concept of product-service system. J Cleaner Prod 10:237–245
7. Manzini E, Vezzoli C (2001) Product service systems and sustainability: opportunities for sustainable solutions. Documentation from a Workshop on Product-Service Systems held by UNEP DTIE in Paris, June 2000. United Nations Environment Programme

8. Williams A (2007) Product service systems in the automobile industry: contribution to system innovation? J Cleaner Prod 15:1093–1103
9. Charter M, Chick A (1997) Welcome to the first issue of the journal of sustainable product design. J Sustainable Prod Des 1(1):5–6
10. Ritzén S (2000) Integrating environmental aspects into product development: proactive measures. Doctoral thesis thesis, Department of machine design, The Royal Institute of Technology, Stockholm
11. International Organization for Standardization (ISO) (2004) Environmental management systems—Requirements with guidance for use. ISO 14001. ISO, Geneva
12. Aloisi de Larderel J (1998) What is a sustainable enterprise? Paper presented at UNEP workshop, 14 October, Paris
13. Schmidt-Bleek F (1997) MIPS and factor 10 for a sustainable and profitable economy. Wuppertal, Institute, Wuppertal
14. Weenen van H (1997) Design for sustainable development: guides and manuals. European Foundation for the Improvement of Living and Working Conditions, Dublin
15. Tischner U, Chmincke SE, Rubik F, Prösler M (2000) How to do Ecodesign? A guide for environmentally and economically sound design. German Federal Environmental Agency, Berlin
16. Robèrt KH, Schmidt-Bleek B, de Aloisi Larderel J, Basile G, Jansen JL, Kuehr R, Price-Thomas P, Suzuki M, Hawken P, Wackernagel M (2002) Strategic sustainable development—selection, design and synergies of applied tools. J Cleaner Prod 10(3):197–214
17. Byggeth SH, Hochschorner E (2006) Handling trade-offs in eco design tools for sustainable product development and procurement. J Cleaner Prod 14(15–16):1420–1430
18. Brezet H, van Hemel C (1997) Eco design: a promising approach to sustainable production and consumption. United Nations Environment Programme (UNEP), France
19. Luttorp C (1997) Design for disassembly. Environmentally adapted product development based on preparing disassembly and sorting. Doctoral thesis, Department of Machine Design, KTH, Sweden
20. Graedel TE (1998) Streamlined life-cycle assessment. Prentice Hall, Upper Saddle River
21. International Organization for Standardization (ISO) (2006) Environmental management—Lifecycle assessment—Principles and framework. ISO 14040. ISO, Geneva
22. Ny H, MacDonald JP, Broman G, Yamamoto R, Robèrt K-H (2006) Sustainability constraints as system boundaries: an approach to making life-cycle management strategic J Ind Ecol 10(1):61–77
23. Robèrt K-H (2000) Tools and concepts for sustainable development, how do they relate to a general framework for sustainable development, and to each other? J Cleaner Prod 8(3):243–254
24. Byggeth SH, Broman GI (2001) Environmental aspects in product development—An investigation among small and medium-sized enterprises. In: Gupta SM (ed) Proceedings of SPIE, environmentally conscious manufacturing, vol. 4193. pp 261–271
25. Byggeth SH, Broman G, Robèrt K-H (2007) A method for sustainable product development based on a modular system of guiding questions. J Cleaner Prod 15(1):1–11
26. Ny H, Hallstedt S, Robèrt K-H, Broman G (2008) Introducing templates for sustainable product development through an evaluation case study of televisions at the Matsushita Electric Group. J Ind Ecol 12(4):600–623
27. Ny H, MacDonald JP, Robèrt K-H, Broman G (2009) Sustainability constraints as system boundaries: introductory steps toward strategic life-cycle management. In: Wang H-F (ed) Web-based green products life cycle management systems: reverse supply chain utilization. IGI Global, Hershey
28. Thompson A, Ny H, Lindahl P, Broman G, Severinsson M (2010) Benefits of a product service system approach for long-life products: the case of light tubes. In: 2nd CIRP, IPS2 conference, Linköping, Sweden, 14–15 Apr 2010

Author Index

A. Chakrabarti (ed.), *CIRP Design 2012*, DOI: 10.1007/978-1-4471-4507-3,